# Applications of Neutron Scattering
# to Soft Condensed Matter

# Applications of Neutron Scattering to Soft Condensed Matter

*Edited by*

**Barbara J. Gabrys**
*Open University*
*Milton Keynes, UK*

## CRC Press
Taylor & Francis Group
Boca Raton  London  New York

CRC Press is an imprint of the
Taylor & Francis Group, an **informa** business

CRC Press
Taylor & Francis Group
6000 Broken Sound Parkway NW, Suite 300
Boca Raton, FL 33487-2742

First issued in paperback 2020

© 2010 by Taylor & Francis Group, LLC
CRC Press is an imprint of Taylor & Francis Group, an Informa business

No claim to original U.S. Government works

ISBN 13: 978-0-367-57885-5 (pbk)
ISBN 13: 978-90-5699-300-9 (hbk)

**Visit the Taylor & Francis Web site at**
**http://www.taylorandfrancis.com**

**and the CRC Press Web site at**
**http://www.crcpress.com**

# Contents

# Preface

From a conservative scientist point of view, neutron scattering is a young discipline. It got a seal of approval in 1994 from the scientific community by awarding the Nobel prize to Clifford G. Shull and Bertram N. Brockhouse. The former got it for 'knowing, where the atoms are', the latter for 'knowing, what they do'. Being young, neutron scattering continues to grow and change. Initially the domain of professional neutron scatterers, at present its techniques are much better known and appreciated, not to forget user friendly. This means that neutron exploitation is broadening, as any member of the peer-review committee at neutron sources would testify. Neutrons are used in engineering, earth sciences, biology, pharmaceutical studies, to name but a few disciplines. This trend, visible in the last decade, implies that neutron use will continue to grow.

Why is this so? Neutron scattering will never be as convenient as a laboratory-based technique, however it has some advantages that make it a very attractive or sometimes the only possible technique to use. Firstly, there is the 'right' wavelength and hence energy of thermal neutrons. While the former is of the order of atomic and intermolecular distances in liquids and solids, the latter corresponds to the range of translational, rotational and vibrational energies of the molecules. Two unique properties of the neutron led to a widespread application of neutron scattering in industry: firstly, it can penetrate matter very deeply due to the weak interaction of the spin of the neutron with the nuclei. Neutrons easily penetrate many metals which means a wide range of pressures and temperatures, which can mimick an industrial process, is accessible. Secondly, the different scattering cross-sections of different isotopes mean that isotopic labelling can be employed to 'select' or 'highlight' different molecules or parts of molecules. Therefore neutrons are ideal probes for studying the shape, organisation and motion of molecules which shape the physical properties of materials.

This book is aimed at present and potential users of neutron sources working in the area of soft condensed matter and materials science, and the emphasis is on the applicability of neutron research to industrial problems. All authors have the experience of working with industry or come from industry, and the materials investigated range from simple halocarbons through homopolymers to drugs.

The opening chapter is a short, tutorial introduction, giving the principles underlying neutron scattering. It provides the necessary background for several techniques explained in more detail in subsequent chapters.

Chapter 2 presents the state-of-the-art in neutron diffraction which is the structure refinement from powders and single crystals as well as *ab initio* structure refinement from powders and single crystals. It also provides comparison of neutron and x-ray single-crystal analyses, and demonstrates how it all works on the example of phase transformations in organic molecular crystals. The point addressed here is polymorphism amongst pharmaceuticals where the structural 'screening' by neutrons of an undesirable polymorph can prevent manufacturing and batch release problems. The first-order phase transition in pyrene is used as a simpler, tutorial example to illustrate the principles of such work, which requires accurate and precise lattice constants determined from rapidly collected data, and sophisticated control of sample environment. This is where neutrons, which can pass easily through a sample cell with thick walls, have a technical advantage over x-rays.

Chapter 3 discusses a combination of diffraction and molecular simulation applied to determination of not only the structure of the liquid but also the internal potential energy associated with this structure. The case study is halocarbons which are a chemically interesting and industrially important set of molecules. They are a model system for studying the influence of changes in the molecular structure, polarizability, and anisotropy on the structure of the liquid. The aim is to develop transferable potentials to predict the properties and structure of these relatively simple liquids; then to apply this approach to more complex cases. In the long term, one hopes to achieve sufficient knowledge to use a desired property as *a starting point* and predict a structure which gives such a property. At present, such an insight into structure-property relationship is rarely possible, most studies being conducted from the other side of the link.

Small angle neutron scattering (SANS) is perhaps the most known neutron scattering technique to industrialists and academics alike. Chapter 4 provides a lucid exposition of basic principles and models. Its application to the studies of adsorbed polymer/surfactant layers in colloidal dispersions and emulsion systems is presented. From the author's viewpoint, the 'structure' is described in terms of the layer thickness, adsorbed amount, bound fraction and segment density distribution. This is the information sought by industrial physical chemists working on sterically stabilised colloidal systems such as paints, agrochemicals and food emulsifiers.

SANS is also the pivotal technique for Chapter 5 where the crystalline polymers are described. It complements and extends Chapter 4 where the principles of SANS and technical know-how (e.g. labelling method) are concerned. Of great interest here is the description of chain conformation in the amorphous and semicrystalline state as well as the structure of polymer networks.

Neutron reflectivity, the youngest of neutron structural techniques, is perhaps the second most widely used technique among industrialists. Chapter 6 describes how, using selectively deuterium labelled polymers, it is possible to obtain a detailed description of the organization of polymers spread at an air-water interface. The spatial distribution of components of amphiphilic copolymers can be specified to length scales of 2 Å. This description is valid under the assumption of kinematic approximation. This approach to data analysis allows for better understanding of the surface visco-elastic data obtained from surface quasielastic light scattering (SQELS) on the same systems. This chapter also highlights the need of further theoretical developments in the SQELS field.

Since neutron reflectivity is a technique capable of probing structural profiles at the interfacial region with resolution of a few Å, it has been used to determine the structure of soluble surfactant monolayers at different interfaces (Chapter 7). In general, soluble monolayers are difficult to study because of their heavy mixing with water. In this chapter, a very sophisticated deuterium labelling of surfactants is used to extract the signal from that layer only, therefore the solvent and substrate can be matched out. Partial labelling of surfactants allows one to obtain information about the internal structure of the surfactant layer with respect to tilting, bending, and intermixing for small fragments. Such information is necessary for understanding the nature of monolayer packing, and is vital to encourage further theoretical efforts.

An important application of SANS is to study the miscibility of polymer blends, i.e. to determine polymer-polymer interaction parameters via measurements of the concentration fluctuations close to the phase boundary. The natural contrast in the blend is enhanced by a judicious deuteration of one of the constituent polymers. Then the interaction parameter and some information about molecular size and shape is obtained. Chapter 8 describes the principles underlying such studies and shows several applications, among others a description of thermodynamical behavior of polyolefins which is of considerable technological importance. In addition, the use of neutron specular reflection to investigate the structure and composition profile of the interface between two polymers is discussed. Such information is again of considerable technological interest: the extent of mixing between the components is one of the factors which determines the mechanical properties of the final product. This chapter is brought to a close with a short review of dynamical investigations of polymer blends by quasi-elastic neutron scattering (QENS). The information sought is the extent of miscibility through the influence of mixing on the local motion.

Dynamics of molecular vibrations is studied by means of inelastic incoherent neutron scattering (INS). Neutron spectroscopy is a technique complementary to more commonly used infra-red and Raman spectroscopies. Again the unique properties of the neutron ensure that INS spectra are different from their optical counterparts. This is because in the neutron scattering from the nuclear potential only the scattering cross-section plays a role, and there are no selection rules to be reckoned with. This means it can easily access the low frequency region where infrared and Raman spectroscopies struggle. Chapter 9 gives a brief introduction to the principles of this technique and instrumentation and several applications. These are advanced materials such as carbon fibre/resin composites, one of the 'big four' polymers (polyethylene) and hydrodesulphurisation catalysts.

The concern of the impact of neutron investigations on wealth creation is visible in all the chapters in this book, and Chapter 10 is no exception. Here the class of commercially important polymers, ionomers, is reviewed. In solution, a few ionic groups attached to a non-polar polymer enhance its surface activity and the ability to control rheological properties. In the bulk, ionic interactions can vastly improve tensile properties or ionomer clarity or miscibility with other polymers. The neutron studies which elucidate the influence of the ionic substitution (sulfonation) on the physico-chemical properties of the starting polymers are SANS for solutions and SANS and wide-angle neutron scattering for the bulk. The model system studied is polystyrene and its ionomers.

The book is brought to an end by a review of the contribution of neutron scattering to pharmaceutical sciences (Chapter 11). This is perhaps the least expected application of neutrons, since pharmaceutics is a very broadly based discipline, and needs input from several diverse areas in order to convert pharmacologically active chemicals to medicines. Increasingly these medicines are not simple concoctions or pills, but are elaborate systems designed to deliver drugs to specific sites in the body. They can also regulate the dose of drug to the patient: consider polymer implants which release hormones over many months or transdermal patches which deliver controlled doses of a drug through the skin. Two areas of application are described in more detail: studies of vesicle-forming surfactants, and perfluorocarbon emulsions and interfacial polymers.

Since this field is one of the most recent applications of neutrons, in my opinion it makes a fitting end to this book which details not only the present developments, but future predictions as well.

I would like to thank the authors who juggled many commitments in order to make this book come to life, and John Lindquist for seeing this project through on behalf of the publisher. A big "thank you" to Jos Schouten for keeping my morale high, and Karolina Pesz for her help with the index. I am grateful to Professor David Brannan, the Dean of Faculty of Mathematics and Computing, the Open University and Dame Fiona Caldicott, the Principle of Somerville College, University of Oxford, for their support and interest in my research.

# Contributors

**V. Arrighi**
Heriot-Watt University
Edinburgh
UK

**D. Barlow**
King's College London
London
UK

**A.N. Burgess**
ICI Technology
Runcorn
UK

**T. Cosgrove**
University of Bristol
Bristol
UK

**A.J. Florence**
University of Strathclyde
Glasgow
UK

**B.J. Gabrys**
The Open University
Milton Keynes
UK

**P.C. Griffiths**
University of Wales Cardiff
Cardiff
UK

**J.S. Higgins**
Imperial College
London
UK

**W.S. Howells**
Rutherford Appleton Laboratory
Oxon
UK

**K.A. Johnson**
University of Liverpool
Liverpool
UK

**K. Kaji**
University of Kyoto
Kyoto
Japan

**S.M. King**
Rutherford Appleton Laboratory
Oxon
UK

**M.J. Lawrence**
King's College London
London
UK

**J.R. Lu**
University of Surrey
Guildford
UK

**K.A. Mort**
University of Keele
Keele
UK

**S.F. Parker**
Rutherford Appleton Laboratory
Oxon
UK

**S.K. Peace**
Unilever Research
Bedford
UK

**R.W. Richards**
University of Durham
Durham
UK

**K. Shankland**
Rutherford Appleton Laboratory
Oxon
UK

**N. Shankland**
University of Strathclyde
Glasgow
UK

**R.K. Thomas**
Oxford University
Oxford
UK

**C. Washington**
University of Nottingham
Nottingham
UK

**C.C. Wilson**
Rutherford Appleton Laboratory
Oxon
UK

**A.M. Young**
Biomaterials Department
Queen Mary and Westfield College
London
UK

**W. Zajac**
H. Niewodniczański Institute of
   Nuclear Physics
Cracow
Poland

# 1. An Introduction to Neutron Scattering

WOJCIECH ZAJAC[1] and BARBARA J. GABRYS[2]

[1] *H. Niewodniczański Institute of Nuclear Physics, Cracow, Poland*
[2] *Faculty of Mathematics and Computing, Open University in the South, Foxcombe Hall, Boars Hill, Oxford OX1 5HR, UK*

Most information about the nature is carried by two kinds of signals:

(a) generated by the objects under study
(b) modified by them

In an everyday experience, this simply means looking at a light emitting object or looking at an object in a reflected light. In physical sciences, where we are dealing with objects such as particles, atoms, molecules and their ensembles, etc. research techniques based on the second type are often referred to as scattering methods. Sometimes they are applied to study "bulk" properties of large objects (radar techniques would be a good example). Most often, however, particles, or radiation is carefully chosen with the aim to explore particular microscopic characteristics of condensed matter. Such radiation then plays the role of a probe that is brought into interaction with the specimen, and this in turn results in a change of the states of the probe and the specimen alike.

In this book we are limited to studying objects ranging from single atoms (at the smallest) up to large molecular aggregates such as emulsion droplets. As constituents of condensed matter, all these objects can raise questions with respect to:

(a) How large and what shape they are (macromolecular or multimolecular units, polymer chains, etc.)

(b) Where they "are", or how they are arranged,

(c) What they "do", or how they move.

Problems of type (a) are somewhat specific, although they belong to the field of structural research (large scale structures), and can be considered a limiting case for the next type, (b). Problems of type (b) and (c) can be grouped together under a common subject of: *"studying space and time correlations among atoms and atomic units"*.

It is therefore the **correlation functions** that will be at the very heart of our interest throughout this book, although they may be hidden behind a different language or formalism. The wave properties of the probe will make it possible to study these correlations.

Most generally speaking, a correlation function refers to the probability of finding object $A$ at coordinates $\mathbf{x}$ at time $t$ given object $B$ is at $\mathbf{x}_0$ at time $t_0$. If $A$ and $B$ are the same, we are referring to autocorrelation or self-correlation functions. Therefore whenever $A$ and $B$ denote particular atomic, submolecular or molecular species and $A \neq B$, we will be looking towards coherent scattering phenomena. On the other hand, stochastic motions, diffusion and migration phenomena, as well as collective motions in polycrystalline samples (seen as the density of phonon states) will all be viewed through **incoherent scattering.**

The probe-specimen interaction may involve the energy transfer between these two. Whenever this happens, the scattering is *inelastic*, if the probe energy (wavelength) is conserved, the scattering is called *elastic*.

Up to this point, we did not say anything about the type of radiation referred to by the word *probe*, although it is clear that in order to prove experimentally useful, such radiation must have wavelength comparable with atomic distances and in order to study collective excitations, comparable with wavelengths corresponding to the associated quasiparticles such as phonons or magnons, as well as it should carry energies comparable with those of atomic or molecular motions. So our experimental expectations can be summarised in the following table:

TABLE 1.

| For solving structural problems | For solving dynamical problems |
| --- | --- |
| Is capable of deeply penetrating the sample Can "see" light atoms equally well as heavy ones | |
| Has got wavelength comparable with interatomic distances. | Exchanges energy with the sample, that is comparable with the energies of molecular and lattice motions under study |
| Does not deposit much energy in the sample when absorbed (sample survives the experiment) | |
| Is capable of interacting with magnetic moments | |
| to investigate magnetic structures | to investigate magnetic dynamics and relaxation |
| | The probe-sample interaction is tractable theoretically with relative ease so that physical models for properties under study can be easily verified (preferably within the linear response approximation) |

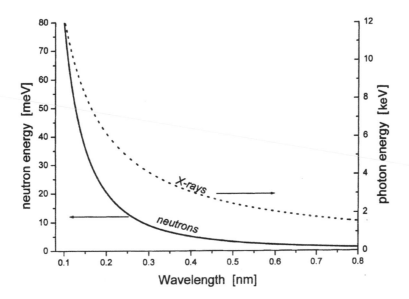

FIGURE 1. Comparison between the energies of a neutron and photon, corresponding to the same wavelength.

With respect to the above, two types of radiation (or associated particles, whichever approach is more suitable in a given case) are considered complementary: electromagnetic radiation (photons) and neutrons, although only neutrons fulfil all the conditions simultaneously. The graph in Figure 1 illustrates this statement.

For quantitative comparison of neutrons and photons, it is in order to recall a few numbers here. Neutron is a chargeless elementary particle with the mass of $m_N = 1.6749286 \cdot 10^{-27}$ kg, and magnetic moment of $\mu_N = 1.04187563 \cdot 10^{-10} \mu_B$ (Bohr magnetons), or in SI units $0.96623707 \cdot 10^{-26}$ J/T. We can produce neutrons whose associated wavelength corresponds to that of X-rays used in crystallography (the characteristic line of Cu-$K_\alpha$ is 0.154 nm), but with energy orders of magnitude smaller than that of X-rays. For example, photon with $\lambda = 0.154$ nm carries the energy of: $E_x = \frac{hc}{\lambda} = 8.048$ keV $= 1.289 \cdot 10^{-15}$ J, while the neutron with the same wavelength has the energy $E_N = \frac{h^2}{2m_N\lambda^2} = 34.467$ meV $= 278$ cm$^{-1}$, i.e. 233500 times smaller. This is the energy of the neutron travelling with the velocity $v = \sqrt{\frac{2E_N}{m_N}} = 2568 \frac{m}{s}$. The above wavenumber value falls within the range of typical molecular vibrational modes as seen by infrared or Raman spectroscopy.

There is a distinct difference between the way neutrons and X-rays are scattered by the matter. X-rays are scattered by the shell electrons. This means that the probability for the scattering event is directly proportional to the atomic number $Z$, hence e.g. hydrogen atoms are hardly "visible". In vibrational spectroscopy, the difference between electromagnetic radiation and neutrons becomes still more important. For neutrons, it is high momentum transfer that enables us to apply a relatively simple theoretical description of the process. On the contrary, in light scattering spectroscopic techniques, where momentum transfers are negligibly small, there is the interaction of electromagnetic radiation with dipole moments that plays important role. This is where the well known selection rules come into play. Neutron scattering is free of such rules.

On the other hand, neutrons interact with matter via two physical processes:

- With atomic nuclei through the strong nuclear interaction
- With magnetic moments of any origin (e.g. electronic orbital magnetic moments).

Because of the nuclear nature of neutron scattering, atoms (via their nuclei) can be considered point scatterers. The probability of a scattering event ("cross section") is a quantity characteristic for a particular isotope. This opens the whole field of isotopic substitution as a method of improving (or even enabling) a great number of experiments. (Hydrogen — deuterium substitution in elastic and quasielastic scattering, "contrast matching" in small angle scattering, being probably the most frequently encountered.)

With respect to their wave properties, neutrons and photons share some general features:

- They both can be polarised with the possible rotation of the plane of polarisation, as it happens with the light passing through optically active media,
- Both exhibit birefringence,
- In both cases there exist refractive indices.

It is worth saying a few words about the last item listed above. For electromagnetic radiation, the refractive indices relative to the vacuum are greater than unity, making it possible for the light to undergo total internal reflection. Neutrons, however, have their refractive indices generally less than one. The corresponding expression for the neutron refractive index is given by (Squires, 1978):

$$n_N = 1 - \frac{1}{2\pi}\rho\lambda^2\bar{b}_{\text{coh}}, \tag{1}$$

where $\rho$ is the number density of atoms (number of scattering nuclei per unit volume), $\lambda$ is the wavelength of the incident neutrons, and $\bar{b}_{\text{coh}}$ is the mean coherent scattering length (see below) for the given nuclei. This refractive index does not depend on the crystal structure, nor does it affect the phase of the incident wave (coherent scattering). It is so because the concept of the refractive index involves forward scattering under small angles (small angle means small momentum transfer, small length of reciprocal space vectors and, consequently, the mentioned properties). This property of the refractive index being less than unity, has far reaching practical consequences: neutrons undergo total "external" reflection (the counterpart of total internal reflection of light). It is therefore possible to build highly effective neutron guides, which are evacuated tubes with their walls covered with appropriate material (usually nickel). An example of a rather more sophisticated use of neutron reflection is a supermirror polariser.

As a result of the neutron-nucleus interaction either of the two events can happen:

- neutron capture, or
- neutron scattering

The first event is important for building neutron detectors because a compound, excited nucleus resulting from neutron capture, can deexcite through the emission of a charged particle or a $\gamma$ quantum, which are easy to count.

In a scattering event, the neutron can exchange both its energy and momentum with the nucleus. In the condensed matter research we typically use neutrons, whose kinetic energy corresponds to the energies of thermal motions of the atoms. Such neutrons are referred to as *thermal*.

For both neutron capture and scattering, we can define measure of probabilities for both events. Let us consider a nucleus which is irradiated with a neutron flux of $I$ particles per second per unit area. Out of this number, $n_a$ get captured, and $n_s$ get scattered. We define coefficients $\sigma_a$ and $\sigma_s$ such that:

$$n_a = I \cdot \sigma_a \quad \text{and} \quad n_s = I \cdot \sigma_s \tag{2}$$

Since $I$ has a dimension of $\frac{1}{s \cdot cm^2}$, and $n$'s have both dimension of $\frac{1}{s}$, it follows that $\sigma$'s are measured in the units of area. Consequently, they are termed *cross sections*. Cross sections are usually expressed in special units, called *barns* (symbol: b). The following definition holds:

$$1b = 10^{-24} \, cm^2, \quad \text{or} \quad 1b = 100 \, fm^2$$

For thermal neutrons, $\sigma_a$ is directly proportional to the neutron wavelength, or inversely proportional to its wavenumber and velocity. This is sometimes referred to as the "$1/v$" law.

In a typical neutron scattering experiment we illuminate the sample with the neutron flux and measure the spatial distribution of the scattered particles, or the change to their energy, or both (some other characteristics may be of interest too, such as the orientation of the neutron spin). In order to be able to talk about probabilities of neutrons being scattered into a solid angle element $d\Omega$ and with the energy transfer of $dE$, the concept of differential cross section or double differential cross section is introduced:

$$\frac{d\sigma}{d\Omega} \quad \text{and} \quad \frac{d^2\sigma}{d\Omega d\omega} = \hbar \cdot \frac{d^2\sigma}{d\Omega d E} \tag{3}$$

It should be pointed out that any spatial distribution of the scattered neutrons results solely from the properties of atomic arrangement within the sample or from their motions. Nuclear scattering of thermal neutrons is isotropic in space. The wavefunction of a scattered neutron of a wavevector **k** at the point **r** can be written as (Squires, 1978):

$$\psi_{sc} = -\frac{b}{r} \exp(ikr) \tag{4}$$

It is characterised by a single parameter $b$, independent of neutron energy or the scattering direction. Indeed, within the Born approximation, such scattering is described by the Fermi pseudopotential (e.g. Squires, 1978; Sköld and Price, 1986):

$$V(\mathbf{r}) = \frac{2\pi \hbar^2}{m} b\delta(\mathbf{r}),$$

where **r** denotes the position of the neutron relative to the nucleus, $m$ is the neutron mass, and $b$ is the same constant, known as the *scattering length*, which — in general — can be a complex number:

$$b = b' - ib''$$

It depends on the spin state of the nucleus-neutron system and on the mass of the nuclide, hence being different for different isotopes of a given element. Consequently, for each element in a real scattering system we refer to a distribution of a scattering lengths. Making use of the Fermi pseudopotential, we can derive the following formula for the double differential cross section (Squires, 1978):

$$\frac{d^2\sigma}{d\Omega dE'} = \frac{k}{k'}\frac{1}{2\pi\hbar}\sum_{j,j'}\overline{b_{j'}^*b_j}\int \langle\exp\left(-i\mathbf{Q}\cdot\mathbf{R}_{j'}(0)\right)\exp\left(i\mathbf{Q}\cdot\mathbf{R}_j(t)\right)\rangle \exp(-i\omega t)dt,$$

(See Fig. 2, *Naming conventions*) where the primed symbols refer to the final state, and $\mathbf{Q} = \mathbf{k} - \mathbf{k}'$ is the scattering vector. Each element of the above sum is weighted by the average product $\overline{b_{j'}^*b_j}$. This comes from the averaging over possible initial and final states that can be encountered in the system. Since the $b$ values of different nuclei are uncorrelated, we have (Squires, 1978):

$$\overline{b_{j'}^*b_j} = |\bar{b}|^2 \text{ for } j' \neq j \quad\text{and}\quad \overline{b_{j'}^*b_j} = \overline{|b|^2} \text{ for } j' = j. \quad (5)$$

With this in mind, we can express the double differential cross section as the sum of two terms, referred to as the coherent and the incoherent one:

$$\left(\frac{d^2\sigma}{d\Omega dE'}\right)_{\text{coh}} = \frac{\sigma_{\text{coh}}}{4\pi}\frac{k}{k'}\frac{1}{2\pi\hbar}\sum_{j,j'}\int \langle\exp\left(-i\mathbf{Q}\cdot\mathbf{R}_{j'}(0)\right)\exp\left(i\mathbf{Q}\cdot\mathbf{R}_j(t)\right)\rangle \exp(-i\omega t)dt$$

(6a)

and

$$\left(\frac{d^2\sigma}{d\Omega dE'}\right)_{\text{inc}} = \frac{\sigma_{\text{inc}}}{4\pi}\frac{k}{k'}\frac{1}{2\pi\hbar}\sum_{j}\int \langle\exp\left(-i\mathbf{Q}\cdot\mathbf{R}_j(0)\right)\exp\left(i\mathbf{Q}\cdot\mathbf{R}_j(t)\right)\rangle \exp(-i\omega t)dt.$$

(6b)

The constants $\sigma_{\text{coh}}$ and $\sigma_{\text{inc}}$ are defined as:

$$\sigma_{\text{coh}} = 4\pi|\bar{b}|^2 \quad\text{and}\quad \sigma_{\text{inc}} = 4\pi\left(\overline{|b|^2} - |\bar{b}|^2\right). \quad (7)$$

The incoherent term accounts for the correlation between the position of the same nucleus at different times, so it is connected with the motion of that nucleus, no matter whether the latter is engaged in some kind of collective vibrations or it undergoes stochastic displacements. On the other hand, the coherent cross section involves correlation between the positions of different nuclei at different times. It therefore includes collective phenomena (as far as dynamics is concerned) and the structural issues. In the language of correlation functions, the latter means spatial correlations at the limit of $t \to \infty$. Obviously, the coherent scattering includes interference effects.

In practice, it is the order of the two operations applied to the scattering lengths, that makes the difference between the coherent and incoherent cross sections: The coherent one results from first taking an average scattering length, and then its square. The incoherent cross section involves averaging of squared scattering lengths.

The quantities commonly used in many formulae, and listed in nuclear data tables, are so called *bound coherent* and *incoherent scattering lengths*. The word *bound* refers to the nucleus being bound, i.e. not free. For free nuclei, the scattering event should be treated in the centre of mass reference frame. If **I** is the spin of the nucleus, and **s** denotes the spin of the neutron, the following holds for the particular spin state of the nucleus-neutron system (e.g. Squires, 1978; Sköld and Price, 1986; see also Sears, 1992):

$$b = b_{\text{coh}} + \frac{2b_{\text{inc}}}{\sqrt{I(I+1)}}\mathbf{I} \cdot \mathbf{s}, \tag{8}$$

$b_{\text{coh}}$ and $b_{\text{inc}}$ being the bound coherent and incoherent scattering lengths.

The scattering lengths and cross sections are related through the following expressions (e.g. Squires, 1978; Sköld and Price, 1986, Sears, 1992):

$$\sigma_s = 4\pi \langle |b|^2 \rangle \quad \text{and} \quad \sigma_a = \frac{4\pi}{k} \langle b'' \rangle \tag{8a}$$

As before, the brackets denote statistical averaging over possible orientations of the nuclear and neutron spins. **k** is the wave vector of the incident neutron. Hence the imaginary part of the scattering length relates to the absorption.

Tables of neutron scattering data can be found in many texts. At the moment of writing, the most up-to-date figures were published in the "Feature section of neutron scattering lengths and cross sections of the elements and their isotopes" in *Neutron News*, 3(3), 29–37, 1992. They are also on the *WWW* available from many sites, e.g. http://ne43.ne.uiuc.edu/n-scatter/n-lengths/.

In Table 2 below we quote nuclear scattering parameters for the most common isotopes.

## 1.0 NAMING CONVENTIONS

A typical neutron scattering experiment can be schematically illustrated in Figure 2 below.

The vector **Q** will be called the *scattering vector*, or the *momentum transfer vector*. The scattering process, in which $|\mathbf{k}_f| = |\mathbf{k}_i|$, is called elastic ($E_f = E_i$), while whenever $E_f \neq E_i$, the process is inelastic. For $E_f > E_i$ it is sometimes called "up-scattering", and for $E_f < E_i$ — "down-scattering". The scattered particles are represented by a spherical wave, with $r_{\text{min}} = b$, $b$ being the scattering length of the nucleus.

## 1.1 STRUCTURE

The term *structural research* refers to the investigation of atomic arrangement within the sample. It covers the whole span of subjects ranging from crystallography to the investigation of large molecular objects such as micelles, clusters or polymer coils. In another words, the quantity being sought is the atomic spatial distribution, or the pair correlation function at

TABLE 2. Bound coherent scattering lengths, $b_{coh}$, bound incoherent scattering lengths, $b_{inc}$, bound coherent scattering cross-sections, $\sigma_{coh}$, bound incoherent scattering cross-sections $\sigma_{inc}$, and absorption cross-sections, $\sigma_{abs}$ (for 2200 m/s neutrons) for selected nuclei.

| Atomic nucleus | Natural abundance | $b_{coh}$ $10^{-15}$ m | $b_{inc}$ $10^{-15}$ m | $\sigma_{coh}$ $10^{-28}$ m$^2$ | $\sigma_{inc}$ $10^{-28}$ m$^2$ | $\sigma_{abs}$ $10^{-28}$ m$^2$ |
|---|---|---|---|---|---|---|
| $^1$H | 99.985 | −3.7406 | 25.274 | 1.7583 | 80.27 | 0.3326 |
| $^2$H | 0.015 | 6.671 | 4.04 | 5.592 | 2.05 | 0.000519 |
| $^3$H | (12.32$a$) | 4.792 | −1.04 | 2.89 | 0.14 | 0 |
| $^{10}$B | 20 | −0.1 − 1.066$i$ | −4.7 + 1.231$i$ | 0.144 | 3 | 3835.(9.) |
| $^{11}$B | 80 | 6.65 | −1.3 | 5.56 | 0.21 | 0.0055 |
| $^{12}$C | 98.9 | 6.6511 | 0 | 5.559 | 0 | 0.00353 |
| $^{13}$C | 1.1 | 6.19 | −0.52 | 4.81 | 0.034 | 0.00137 |
| $^{14}$N | 99.63 | 9.37 | 2.0 | 11.03 | 0.5 | 1.91 |
| $^{15}$N | 0.37 | 6.44 | −0.02 | 5.21 | 0.00005 | 0.000024 |
| $^{16}$O | 99.762 | 5.803 | 0 | 4.232 | 0 | 0.0001 |
| $^{17}$O | 0.038 | 5.78 | 0.18 | 4.2 | 0.004 | 0.236 |
| $^{18}$O | 0.2 | 5.84 | 0 | 4.29 | 0 | 0.00016 |
| F | 100 | 5.654 | −0.082 | 4.017 | 0.0008 | 0.0096 |
| Na | 100 | 3.63 | 3.59 | 1.66 | 1.62 | |
| $^{28}$Si | 92.23 | 4.107 | 0 | 2.12 | 0 | 0.177 |
| $^{29}$Si | 4.67 | 4.70 | 0.09 | 2.78 | 0.001 | 0.101 |
| $^{30}$Si | 3.1 | 4.58 | 0 | 2.64 | 0 | 0.107 |
| $^{46}$Ti | 8.2 | 4.93 | 0 | 3.05 | 0 | 0.59 |
| $^{47}$Ti | 7.4 | 3.63 | −3.5 | 1.66 | 1.5 | 1.7 |
| $^{48}$Ti | 73.8 | −6.08 | 0 | 4.65 | 0 | 7.84 |
| $^{49}$Ti | 5.4 | 1.04 | 5.1 | 0.14 | 3.3 | 2.2 |
| $^{50}$Ti | 5.2 | 6.18 | 0 | 4.8 | 0 | 0.179 |
| $^{106}$Cd | 1.25 | 5.(2.) | 0 | 3.1 | 0 | 1 |
| $^{108}$Cd | 0.89 | 5.4 | 0 | 3.7 | 0 | 1.1 |
| $^{110}$Cd | 12.51 | 5.9 | 0 | 4.4 | 0 | 11 |
| $^{111}$Cd | 12.81 | 6.5 | − | 5.3 | 0.3 | 24 |
| $^{112}$Cd | 24.13 | 6.4 | 0 | 5.1 | 0 | 2.2 |
| $^{113}$Cd | 12.22 | −8.0 − 5.73$i$ | − | 12.1 | 0.3 | 20600.(400.) |
| $^{114}$Cd | 28.72 | 7.5 | 0 | 7.1 | 0 | 0.34 |
| $^{116}$Cd | 7.47 | 6.3 | 0 | 5 | 0 | 0.075 |
| $^{152}$Gd | 0.2 | 10.(3.) | 0 | 13.(8.) | 0 | 735.(20.) |
| $^{154}$Gd | 2.1 | 10.(3.) | 0 | 13.(8.) | 0 | 85.(12.) |
| $^{155}$Gd | 14.8 | 6.0 − 17.0$i$ | (+/−)5.(5.) − 13.16$i$ | 40.8 | 25.(6.) | 61100.(400.) |
| $^{156}$Gd | 20.6 | 6.3 | 0 | 5 | 0 | 1.5(1.2) |
| $^{157}$Gd | 15.7 | −1.14 − 71.9$i$ | (+/−)5.(5.) − 55.8$i$ | 650.(4.) | 394.(7.) | 259000.(700.) |
| $^{158}$Gd | 24.8 | 9.(2.) | 0 | 10.(5.) | 0 | 2.2 |
| $^{160}$Gd | 21.8 | 9.15 | 0 | 10.52 | 0 | 0.77 |

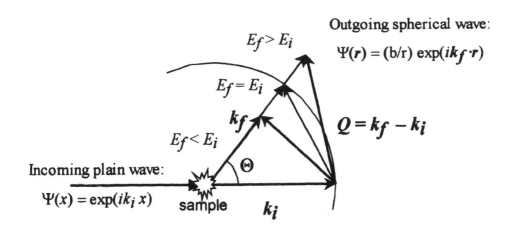

FIGURE 2. Schematic illustration of a neutron scattering geometry and vector relations.

$t \rightarrow \infty$. In a coherent scattering process, the outgoing spherical waves can interfere, thus "translating" the pair correlation function into a measurable function of $\mathbf{Q}$. Indeed, let us assume that the spatial correlation function has the general form of:

$$G(\mathbf{r}) = \int \rho(\mathbf{r})\rho(\mathbf{x}+\mathbf{r})d\mathbf{r}, \tag{9}$$

where for discrete atom distribution we have: $\rho(\mathbf{r}) = \sum_n \delta(\mathbf{r} - \mathbf{r_n})$. Hence the Fourier transform becomes:

$$G'(\mathbf{Q}) = \frac{1}{2\pi} \sum_{n,m} \exp(-i\mathbf{Q} \cdot (\mathbf{r}_m - \mathbf{r}_n)) \tag{10}$$

On the other hand, if the spherical wave, scattered off the nucleus positioned at the point $\mathbf{r}_n$ can be expressed as:

$$\Psi^{sc}(\mathbf{r}) = -b\frac{e^{ikr}}{r} \sum_n \exp(i\mathbf{Q}\mathbf{r}_n), \tag{11}$$

then it may be proved (e.g. Squires 1978, Sköld and Price, 1986) that the elastic coherent differential cross section is:

$$\frac{d\sigma}{d\Omega}(\Theta, \varphi) = (2\pi b)^2 \sum_{n,m} \exp\left(i\mathbf{Q} \cdot (\mathbf{r}_n - \mathbf{r}_m)\right) \tag{12}$$

By comparing equation (10) with (12) we see that by measuring the differential cross section we in fact access the Fourier transform of the spatial correlation function:

$$\frac{d\sigma}{d\Omega}(\Theta, \varphi) = (2\pi b)^2 G'(-\mathbf{Q}) \tag{13}$$

Equation (13) opens up the area of neutron crystallography, the field of perhaps the most successful use of thermal neutron scattering. Through their interaction with magnetic moments neutrons remain the only tool for solving complex magnetic structures. Along with the progress in high resolution instrumentation, we have been witnessing development of data analysis methods. A very popular one dates back to late sixties and has become a standard in the area of structure refinement: the Rietveld method (Rietveld, 1969). These subjects are extensively covered in numerous texts on crystallography, and real life examples are presented in Chapter 2 where further references can be found.

Neutron diffractometers can be built on both steady and pulsed sources. In the first case the conventional $I(2\Theta)$ spectrum is collected. On a pulsed source, on each detector position we measure the time-of-flight spectrum, which for a known flightpath translates into $I(\lambda)$. To increase resolution, very long flightpaths are used, for example the instrument HRPD (High Resolution Powder Diffractometer) at ISIS, Rutherford Appleton Laboratory, stands 100m away from the neutron source.

### 1.1.1 Non-Crystalline Matter: Small Angle Scattering

For elastic, coherent scattering, where $|\mathbf{k_f}| = |\mathbf{k_i}|$, we can solve the $\mathbf{k_i}$–$\mathbf{k_f}$–$\mathbf{Q}$ triangle (cf Figure 2):

$$Q = |\mathbf{Q}| = \frac{4\pi}{\lambda} \sin \frac{\Theta}{2}.$$

Despite the fact that we no longer deal with crystallographic planes, we may still write down the condition for constructive interference in terms of the Bragg's law:

$$2d \sin \left( \frac{\Theta}{2} \right) = n\lambda \quad \text{and we obtain:} \quad d = \frac{2\pi}{Q} \qquad (14)$$

The latter formula expresses the relationship between the "size" of objects under study (be it crystalline interplanar distances or average diameters of macromolecular aggregates) and the corresponding neutron momentum transfer in a diffraction experiment. The larger this size is the smaller momentum transfers are required. This, again, translates into the scattering angles. Very large objects require investigation in an almost direct beam. This field of neutron study is called *small angle neutron scattering (SANS)* or *low-Q scattering*. By convention, SANS means diffraction experiments with $0.05 \text{ nm}^{-1} < Q < 2 \text{ nm}^{-1} (0.005 \text{ Å}^{-1} < Q < 0.2 \text{ Å}^{-1})$.

There exists a technique called *Ultra Small-Angle Neutron Scattering (USANS)* that pushes the above lower limit towards still smaller momentum transfers. A recent development in this field, a Double-Crystal Diffractometer for Ultra Small-Angle Neutron Scattering built at High Flux Isotope Reactor, Oak Ridge National Laboratory allows the minimum measurable scattering vector to be $Q \min = 3.5*10^{-5} \text{ Å}^{-1}$ which corresponds to the distance in real space of $2\pi/Q_{min} \sim 1.8*10^5 \text{ Å}$ (Agamalian, 1997; Drews, 1997).

Typical interatomic distances are too short to be "seen" in a small angle experiment. Instead, we will be speaking of a scattering length density $\delta(\mathbf{r})$, or the average scattering length $\langle b \rangle$ defined for the volume of a size distinguishable in terms of (14). Thus the (complex) scattering amplitude can be written down as:

$$A(\mathbf{Q}) = \int \delta(\mathbf{r}) \exp(-i\mathbf{Q} \cdot \mathbf{r}) d^3\mathbf{r}, \tag{15}$$

The scattering intensity (the one we actually measure in an experiment) is a squared modulus of (15), i.e.

$$I(\mathbf{Q}) = |A(\mathbf{Q})|^2 = \frac{1}{V} \int G(\mathbf{r}) \exp(-i\mathbf{Q} \cdot r) d^3\mathbf{r}, \tag{16}$$

where

$$G(\mathbf{r}) = \frac{1}{V} \int \delta(\mathbf{x}) \delta(\mathbf{x} + \mathbf{r}) d^3\mathbf{x} \tag{17}$$

is the spatial correlation function between average scattering length densities measured at two points, a distance $\mathbf{r}$ apart.

The derivation of equations (16) and (17) was necessary to make a point essential for small angle scattering work. If the objects of interest $O$ are embedded (suspended, dissolved) in medium $M$, then it is the **difference** between average scattering densities of the two, that will provide useful experimental information. This is known as the *contrast*. Were the average scattering length densities $\langle b \rangle_O$ and $\langle b \rangle_M$ equal, we would see nothing in a small angle scattering. Given a complex scattering system, whose scattering intensity could be difficult to interpret, we can therefore sometimes use suitable isotopic substitution to "play around" with the contrast in order to make some parts of the sample invisible.

The integration in (17) is a three-dimensional one, i.e. $G(\mathbf{r})$ contains two types of information: on the shape of scattering object (as defined by the concept of contrast), and on the spatial arrangement of such objects.

In practical terms, the scattered neutron intensity recorded by a particular detector element, (there may be large arrays of position-sensitive detectors), will be proportional to the differential scattering cross section. The proportionality constant will depend on the neutron wavelength, detector efficiency, sample size and its transmission, and are specific to a given experimental setup. Therefore, the quantity being sought is the differential cross section. In view of the above discussion, it can be written (Guinier and Fournet, 1955; Feigin and Svergun, 1987; King, 1999):

$$\frac{d\sigma_{\text{coh}}}{d\Omega}(Q) \propto P(Q)S(Q) \tag{18}$$

For simplicity, cylindrical symmetry is assumed in the above equation. The first term in this expression is known as the *form factor*. It carries the information on the size of the scattering objects. For typical shapes such as a sphere, a disc, a thin rod, or a Gaussian random coil, simple theoretical expressions have been derived (see e.g. King, 1999 and references therein). The second term, $S(Q)$, is *the structure factor*, and is interparticle in nature.

Further details on small angle scattering can be found in many text books, for example (Feigin and Svergun 1987; Higgins and Benoit 1994) or very useful reports and reviews such as (King, 1999; Williams *et al.*, 1994). In this book Chapters 4 and 5 give more detail about models used in SANS and present several examples.

### 1.1.2 Non-Crystalline Matter: The Radial Distribution Functions and Short Range Order

No matter how immensely large the field of neutron crystallography is and how popular is the small angle scattering among biochemists, material technology specialists and physicists engaged in fundamental studies, the field of structural research would be far from complete without methods especially designed for investigating disordered systems. By word "disordered" we mean liquid and amorphous materials, as well as condensed matter lacking long range ordering but exhibiting a short range one, such as some polymers do. This is a very interesting field of research, often producing highly disputed or even controversial findings. Even "model" structureless matter such as amorphous metallic alloys, prove to exhibit short range "preferential ordering" which gives rise to the so called quasi-crystalline structures of otherwise non-existing five-fold symmetry (Wolny, 1993).

Let us re-consider the pair correlation function $G(\mathbf{r})$. Given the scattering centre is positioned at the origin, we can write down the structure factor $S(\mathbf{Q})$. In the absence of any side-effects such as inelastic scattering or attenuation, it is the structure factor that we are measuring, identical to the differential cross-section. If $\rho$ is the number density of atoms, then (Soper *et al.*, 1989):

$$S(\mathbf{Q}) = 1 + \rho \int d\mathbf{r}\big(G(\mathbf{r}) - 1\big) \exp(i\mathbf{Q} \cdot \mathbf{r}), \tag{19}$$

Assuming for simplicity that the sample is isotropic, without any preferred direction, we can reduce equation (19) to the discrete sum in one dimension. Let the indices $\alpha, \beta, \ldots$ denote atomic species present in the sample, and let $f_\alpha$ denote the corresponding atomic fraction, then the total structure factor becomes (Soper *et al.*, 1989; Dore, 1984):

$$S_{\text{total}}(Q) = \sum_\alpha f_\alpha \langle b_\alpha^2 \rangle + \sum_{\alpha,\beta} f_\alpha \langle b_\alpha \rangle f_\beta \langle b_\beta \rangle \big(S_{\alpha,\beta}(Q) - 1\big) \tag{20}$$

As before, angular brackets mean averaging over isotopes and spin states of nucleus-neutron systems. The first term in (20) refers to "single atom" scattering, whereas the second one, called "distinct", is that which actually produces interference effects.

Another difficulty commonly encountered in structural research of hydrogen-rich materials such as polymers, is connected with incoherent scattering on hydrogen nuclei. Substitution with deuterium is not always possible and, strictly speaking, it always renders a *different* scattering system. There does exist an elegant remedy, although experimentally difficult and expensive (both in terms of instrumentation and beam time). This is through the use of polarised neutrons with polarisation analysis of outgoing beam. During the experiment (Schärpf *et al.*, 1990), one makes use of a spin flipper which, with the fixed orientation of both polariser and analysers, allows separate counting of neutrons whose spin has been flipped in a scattering event and those which preserved their spin state. The following relationships hold:

$$I_{\uparrow\downarrow} = \frac{2}{3}I_{\text{spin incoh}}$$

$$I_{\uparrow\uparrow} = I_{\text{coh}} + \frac{1}{3}I_{\text{spin incoh}} + I_{\text{isotope incoh}} \tag{21}$$

Spin-incoherent cross-sections are tabulated and can be made use of, so that simple arithmetics yields pure coherent spectrum. In fact, in actual experimental practice a number of corrections must be carried out, of which the treatment of multiple scattering is by no means a simple task (as double spin-flip is recorded as no spin flip). Nevertheless some new results on short-range order in polymers have been obtained in this way (Schärpf *et al.*, 1990; Lamers *et al.*, 1992; Gabryś *et al.*, 1996).

### 1.1.3  Neutron Specular Reflection — Investigation of Surfaces and Interfaces

Physics of surfaces and interfaces can also be studied by a specific tool among neutron scattering techniques, the *neutron specular reflection*. A smooth surface, illuminated under a grazing angle, will produce a reflected beam (always through a purely coherent scattering). This can be used to detect changes of the refractive index, normal to the interface.

This technique is particularly useful for studying surface adsorption phenomena (e.g. surfactant adsorption, hydrogen adsorption on metallic surfaces), polymer thin films, polymer-polymer interfaces, biological membranes, etc.

Magnetic properties of thin layers can also be studied with neutron reflectometry. Using polarised beam with polarisation analysis allows one to reconstruct the arrangement of magnetic moments in a thin film or surface layer (Fermon, 1995).

## 1.2  DYNAMICS

Investigation of the structure of liquid matter as well as that exhibiting a rich variety of internal motions, inevitably rises questions about dynamic contributions to the scattering, since the scattering will then no longer remain elastic (Soper *et al.*, 1989; Dore, 1994).

Atoms and their ensembles (such as submolecular units or atomic groups) undergo motions of different types. These can be collective or stochastic. Investigation of such motions involve energy exchange between the sample and the probe, i.e. inelastic (or quasielastic) scattering. Stochastic motions are seen in incoherent scattering processes, whereas collective phenomena (quasiparticles such as phonons or magnons) give rise also to coherent scattering.

### 1.2.1  Coherent Inelastic Scattering

In order to understand the principle of coherent scattering from collective motions let us recall a textbook example of light diffraction from an irregularly ruled grating (the distance between lines varies sinusoidally). In such a case the main diffraction pattern is accompanied by a weak signal known as *ghosts* (cf. Squires, 1978). A somewhat similar situation occurs when there is a regular wave of collective displacements of atoms in a crystal lattice. Such a wave, characterised by a wavevector $\mathbf{q}$, represents a phonon travelling in the direction $\mathbf{q}$ with the velocity $\omega/q$. The condition for constructive interference can then be obtained from the momentum conservation rule in a way analogous to that used in Bragg scattering:

$$\mathbf{k} - \mathbf{k}' = \tau \pm \mathbf{q}, \tag{22}$$

where $\mathbf{k}$ and $\mathbf{k}'$ refer to the neutron momentum before and after scattering, and $\tau$ is the reciprocal lattice vector. The plus or minus signs correspond to the creation or annihilation of a phonon in the particular normal mode. The above qualitative description applies to the so called one-phonon scattering. It can be further generalised to the multiphonon case. With this concept in mind, the expression (6) for the coherent cross section can be transformed to represent the double differential coherent cross section for the creation or annihilation of a phonon in a general crystal lattice. Derivation of these quite complex formulae is beyond the scope of this chapter and can be found in many texts on neutron scattering (see e.g. Lovesey, 1984).

Collective motions in crystals were studied theoretically long ago, but could not be measured experimentally before the advent of neutron scattering methods. Although some integrated information could be obtained by e.g. adiabatic calorimetry, neutrons proved unique in measuring the *phonon dispersion relations*. The dispersion relation is the dependence of the phonon energy (or frequency) upon its momentum. Such curves are measured down typical characteristic directions in the Brillouin zone. The experiment is performed on an oriented single crystal by illuminating it with a monoenergetic neutron beam. Then the energy distributions of scattered neutrons are measured in various directions. Many elegant and important experiments of this type were already performed in mid-sixties. An example of phonon dispersion relations is given in Figure 3.

### 1.2.2 *Incoherent Inelastic Neutron Scattering (IINS)*

Molecular and lattice vibrations are more often studied through incoherent inelastic scattering. This relatively simple experimental technique, sensitive to correlated and uncorrelated vibrations alike, relies on the following principle: Imagine the sample is irradiated with a beam of monoenergetic (a few eV) neutrons. Upon interaction with the sample neutrons gain energy (become "heated up") from intramolecular and lattice vibrations. In this way the neutron energy gain spectrum will reflect the phonon density of states function $g(\omega)$. The more vibration states are occupied in the vicinity of a given frequency, the more likely is the corresponding neutron energy gain. In practice, whenever hydrogenous samples are being investigated, we speak of *hydrogenous* phonon density of states, since the most information comes from these nuclei due to their large inelastic scattering cross section. Of course, a symmetrical picture can be drawn for the case of the neutron energy loss, when the lattice is heated up.

For a given crystal lattice, an expression can be derived to link the double differential incoherent cross section to the phonon density of states. This derivation is "lattice-symmetry dependent" because it includes averaging of the squared products $(\kappa \cdot e_n)^2$ for each normal mode $n$, where $\kappa = \mathbf{k}' - \mathbf{k}$, and $e_n$ is the polarisation vector.

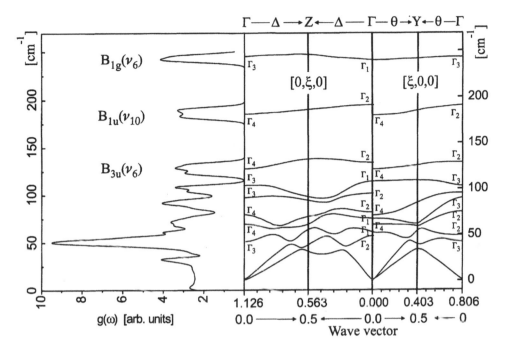

FIGURE 3. Phonon dispersion relations and calculated phonon density of states for crystalline biphenyl. (Courtesy of Dr I. Natkaniec)

In the simple case of a one phonon up-scattering (neutron energy gain) in a cubic crystal, and under harmonic approximation we have:

$$\left(\frac{d^2\sigma}{d\Omega dE'}\right)_{\text{inc}} = \frac{1}{4\pi}\frac{k'}{k}\frac{\kappa^2}{4M}\sum_j \sigma_j^{\text{inc}} \exp(-2W_j)\frac{g_i(\omega)}{\omega}\left(\coth\left(\frac{1}{2}\hbar\omega\beta\right)-1\right) \quad (23)$$

The sum runs over all atoms in the molecule. $M$ stands for the mass of the molecule, $W_j$ is the Debye-Waller factor, $\beta$ is Boltzmann reduced temperature. This formula is derived in many texts on neutron scattering. Figure 3 illustrates the relationship between the phonon dispersion relations, phonon density of states, and the actually measured IINS spectrum in a well known molecular crystal of biphenyl.

The phonon density of states can be obtained from the dispersion relations by their integration, thus losing any information on the phonon momenta. Note a distinct separation of the "phonon area" or "external vibrations range" and the "internal vibrations range" in the IINS spectrum. Examples of applications of the IINS to advanced materials and polymers can be found in Chapter 9.

Lattice and intramolecular vibrations are by no means the only "dynamical phenomena" accessible to the IINS method. With a high energy resolution instrument we can explore the spectrum area close to the elastic line. This enables us to study tunnelling phenomena (an oscillator vibrating in a double-well potential with the wells separated with a not-so-high barrier has got its energy levels split. The lower the barrier is the greater the splitting).

Through inelastic scattering with very small energy transfers we entered the world of *"almost-elastic scattering"*. Now we are going to take a closer look at it.

### 1.2.3 Quasielastic Neutron Scattering (QENS)

Let us recall formula (6b) for the double differential cross section for incoherent scattering. As said above, this quantity depends on the spatial correlations between the positions of the same nucleus at different times, i.e. on the self-pair correlation function. Quite often these positions are known, and the probability can be calculated that a given nucleus jumps from one position to another, viz. instantaneous or stochastic rotation of the $CH_3$ group about its 3-fold axis. Similarly, the diffusion is theoretically tractable, either linear (Brownian motion) or rotational (atoms on a circle or sphere).

The subject of quasielastic neutron scattering is covered in the comprehensive text by M. Bée (1994). Here we only sketch the main concepts behind this method.

Since we are dealing with stochastic motions, there is no correlation between the positions of a given nucleus at $t = 0$ and $t = \infty$. Hence the correlation function in (6b) becomes (Bée, 1994):

$$G(r, \infty) = \frac{1}{N} \sum_j \int \langle \delta(\mathbf{r} - \mathbf{r}') + \mathbf{R}_j(0) \rangle \langle \delta(\mathbf{r}') - \mathbf{R}_j(0) \rangle d\mathbf{r}' \qquad (24)$$

The pair correlation function is a space-Fourier transform of the so called *intermediate scattering function* $I_{inc}(\mathbf{Q}, t)$. The latter, in turn, is a time-Fourier transform of the familiar scattering function $S_{inc}(\mathbf{Q}, \omega)$. The intermediate scattering function can be separated into its time-dependent and time-independent parts (Bée, 1994):

$$I_{inc}(\mathbf{Q}, t) = I_{inc}(\mathbf{Q}, \infty) + I'_{inc}(\mathbf{Q}, t) \qquad (25)$$

Making use of (24) and taking both Fourier transforms, we arrive at:

$$I_{inc}(\mathbf{Q}, \infty) = \frac{1}{N} \sum_j |\langle \exp(i\mathbf{Q} \cdot \mathbf{R}_j) \rangle|^2 \qquad (26)$$

and

$$S_{inc}(\mathbf{Q}, \omega) = I_{inc}(\mathbf{Q}, \infty)\delta(\omega) + S_{inc}^{qel}(\mathbf{Q}, \omega) \qquad (27)$$

In this way the scattering function has been separated into its elastic part and a second, called *quasielastic*, superimposed upon it. The elastic part (26) in (27) vanishes for translational (Brownian) diffusion, and it is a characteristic feature for this type of motion. If the reorientation of a given nucleus is confined to a circle or sphere (be it instantaneous jumps over a discrete set of sites or rotational diffusion), then we have an non-zero elastic component. The type of motion under study can then be determined by measuring the relative contribution of the elastic line, the Elastic Incoherent Structure Factor (EISF), as a function of momentum transfer, and by detailed investigation of the shape of quasielastic broadening. This is typically done by constructing a theoretical model for the motion, calculating the self-pair correlation function, and then the model scattering function. The latter is convolved with the instrumental function and least-squares fitted to the neutron scattering spectra.

This is an elegant method of studying stochastic motions of molecules and molecular groups, although by no means experimentally easy or fast in data processing. Real-life QENS spectra are often contaminated with Bragg reflections from the sample or affected by multiple scattering. Purposefully applied, it can answer many questions on reorientation processes in molecular crystals, liquid crystals, crystals with isostructural phase transitions. It proved useful in studying migration processes in e.g. porous materials, superionic conductors, other non-metallic conductors, or of hydrogen in metals. In this book, some application of QENS to polymers is presented in Chapter 8.

### 1.2.4  Neutron Sources and Instrumentation

Neutrons are elementary particles that "live" inside the atomic nuclei. A free neutron (i.e. one that is not incorporated in a nucleus) has a mean lifetime of $\tau_{1/2} = 886.7 \pm 1.9$ s and dies through $\beta$ decay: $n \rightarrow p + e^- + \bar{v}_e$. For scattering experiments, neutrons must be therefore produced by carrying out appropriate nuclear reactions. A few such reactions are known, of which two are most frequently used: nuclear fission, schematically illustrated in Figure 4, and spallation, sketched in Figure 5.

To date, intense beams of low energy neutrons have been most commonly produced in *moderated fission reactors*, optimised for this purpose (high neutron brightness). Nuclear reactors differ greatly in design, so we shall limit ourselves to mere mentioning some issues they share. A fissile nucleus is capable of capturing a stray neutron (incrementing its mass number by one), and subsequently breaking into highly excited *fission fragments*. Neutrons of the energy in the range of megaelectronvolts are evaporated from these fragments. Uranium $^{235}$U is probably the most popular fissile isotope, $^{233}$U, $^{239}$Pu are other examples. The cross section for the neutron capture with subsequent fission, is a steep function of the energy of the impinging particle. In the case of uranium $^{235}$U, this cross section $\sigma_f = 4$b for $E_n = 10$ keV, whereas $\sigma_f = 579$b for $E_n = 0.0253$ eV.

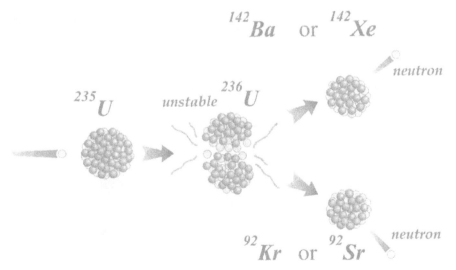

FIGURE 4. Schematic illustration of neutron production in a $^{235}$U fission reaction.

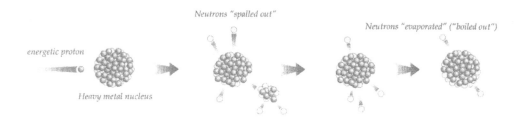

FIGURE 5. Schematic illustration of neutron production in a nuclear spallation reaction.

(Thorium $^{232}$Th is a contrary example, it fissions when hit by a fast neutron, and $^{238}$U requires neutrons above the threshold energy of 1.1 MeV, below which it is not fissile.) The neutron released following the fission reaction of $^{235}$U, has the energy of about 2 MeV. If the particular reactor operates on thermal or epithermal neutrons, a *moderator* is required to slow down the neutrons. There exist reactors operating on fast neutrons. Such reactors are called *fast reactors*.

Research reactors are typically *steady neutron sources*: they provide a constant flux of neutrons. An interesting exception is the fast pulsed reactor IBR-2 at Joined Institute for Nuclear Research in Dubna, Russia. This machine is equipped with rotating neutron reflectors and becomes critical only when the reflectors meet against the nuclear fuel.

There exists an alternative way of producing neutrons that can win the performance competition with the best research reactors: **nuclear spallation**. When a high energy proton (> 800 MeV, typically around 1 GeV) hits a heavy metal nucleus (such as tungsten, tantalum, lead or uranium) some neutrons become "knocked out" of the nucleus. Moreover, some more neutrons may be evaporated from a highly excited nucleus while the latter deexcites (see Figure 5).

The spallation reaction can produce between 20 and 30 neutrons from a single nucleus. More precisely, the number $N_n$ of neutrons with mean energy $T$ produced from a single nucleus after a collision with proton, whose energy is $E_p$ equals:

$$N_n + \frac{E_p}{S_n + T},\tag{28}$$

where $S_n$ is the reparation energy. For a given target, a semi-empirical formula is used:

$$N_p = a \cdot (A + 20) \cdot (E_p - b)\tag{29}$$

where $A$ is the mass munber of the target nucleus, and $a$ and $b$ are constants dependent on the shape the target. The tantalum target at ISIS (Rutherford Appleton Laboratory) gives 25 neutrons per 800 MeV proton.

A schematic layout of a spallation neutron source is shown in Figure 6. H$^-$ ions are preaccelerated in a linear accelerator. Then they are directed onto a stripper foil where electrons are stripped off. Remaining protons are injected into a synchrotron (which may sometimes also work as a storage ring). When they reach their final energy, they are

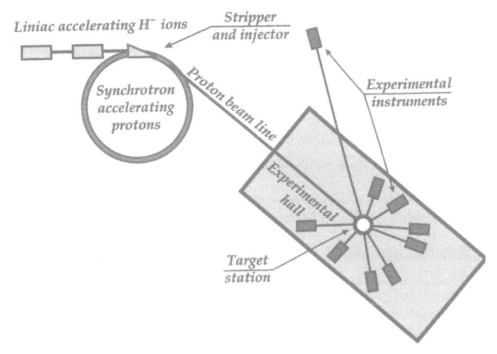

*Liniac accelerating H⁻ ions*

*Stripper and injector*

*Synchrotron accelerating protons*

*Proton beam line*

*Experimental instruments*

*Experimental hall*

*Target station*

FIGURE 6. A typical layout of a pulsed (accelerator) neutron scattering facility.

extracted from the ring and directed onto the *target station*, where the neutrons are actually produced. Typically, a neutron spallation source, alternatively called an *accelerator source*, as described above, is a pulsed one. Emitting neutrons in very short bursts naturally favours instruments using time-of-flight technique of analysing energy (or neutron wavelength). However, the world's first steady spallation source has already been built. It is the *Swiss Spallation Neutron Source*, or *SINQ*, located at the Paul Scherrer Institute, capable of supplying a flux of about $10^{14} ns^{-1}$ cm$^{-2}$.

In any case, neutrons of desired energy (wavelength) are obtained by bringing them to thermal equilibrium with the appropriate moderator (such as water, liquid methane or liquid hydrogen (deuterium)). For the given moderator temperature $T$ the wavelengths or energies of the neutrons are governed by the Maxwellian distribution. If $N$ denotes the number of neutrons in the unit volume, passing through the moderator in the unit time, then the number of neutrons with the velocity between $v$ and $v + dv$ is given by:

$$N(v) = \left(\frac{m_n}{2\pi k_B T}\right)^{\frac{3}{2}} \exp\left(-\frac{m_n v^2}{2 k_B T}\right) v^2 \tag{30}$$

This distribution has a maximum value $v_0$ and root mean squared value $v_{\text{rmsq}}$ at:

$$v_0 = \sqrt{\frac{2 k_B T}{m_n}} \quad \text{and} \quad v_{\text{rmsq}} = \sqrt{\frac{3 k_B T}{m_n}} \tag{31}$$

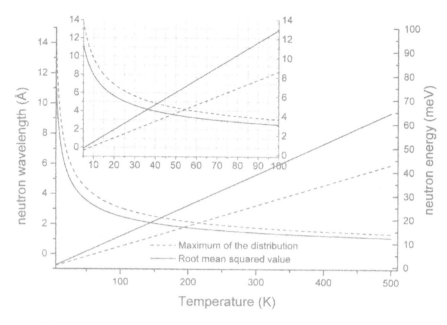

FIGURE 7. Neutron wavelength and energy corresponding to the maximum (dashed line) and root mean squared value of the Maxwellian distribution of neutron velocities at typical temperatures used in neutron scattering experiments. Low temperature range is expanded in the inset.

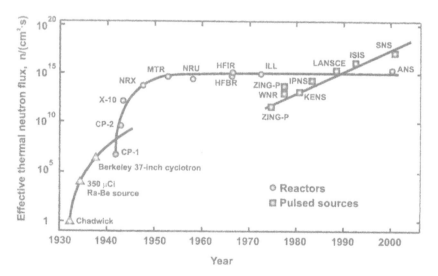

FIGURE 8. Performance of most important research neutron sources. Data published in the WWW (http://www.ornl.gov/sns/sns_brochure/sns_brochure.pdf), Spallation Neutron Source Brochure. (This figure, previously published in *Spallation Neutron Source: The Next-Generation Neutron Scattering Facility for the United States*, ORNL/M-6369, Oak Ridge National Laboratory, 1988, was adapted from *Neutron Scattering*, ed. K. Sköld and D.L. Price, Academic Press, 1986.)

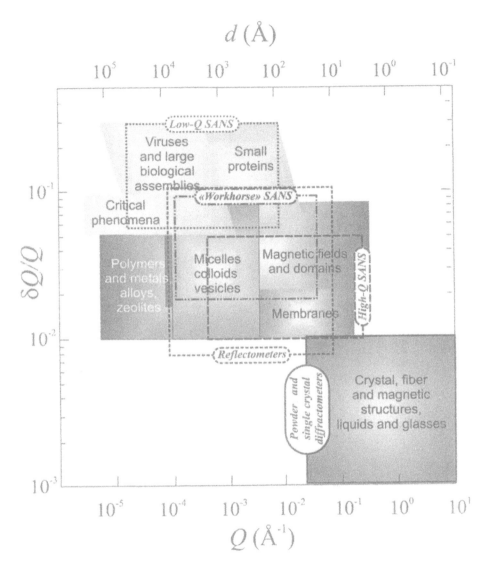

FIGURE 9. Schematic map of structural research interest areas and corresponding coherent neutron scattering methods. A compilation of pictures published in the WWW in: http://www.ornl.gov/jins/lss.pdf

Neutron wavelengths and energies corresponding to the above $v_0$ and $v_{rmsq}$ are shown in Figure 7.

The performance of the existing research neutron sources is compared in Figure 8.

Contemporary neutron scattering facility contains a collection of highly optimised instruments of considerable complexity as well as ancillary equipment. At the receiving end of the instrument, these are: neutron guides, choppers, filters, collimators and counters.

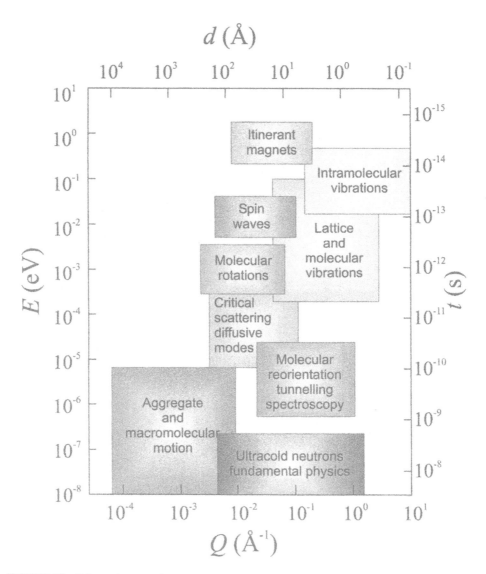

FIGURE 10. Schematic map of dynamical research interest areas, pictures published in the WWW in: http://www.ornl.gov/jins/lss.pdf

Sample environment equipment comprises cryostats, furnaces, magnets, etc. Data acquisition electronics and dedicated computer support conclude the list.

Most important research fields with their characteristic areas of momentum transfer and relative resolution or energy transfer are schematically shown in Figures 9 and 10.

Figure 9, which illustrates structural research (diffraction techniques), also tells us what type of instrument to look for in the particular research field. Issues covered by Figure 10

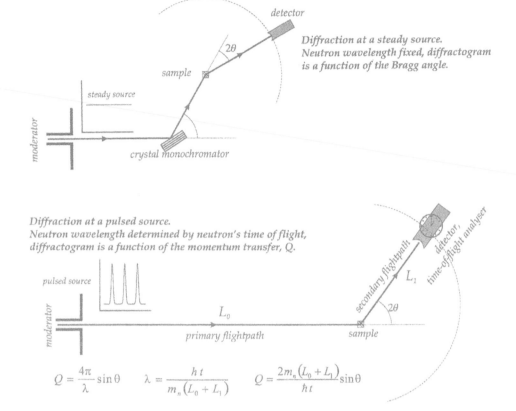

FIGURE 11. Comparison of diffractometers operating at a steady and pulsed sources. A schematic illustration.

are studied on inelastic scattering instruments, and devices especially built to the specification of a given energy transfer/momentum transfer area should be looked for.

Neutron scattering instruments differ with respect to the type of neutron flux they receive, i.e. whether they operate at a steady or pulsed source. To illustrate the difference, we present a comparison of two diffractometers: a classical one, with the fixed wavelength incident beam (steady source), and the one operating on a pulsed source, where neutron wavelength is determined by its time of flight. A schematic, very simplified picture is shown in Figure 11.

A similar comparison of two instruments for the dynamical analysis of single crystals (inelastic scattering) can be found in Figure 12. A detailed review of neutron scattering instrumentation can be found in the text *Neutron Scattering* by Sköld and Price, 1986.

In order to get a feeling of how the instruments are optimised for particular needs, in Figure 13 we compare three designs of inelastic scattering spectrometers. The well known IN10 at ILL, Grenoble is a high resolution backscattering machine operating on 2.080 eV neutrons ($\lambda = 6.271$ Å), with energy transfers $\varepsilon = \pm 1.5\mu$ eV. Incident neutron wavelength can be modified with a Doppler drive or cryofurnace.

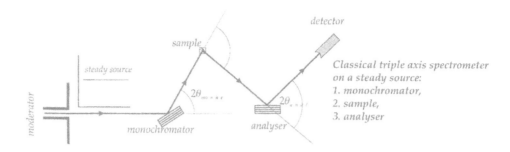

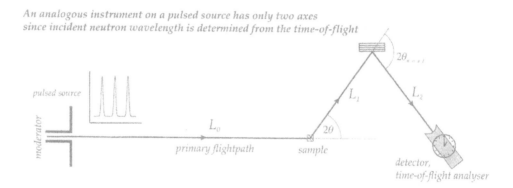

FIGURE 12. Comparison of inelastic scattering instruments (crystal analysers) operating at a steady and pulsed sources. A schematic illustration.

A general purpose inelastic and quasielastic spectrometer, NERA-PR, is installed at the pulsed reactor IBR-2 at Dubna. It is an *all-in-one* instrument, specially designed for the study of molecular motions in orientationally disordered phases (molecular crystals, liquid crystals etc.) with simultaneous control of crystal phase parameters (diffraction). This is an inverted geometry spectrometer that can operate in the inelastic or quasielastic scattering modes.

Finally, a schematic diagram of a new high sensitivity inelastic scattering instrument, TOSCA (a **T**hermal **O**riginal **S**pectrometer with **C**ylindrical **A**nalysers), recently installed at the ISIS spallation source at the Rutherford Appleton Laboratory. This machine replaces TFXA (**T**ime-**F**ocused **C**rystal **A**nalyser), an inelastic scattering spectrometer, inheriting positive features of its predecessor and significantly improving sensitivity.

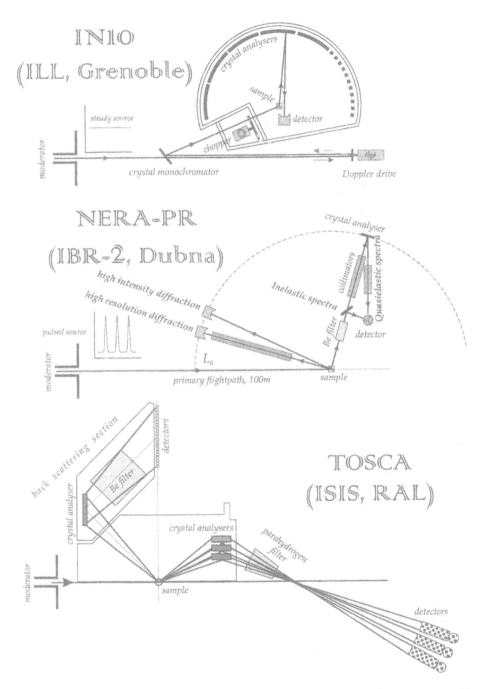

FIGURE 13. The principles of operation of three different inelastic scattering instruments. Note, how they are optimised for resolution (IN10), versatility (NERA-PR) and efficiency (TOSCA).

## 1.3 SUMMARY

In this briefest of introductions, we intended to outline the fundamentals of neutron scattering and provide some sign-posts to the subject matter. This accounts for a short reference list, since it was not possible to do justice to such wide and burgeoning field as neutron scattering. More choice of literature can be found in other chapters in this book. Hence we included either the celebrated textbooks or most recent reviews or simply our favorite references, for which we do not apologize.

## References

Agamalian, M., Wignall, G.D. and Triolo, R. (1997) *J. Appl. Cryst.* **30**, 345.
Bée, M. (1994) *Quasielastic Neutron Scattering*, Adam Hilger, Bristol and Philadelphia.
Dahlborg, U., Lovesey, S.W. (1997) Physics and Chemistry of Materials from Neutron Diffraction and Spectroscopy, *RAL Report* **RAL-90-077**.
Dore, J.C. (1984) Structural Studies of Molecular Liquids by Neutron and X-Ray Diffraction in *Molecular Liquids — Dynamics and Interactions*, Barnes, A.J., 383–409.
Drews, A.R, Barker, J.G., Glinka, C.J., Agamalian, M. (1997) *Physica B*, **241**, 189.
Feigin, L.A., Svergun, D.I. (1987) *Structure Analysis by Small-Angle X-Ray and Neutron Scattering*, Plenum Press, New York and London.
Fermon, C. (1995) *Physica B*, **213**, 910–913.
Guinier, A., Fournet, G. (1995) *Small Angle Scattering of X-rays*, John Wiley, New York, 1955.
Higgins, J.S., Benoit, H.C. (1994) *Polymers and Neutron Scattering*, Clarendon Press, Oxford.
Higgins, J.S., Bucknall, D.G. (1997) Neutron reflection studies of polymer-polymer interfaces, *RAL Report*.
King, S.M. (1999) Small-angle neutron scattering Chapter 7 in *Modern Techniques for Polymer Characterisation*, Pethrick, R.A. and Dawkins, J.V. (Eds) John Wiley & Sons Ltd, New York.
Lovesey, S.W. (1984) *Theory of Neutron Scattering from Condensed Matter*, Clarendon Press, Oxford.
Lovesey, S.W. (1993) Basic Concepts for the Interpretation of Neutron Scattering Experiments at an Atomic Level of Description, *RAL Report*, **RAL-93-058**.
Sears, V.F. (1992) Neutron Scattering Lengths and Cross Sections, *Neutron News*, **3**, 29.
Schärpf, O., Gabrys, B. and Peiffer, D.G. (1990) Short-Range Order in Isotactic, Atactic and Sulphonated Polystyrene Measured by Polarised Neutrons (Parts A and B) *ILL Report* **90SC26T**.
Sköld, K. and Price, D.L. (1986) *Methods in Experimental Physics: Neutron Scattering* (Celotta, R. and Levine, J.), Academic Press, London and New York.
Soper, A.K., Howells, W.S. and Hannon, A.C. (1989) ATLAS — Analysis of Time-of-Flight Diffraction Data from Liquid and Amorphous Samples, *RAL Report* **RAL-89-046**.
Squires, G.L. (1978) *Introduction to The Theory of Thermal Neutron Scattering*, Cambridge University Press.
Rietveld, H.M. (1969) *J. Appl. Cryst.*, **2**, 65.
Williams, C.E., May, R.P. and Guinier, A. (1994) Small-angle scattering of x-rays and neutrons in *Material Science and Technology*, Cahn, R.W. (Ed.), VCH Verlagsgesellschaft, Weinheim, **2B**, 613653.
Wolny, J. (1993) Why do quasicrystals have sharp diffraction peaks? *J. Non-Crystalline Solids*, **153**, 293 and references therein.

# 2. Molecular Structures Determined By Neutron Diffraction

NORMAN SHANKLAND[1], ALASTAIR J. FLORENCE[1], CHICK C. WILSON[2] and KENNETH SHANKLAND[2]

[1] *Department of Pharmaceutical Sciences, University of Strathclyde, Glasgow G4 0NR, UK*
[2] *ISIS Facility, Rutherford Appleton Laboratory, Chilton, Didcot, Oxon, OX11 0QX, UK*

## 2.1 INTRODUCTION

### 2.1.1 General Background

Neutron and X-ray diffraction techniques are widely used to study the molecular and crystal structure of organic compounds. Their power stems from their ability to 'see' atoms in 3-dimensional space, and the techniques are complementary in the sense that X-ray diffraction is used routinely for structure *solution*, while neutron diffraction has clear advantages in the case of structure *refinement*. The fundamentals and applications of neutron diffraction have already been covered comprehensively (Bacon, 1975a) and concisely (Biggin, 1986).

This chapter concerns crystal structures determined from measurements of diffracted Bragg intensities, where the molecules are of a size typically less than *ca.* 100 atoms. Only organic structures are covered specifically.

We start with the following question: if experimental measurement of accurate structural parameters is required, why should we consider neutron diffraction?

There is a good, fundamental answer to this question — neutrons are scattered by atomic nuclei, and therefore reveal directly the positions and displacement amplitudes of the nuclei in a molecule. Since the coherent neutron scattering lengths, $b$, for the elements commonly encountered in organic molecules fall in a relatively narrow range, neutron diffraction is very effective for determining the positions of $^1$H- and $^2$H-atoms ($b = -3.7406$ and 6.671, respectively) in the presence of heavier atoms such as $^{12}$C ($b = 6.6511$), $^{14}$N ($b = 9.37$), $^{16}$O ($b = 5.803$) and $^{32}$S ($b = 2.804$) ($b$ in units of $10^{-15}$ m; data taken from Sears (1992)).

The sign of the neutron scattering length of $^1$H (negative) can, in itself, be useful for distinguishing $^1$H-atoms from other atoms in the same molecule which have a positive scattering length, as the following three examples illustrate: (a) assignment of the absolute configuration of $\alpha$-monodeuterioglycolic acid (Johnson *et al.*, 1965); (b) refinement of the site occupancy factors of $^1$H- and $^2$H-atoms in deuterated dopamine (Table 1); (c) interpretation of the complex, disordered hydration state in vitamin B$_{12}$ coenzyme (Finney, 1995).

It is also worth stating at this point that if non-ambient sample environments are required, variable temperature and pressure experiments are easier to perform on a neutron beamline than they are on an X-ray beamline. For single-crystal neutron diffraction, the SXD instrument at ISIS (Keen and Wilson, 1996) offers 5 K to 1300 K. Instrument D10 at the Institut Laue-Langevin at Grenoble offers 0.1 to 1000 K with full 4-circle geometry, and pressures up to 27 kbar in 2-axis mode. The application of non-ambient sample environments in powder diffraction analysis of phase transformations is particularly useful, and is covered in Section 2.4.4.

### 2.1.2   Do We Need Accurate Molecular Crystal Structures?

"It is a striking observation that the level of understanding of biological structure and function has, over the past three decades, been directly proportional to the level of crystallinity of the molecules involved" (Jeffrey and Saenger, 1991a).

In other words, crystal structure solutions enable a high level of understanding of molecular structure. In the case of small molecules, this understanding comes from accurate determination of geometry, atomic displacements and electron density distribution (see Section 2.4.).

It has been argued that studying the crystal structures of small molecules can shed insights into the mechanism of ligand-macromolecule interactions in the body, when direct evidence on the mechanism is not readily available (Glusker *et al.*, 1994b). One way to do this is to use the Cambridge Structural Database (CSD, Allen, 1992; Allen and Kennard, 1993; Allen, Kennard and Watson, 1994) to retrieve the crystal structures and analyse them with a view to identifying preferred directions for non-covalent interaction between functional groups of interest (Glusker *et al.*, 1994b). For example, the positive charge on the quaternary ammonium neurotransmitter acetylcholine (ACh) is known to be important to its action at receptor sites in the body. The implication is that the cation interacts with complementary electronegative moieties on the receptor. Rosenfield and Murray-Rust (1982) chose to model ACh-receptor interactions by analysing the distribution of halide ions (Cl$^-$, Br$^-$, I$^-$) in 34 crystal structures of molecules containing a quaternary ammonium group. The analysis clearly revealed preferred cation-anion orientations, leading the authors to advance the hypothesis that preferred orientation is also likely in ACh-receptor binding.

A related approach to analysing directional preferences utilises the 'IsoStar' database (Bruno *et al.*, 1997). IsoStar is a 'secondary source' in the sense that it is a comprehensive compilation of information on non-covalent interactions derived from three 'primary sources': the CSD, the Brookhaven Protein Data Bank, and molecular orbital calculations.

**TABLE 1.** Refined site occupancy factors for $^1$H and $^2$H in deuterated$^\dagger$ dopamine deuteriobromide (see Section 2.3.2. for refinement details).

| atom | site occupancy |
|------|----------------|
| $^2$H$_A$ ($^2$H$_B$) | 0.559(4) |
| $^1$H$_A$ ($^1$H$_B$) | 0.441(4) |
| $^2$H$_C$ ($^2$H$_D$, $^2$H$_E$) | 0.918(3) |
| $^1$H$_C$ ($^1$H$_D$, $^1$H$_E$) | 0.082(3) |
| $^2$H$_F$ | 0.929(7) |
| $^1$H$_F$ | 0.071(7) |
| $^2$H$_G$ | 0.878(5) |
| $^1$H$_G$ | 0.122(5) |

$^\dagger$Note that the molecule is not perdeuterated. As a consequence of the synthetic pathway employed, there is a 50% probability of either H$_A$ or H$_B$ being an $^1$H-atom. This does not pose a problem for the structure refinement, and the refined occupancies for these positions are close to the expected 50% value. There is also evidence of exchange of $^1$H for $^2$H on the N- and O-atoms. Strict constraints were used to reduce the number of independent variables in the least-squares refinement, and details are given in Shankland *et al.* (1996).

An example of the application of IsoStar is provided by Cole *et al.* (1998) in their analysis of directional preferences in the intermolecular interactions of hydrophobic groups. The CSD has also proved valuable in molecular conformational analysis (see Shankland *et al.* (1998a) for a summary of references), and in the study of chemical reaction paths (see Glusker *et al.* (1994c) for a summary of references).

In addition to what has just been said in relation to the importance of crystal structures, it is also true to say that diffraction studies contribute to the development of theoretical methods for structure determination in two ways: (a) accurate experimental structures enable validation of *ab initio* calculations (Section 2.4.2.); (b) molecular mechanics and semi-empirical methods are parameterised using experimental data, much of which is derived from crystal structures. As the number and quality of experimental structures increases, so the theoretical methods become more widely applicable and reliable. X-ray diffraction is longer established than neutron diffraction, and provides by far the greatest source of crystal structure data (compare Figures 1a and 1b). The fact that less than 1% of the structures in Figure 1a are derived from neutron data also reflects the fact that X-rays are preferred for routine structure solutions, because X-ray diffractometers have a size and operational requirements which allow them to be run within the laboratory environment.

## 2.2 SINGLE-CRYSTAL NEUTRON DIFFRACTION

### 2.2.1 General Background

Diffraction of neutrons was first demonstrated in 1936, and the first neutron diffractometer was built in 1945. In the first decade of neutron diffraction studies, reactors were delivering *ca.* $2 \times 10^{12}$ neutrons $cm^{-2}$ $s^{-1}$ maximum. By 1960 this figure had increased by a factor of 20, and by 1974 fluxes in excess of $10^{15}$ were available (Bacon, 1975b). These increases in reactor output were significant to the development of the single-crystal method, because neutrons are scattered relatively weakly, and high fluxes enabled adequate counting rates to be achieved from single-crystals with a volume sufficiently small to avoid large errors arising from secondary extinction (Bacon, 1975e; for a description of extinction effects see, for example, Glusker *et al.*, 1994a and Bacon, 1975c).

On modern-day instruments, crystals of volume 2–5 $mm^3$ (very approximately) will suffice for a neutron single-crystal diffraction experiment (*cf.* 180 $mm^3$ in the first ever study of an aromatic molecule, $\alpha$-resorcinol, by neutron diffraction (Bacon and Curry, 1956)). Encouraging a crystal to grow to this size is often straightforward, and common ways to go about this have been summarised by Etter *et al.* (1986) and Shankland *et al.* (1998b). When difficulties are encountered, it is rarely obvious why, and the usual way forward is an empirical sampling of the different techniques and conditions. If compound is in short supply, the small-scale recrystallisation apparatus shown in Figure 2 is worth considering. The method is suited to recrystallising very small amounts of material (0.005 g), and Watkin (1972) has reported growing crystals of *ca.* 1 mm edge in one to seven days.

One reason for the small number of neutron structures containing >100 atoms (Figure 1b) is the difficulty often encountered in growing suitable crystals of larger molecules such as oligonucleotides, oligopeptides and oligosaccharides (Jeffrey and Saenger, 1991b). The cyclodextrins are the notable exception in that they crystallise readily as hydrates and inclusion compounds, and much of what is known about their complex hydrogen bonding patterns has been derived from single-crystal neutron refinements (Ding *et al.*, 1991; Jeffrey and Saenger, 1991c; Steiner and Saenger, 1991; Steiner *et al.*, 1991; Steiner and Saenger, 1992a,b,c; Steiner and Saenger, 1993a,b; Steiner and Saenger, 1994). For example, single-crystal neutron diffraction analysis of partially deuterated $\beta$-cyclodextrin ethanol octahydrate ($(C_6[^1H]_7[^2H]_3O_5)_7.C_2[^2H]_5O[^2H].8[^2H]_2O$; space group = $P2_1$; $a = 21.125(2)$, $b = 10.212(1)$, $c = 15.215(2)$Å, $\beta = 111.47(1)°$ at $T = 295$ K) (Steiner *et al.*, 1991) located all $^1H$ and $^2H$ atoms, a significant achievement in a problem of this complexity.

### 2.2.2 Sample Deuteration

Isotopes are generally considered to be indistinguishable by X-ray crystal structure analysis (although, see ammonium malate, below), whereas coherent neutron scattering lengths $b$ for isotopes tend to differ significantly. The difference between $b$ for $^1H$ and $^2H$, for example, (Section 2.1.1.) means that there are advantages to be gained in collecting neutron data

(a)

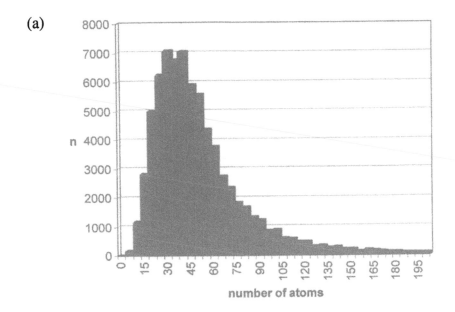

(b)

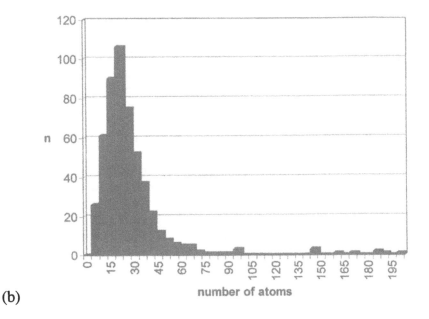

FIGURE 1. Distribution of the number of crystal structures, $n$, of organic (not organometallic) molecules in the Oct '98 CSD *versus* the number of atoms reported for each structure. (a) X-ray structures plus neutron structures; (b) neutron structures only (of which *ca.* 8% are derived from powder diffraction data). Note that the 'number of atoms' includes hydrogen atoms, if reported for a structure.

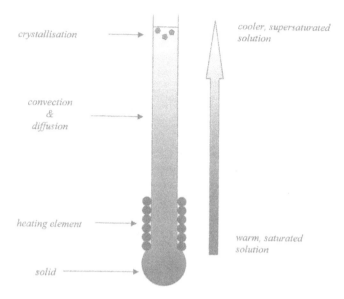

FIGURE 2. Small-scale recrystallisation apparatus (after Watkin, 1972). A saturated solution is prepared using a suitable solvent over a reservoir of excess solid solute. The heating element creates a thermal (and concentration) gradient, such that solute is transferred by convection and diffusion to the cooler upper region, where crystallisation takes place.

from a crystal of a deuterated analogue of a compound under study. $^2$H has a coherent scattering length of significantly greater magnitude than $^1$H, so that while single-crystal neutron diffraction will determine the positions of $^1$H atoms with good accuracy and precision, the situation is better still for $^2$H. In addition, deuteration reduces the incoherent background scattering observed with samples containing $^1$H, so that sample deuteration is a useful route to further improve the accuracy and precision of a single-crystal neutron experiment.

It should be borne in mind that $^1$H and $^2$H differ with respect to their vibrational behaviour. The isotropic displacement parameter $U$ (representing the mean-square displacement of an atom from its average position) for an $^2$H-atom is smaller than that of an $^1$H-atom by an estimated $30$–$50 \times 10^{-4}$ Å$^2$ (Seiler *et al.*, 1984). This difference was just enough for $^1$H and $^2$H to be distinguished in a careful X-ray analysis of the vibrational parameters in mono-ammonium $2S$-$3$-$^2$H-malate (Seiler *et al.*, 1984):

**Mono-ammonium 2$S$-3-$^2$H-malate**
(H* = $^2$H; all other H = $^1$H)

The H-atom displacement parameters were refined isotropically. Although the solitary $^2$H-atom on C3 could be identified by its smaller $U_H$, the authors emphasised that, "Even with large and accurately measured data sets, the X-ray method cannot be expected to match neutron diffraction in its ability to distinguish between protium (i.e. $^1$H) and deuterium" (Seiler *et al.*, 1984).

Exchanging $^2$H for $^1$H, where the H-atom is hydrogen bonded, can have a small influence on hydrogen bond lengths (Pollock *et al.*, 1956; Delaplane and Ibers, 1969; Sabine *et al.*, 1969; Tellgren and Olovsson, 1971; Thomas, 1973; Albinati *et al.*, 1978a,b; Koetzle *et al.*, 1978). Small isotopic effects have also been observed in studies of solid state phase transformations, for example at the 107 K phase transition in N-nitrodimethylamine (Filhol *et al.*, 1980), at the $\delta$–$\alpha'$ phase transition in methylammonium iodide (Yamamuro *et al.*, 1992) and at the 127 K transition in acetone (Ibberson *et al.*, 1995; Section 2.4.4.).

### 2.2.3 An Example of Structure Refinement

Creatine phosphate acts as an energy reservoir in vertebrate muscle. When energy is required, the phosphate is hydrolysed and a relatively large amount of energy is released in the process. The hydrolysis product creatine is a resonance-stabilised zwitterion:

**Creatine**

In the crystal structure, creatine also exists as a zwitterion, and the observed molecular structure is, in effect, a hybrid of the three resonance structures shown above. A single-crystal neutron diffraction analysis of creatine was carried out to determine atomic positional and displacement parameters for input into an X-ray analysis of electron density and chemical bonding in the compound (the logic behind the use of neutron diffraction data in studies of electron density distribution is explained in Section 2.4.1.).

Diffraction data were collected from a crystal of creatine monohydrate on the SXD instrument at ISIS using the time-of-flight Laue diffraction method (Wilson, 1990, 1997a). Collection of a complete structure factor set in this diffraction geometry consists of the accumulation of a series of data frames, each containing a large volume of reciprocal space. Data for two refinements were collected (Frampton *et al.*, 1997); (a) 43 frames of *ca.* 3 h each at 123 K (3709 reflections unique) and (b) 18 frames of *ca.* 20 to 40 min. each at 20 K (3155 reflections unique).

Using the X-ray structure as a starting model, the corrected and merged data were used to refine a total of 190 parameters (including 3 positional plus 6 anisotropic displacement parameters per atom) by full-matrix least-squares methods within the program *GSAS*

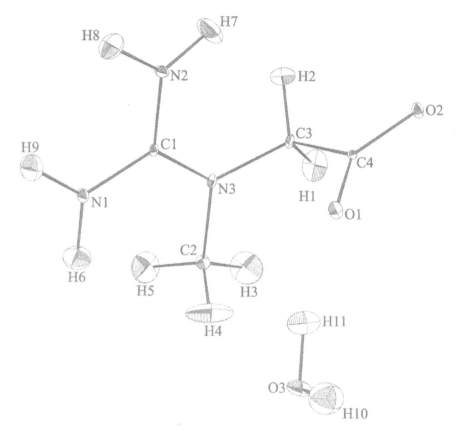

FIGURE 3.  ORTEP (Johnson, 1971) plot of creatine monohyrate at 20 K, with atomic displacement ellipsoids plotted at 50% probability. The structure was refined from data collected in only *ca.* 9 hours, albeit with a lower precision than the longer data collection period at 123 K. Selected bond lengths (in Å) at 20 K [123 K] : C(1)–N(1) = 1.338(2)[1.332(1)]; C(1)–N(2) = 1.329(3)[1.330(1)]; C(1)–N(3) = 1.343(3)[1.338(1)]; C(2)–N(3) = 1.461(3) [1.455(1)]; C(3)–N(3) = 1.452(2)[1.451(1)]; N(1)–H(6) = 1.023(5)[1.010(3)]; N(1)–H(9) = 1.029(5)[1.028(3)]; C(2)–H(3) = 1.079(7)[1.081(3)]; C(2)–H(4) = 1.088(6)[1.085(4)]; C(2)–H(5) = 1.091(5)[1.091(3)]; O(3)–H(10) = 0.976(8)[0.975(3)]; O(3)–H(11) = 0.975(5)[0.979(3)]. The mean-square torsional amplitude ($\langle \phi^2 \rangle$) of the –CH$_3$ motion is *ca.* 167 deg$^2$ at 20 K and 197 deg$^2$ at 123 K.

(Larsen and Von Dreele, 1994), to give an agreement factor $wR(F)$ (as defined in the *GSAS* manual) equal to 0.045 at 123 K and 0.055 at 20 K (Figure 3).

With regard to the positional parameters: (a) bond lengths and angles involving C-, N- and O-atoms derived from the neutron refinement are in good agreement with the previously determined X-ray structure; (b) the weighted mean C–H, N–H and O–H bond lengths in the neutron refinement are 1.089(1), 1.021(1) and 0.977(2) Å respectively, averaged over the 20 K and 123 K structures, which represents a significant improvement upon the X-ray structure.

Determining anisotropic H-atom displacement parameters is routine with neutron (but not X-ray) diffraction, and the large, anisotropic displacements observed in Figure 3 for atoms

H3, H4 and H5 of creatine suggest significant torsional motion of the methyl group at 20 K. The principal axes of the ellipsoids drawn at each atomic position represent the mean-square displacement amplitudes along each of three mutually perpendicular directions. It is not uncommon to observe torsional motion in crystal structures, and torsional amplitudes can be estimated using atomic positional and anisotropic displacement parameters determined from crystal diffraction data (Trueblood and Dunitz, 1983). Figure 4 shows the effect of temperature on the torsional motion of the $-CH_3$ group in the analgesic drug paracetamol.

**Paracetamol**

Figure 5 shows a structure of benzoic acid which illustrates a related aspect of structural chemistry where anisotropic H-atom displacement parameters can be revealing, namely disorder in the position of H-atoms.

## 2.3 POWDER NEUTRON DIFFRACTION

### 2.3.1 *General Background*

Early neutron diffraction experiments were conducted mostly with powders. The quality of the data was somewhat limited by the low neutron fluxes available at the time, but soon improved with the advent of high flux beam reactors and spallation sources, and diffractometers with high angular resolution. This, alongside the introduction of the Rietveld (1969) method, opened the door to routine crystal structure refinement, and offered the possibility of the application of direct methods of structure solution via the closely related Pawley (1981) type refinement.

Measuring the single dimension of a powder diffraction pattern is simpler than having to collect three dimensional single-crystal data, and high precision structure refinements can be obtained with 6 to 12 hours of data collection on an instrument such as HRPD at ISIS (Ibberson *et al.*, 1992a). However, this experimental simplicity comes at a price, namely the inevitable loss of intensity information that arises from the collapse of three dimensions of reciprocal space onto the single dimension of the powder diffraction pattern and ultimately limits the complexity of structural problem that can be tackled (see Section 2.3.2. for examples).

There are some other practicalities to be observed. Several grams of powdered sample are required to ensure good counting rates, and any $^1H$ atoms in the molecule must be fully (or almost fully) substituted by $^2H$ to eliminate (or reduce substantially) the incoherent background scattering associated with $^1H$. This incoherent background is a problem with powders because the whole of the irradiated powder sample contributes to the background, while only a (correctly oriented) fraction of the sample contributes to the Bragg peak intensities (Bacon, 1975d). It is much less of a problem with single crystals because the

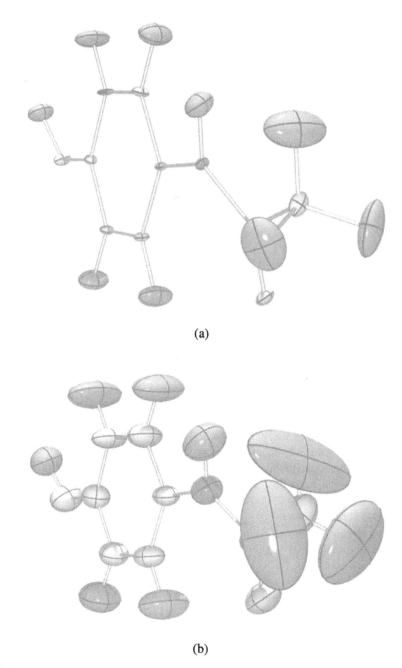

(a)

(b)

FIGURE 4. The neutron single-crystal structure of paracetamol at (a) 20 K; (b) 330 K, with atomic displacement ellipsoids plotted at 50% probability. The size of the ellipsoids increases with increasing temperature, and torsional motion of the $-CH_3$ is apparent from the large (relative to other atoms in the molecule), anisotropic ellipsoids of the methyl H-atoms. The mean-square torsional amplitude ($\langle \phi^2 \rangle$) of the motion increases with increasing temperature, from *ca.* 170 deg$^2$ at 20 K to 930 deg$^2$ at 330 K (Wilson, 1997b). Image generated with POV-Ray$^{TM}$ using output from Ortep-3 for Windows (ver. 1.05, Louis J. Farrugia, Glasgow University).

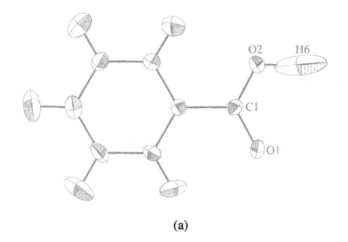

(a)

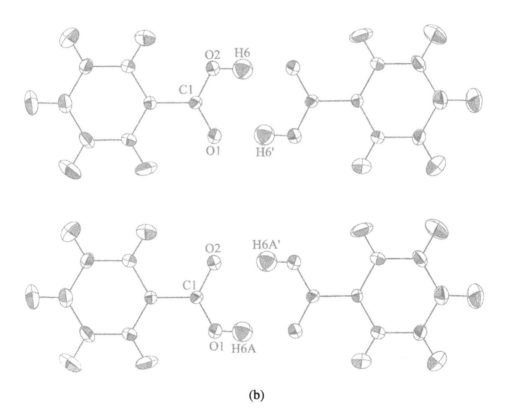

(b)

FIGURE 5. The neutron single-crystal structure of benzoic acid at 175 K with atomic displacement ellipsoids plotted at 50% probability and atom H6: (a) restricted to a single site; (b) refined over two sites. The large, anisotropic displacement ellipsoid of H6 in refinement (a) suggests that this atom is disordered, and this is confirmed in refinement (b), which resulted in site occupancy factors of *ca.* 62% for H6 (H6′) and 38% for H6A (H6A′) (Wilson *et al.*, 1996).

whole volume of the crystal contributes to a Bragg peak, and deuteration is by no means a prerequisite for single-crystal work. The relatively small number of neutron powder diffraction studies of molecular structure is linked in part to the practicality of obtaining deuterated analogues. If a deuterated analogue is not available, single-crystal diffraction may be a more practical route to a high precision structure refinement. On the other hand, powder diffraction may be more appropriate in situations where: (a) it is more convenient to work with a polycrystalline powder, rather than grow a single crystal of a particular phase (the volatile organic molecules in Table 2 are a good example of this), or (b) precise lattice parameter measurements are required, for example, as part of a survey of solid phases (Section 2.4.4.). The field of inorganic chemistry, where powder neutron diffraction has made a bigger impact (see, for example, Hewat (1986)), is not covered here.

### 2.3.2  An Example of Structure Refinement

Dopamine is a neurotransmitter with peripheral and central actions, and disorders of dopaminergic function are implicated in a number of neurological disorders, including Parkinson's disease.

**Dopamine** (cation form)

Amongst the key elements in the structure-activity relationship, interaction of -$NH_3^+$ with aromatic moieties on dopamine receptors is thought to be important (Dougherty, 1996). Accordingly, the following study was undertaken to test the ability of powder neutron diffraction to improve upon the previously reported X-ray study.

Diffraction data were collected from *ca.* 1 g of [$\beta$-$^1H_1$]decadeuteriodopamine deuteri-obromide on HRPD for a period of 11 hours over a time-of-flight range 30 to 230 ms corresponding to a $d$-spacing range 0.6 to 4.8 Å (Shankland *et al.*, 1996). After reduction, data in the range 42 to 210 ms (0.9 to 4.4 Å) were binned in time channels of $\Delta t/t = 3 \times 10^{-4}$ and a full profile refinement performed using the Rietveld (1969) method as implemented in the program *TF12LS* (David *et al.*, 1992a), using the isomorphous single-crystal X-ray structure of dopamine hydrochloride as the starting model.

At 293 K, the refined lattice parameters are $a = 10.66714(4)$; $b = 11.47801(5)$; $c = 7.94104(2)$ Å (space group = $Pbc2_1$). The refinement, which included a total of 93 positional plus atomic displacement parameters, plus site occupancy factors (Table 1), converged to give a $\chi^2$ of 1.99 for the fit, with 5366 observations and 111 basic variables. The high quality of the profile fit (Figure 6) is indicated by the agreement factors $R_{wp} = 2.47\%$, $R_p = 2.74\%$, $R_I = 5.70\%$, $R_E = 1.75\%$ (agreement factors as defined in Young (1995)).

TABLE 2. Crystal structures of volatile organic molecules determined from powder neutron diffraction data.

| Structure[†] | m.p./K | Reference |
|---|---|---|
| Methane | 91 | Arzi and Sándor, 1975 |
| –tribromo | 281 | Myers et al., 1983 |
| –diiodo | 279 | Prystupa et al., 1989 |
| –nitro[‡] | 245 | Trevino et al., 1980; |
| | | David et al., 1992b |
| –dibromo | 221 | Prystupa et al., 1989 |
| –fluorotribromo | 200 | Fitch and Cockcroft, 1992 |
| –bromo | 179 | Gerlach et al., 1986 |
| –fluorotrichloro | 162 | Cockcroft and Fitch, 1994 |
| –fluoro | 131 | Ibberson and Prager, 1996 |
| –dichlorodifluoro | 115 | Cockcroft and Fitch, 1991 |
| –bromotrifluoro | 101 | Jouanneaux et al., 1992 |
| –chlorotrifluoro | 92 | Pawley and Hewat, 1985 |
| –tetrafluoro[‡] | 90 | Fitch and Cockcroft, 1993 |
| –chloroiodo | | Torrie et al., 1993 |
| –iodotrifluoro | | Clarke et al., 1993 |
| Acetic acid[‡] | 290 | Albinati et al., 1978a |
| 1,4-Dimethylbenzene[‡] | 286 | Prager et al., 1991 |
| Formic acid[‡] | 281 | Albinati et al., 1978b |
| Cyclohexane | 280 | Wilding et al., 1993 |
| Benzene[‡] | 279 | David et al., 1992b |
| 4-Methyl pyridine[‡] | 277 | Carlile et al., 1990 |
| Dimethylacetylene | 241 | Ibberson and Prager, 1995 |
| Acetonitrile | 229 | Antson et al., 1987 |
| Acetylene | 192 | Koski and Sandor, 1975 |
| Toluene[‡] | 178 | Ibberson et al., 1992b |
| Methanol | 176 | Torrie et al., 1989 |
| Tetrahydrofuran[‡] | 165 | David and Ibberson, 1992 |
| 1,2-Dibromotetrafluoroethane | 163 | Pawley and Whitely, 1988 |
| n-Butane | 135 | Refson and Pawley, 1986 |
| Ethylene | 104 | Press and Eckert, 1976 |

[†]Identified primarily via CSD. Polycrystalline powders were produced using liquid nitrogen or other refrigerant.
[‡]Refinement of previously known crystal structure. The others are examples of structure solutions obtained either by direct methods (dimethylacetylene, fluoromethane, bromotrifluoromethane, fluorotribromomethane) or by refining trial structures.
Melting points taken from CRC Handbook of Chemistry and Physics, 79th ed.

As expected: (a) good agreement is observed between neutron and X-ray structures for bond lengths and angles involving C-, N- and O- atoms; (b) the mean C–$^2$H bond length observed in the neutron refinement (1.075(12) Å) is a significant improvement on the X-ray mean of 0.94 Å, with a precision ca. 5 times better.

Whilst this is a complicated refinement for a powder pattern, the observation-to-parameter ratio (number of independent reflections/number of refined parameters) is very low

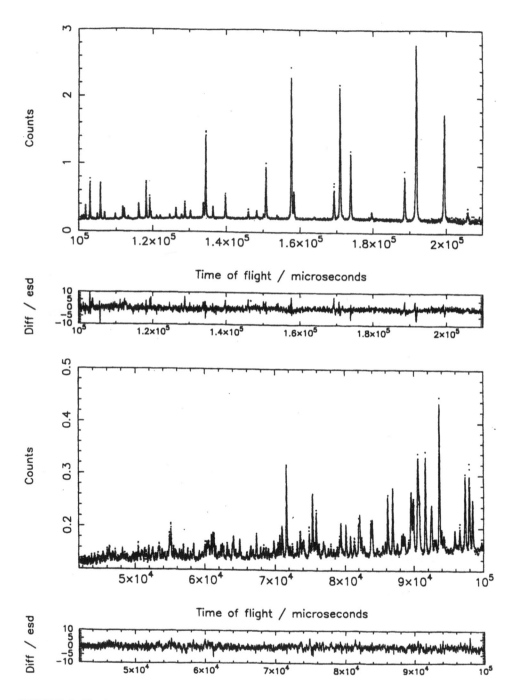

FIGURE 6. Final observed (points) and calculated (line) profiles for Rietveld refinement of deuterated dopamine deuteriobromide at 293 K. The difference plot is also shown.

compared to a single-crystal data set. Nonetheless, this and other examples such as deuterated oxalic acid dihydrate [63][†] (Lehmann *et al.*, 1994), tetrahydrofuran [59] (David and Ibberson, 1992) and toluene [270] (Ibberson *et al.*, 1992b) show that, provided intrinsic line width is well matched to the instrumental resolution, time-of-flight neutron powder data can give an observation-to-parameter ratio good enough to refine moderately complex organic molecular structures to a high degree of accuracy and precision.

## 2.4 MOLECULAR STRUCTURES BY NEUTRON DIFFRACTION

Sections 2.4.1.–2.4.3. highlight analyses of molecular structure in which particular care and attention has been given to scrutinising the level of agreement between state-of-the-art single-crystal neutron refinement and: (a) single-crystal X-ray refinement; (b) high-level *ab initio* calculation; (c) powder neutron refinement. Section 2.4.4. goes on to illustrate the application of powder neutron diffraction to the investigation of phase transformations in organic molecular solids.

### 2.4.1 Comparisons of Neutron and X-ray Single-Crystal Analyses

The positional and atomic displacement parameters obtained from four X-ray and five neutron diffraction low-temperature single-crystal analyses of oxalic acid dihydrate have been compared in a project set up under the auspices of the IUCr Commission on Charge, Spin and Momentum Densities (Coppens *et al.*, 1984). The comparisons included C- and O-atoms (X-ray) and C-, O- and H-atoms (neutron) and showed that:

(a) Atomic displacement parameters were obtained more reproducibly from the neutron data.

(b) Significant differences occurred between X-ray and neutron atomic displacement parameters, although very good agreement was obtained between one of the neutron data sets and one of the high-order X-ray refinements (average of 0.995 for the $U_{11}$, $U_{22}$ and $U_{33}$ ratios).

(c) Non-H atom positional parameters determined from the X-ray refinements agreed to within several $10^{-4}$ Å.

(d) There were no significant differences between positional parameters of all 7 atoms determined from the neutron data sets.

(e) The significance of differences between X-ray and neutron positional parameters for non-H atoms was much reduced in high-order X-ray refinements (average differences of 0.0004 to 0.0006 Å, comparing two high-order X-ray refinements with three neutron data sets).

These findings are essentially echoed elsewhere (see, for example, Hamilton, 1969; Coppens and Vos, 1971; Stevens and Hope, 1975; Bats *et al.*, 1977; Allen, 1986; Hirshfeld, 1992) and show that, provided appropriate care is taken in the experimental measurements and data analysis, good agreement can be achieved between neutron and X-ray single-crystal

---

[†]Figure in square brackets equals number of positional plus atomic displacement parameters refined.

TABLE 3. Selected X-ray studies of the electron density distribution in organic molecules which utilise data collected from a neutron diffraction analysis of the compound.

| Structure | Molecular formula | Reference |
|---|---|---|
| *s*-Triazine[1] | $C_3H_3N_3$ | Coppens, 1967 |
| Oxalic acid dihydrate[2] | $C_2H_2O_4.(H_2O)_2$ | Coppens *et al.*, 1969 |
| Hexamethylenetetramine[3] | $C_6H_{12}N_4$ | Duckworth *et al.*, 1970 |
| Cyanuric acid[1] | $C_3H_3N_3O_3$ | Coppens and Vos, 1971 |
| Tetracyanoethylene oxide | $C_6N_4O$ | Matthews and Stucky, 1971 |
| Ammonium oxalate monohydrate | $(NH_4^+)_2.C_2O_4^{2-}.H_2O$ | Taylor and Sabine, 1972 |
| $\alpha$-Glycine | $C_2H_5NO_2$ | Almlöf *et al.*, 1973 |
| Tetracyanoethylene[4] | $C_6N_4$ | Becker *et al.*, 1973 |
| Sucrose | $C_{12}H_{22}O_{11}$ | Hanson *et al.*, 1973 |
| 2,4,6-Trinitrobenzene-sulphonic acid tetrahydrate | $C_6H_3N_3O_9S.4H_2O$ | Lundgren and Tellgren, 1974 |
| $\alpha$-glycylglycine[5] | $C_4H_8N_2O_3$ | Griffin and Coppens, 1975 |
| Lithium formate monohydrate[6] | $Li^+.HCOO^-.H_2O$ | Thomas *et al.*, 1975 |
| 2-Amino-5-chloropyridine | $C_5H_5N_2Cl$ | Kvick *et al.*, 1976 |
| *p*-Nitropyridine *N*-oxide[7] | $C_5H_4N_2O_3$ | Wang *et al.*, 1976 |
| Sodium hydrogen diacetate | $Na^+.H(C_2H_3O_2)_2^-$ | Stevens *et al.*, 1977 |
| Sodium hydrogen oxalate monohydrate | $Na^+.C_2HO_4^-.H_2O$ | Tellgren *et al.*, 1977 |
| Dimethylammonium hydrogen oxalate | $(CH_3)_2NH_2^+.C_2HO_4^-$ | Thomas, 1977 |
| Pyridine-2,3-dicarboxylic acid | $C_7H_5NO_4$ | Koetzle *et al.*, 1978 |
| Urea[8] | $CO(NH_2)_2$ | Scheringer *et al.*, 1978 |
| Hydrazinium hydrogen oxalate | $N_2H_5^+.C_2HO_4^-$ | Thomas and Liminga, 1978 |
| Parabanic acid[9] | $C_3H_2N_2O_3$ | Craven and McMullan, 1979 |
| Putrescine diphosphate | $C_4H_{14}N_2^{2+}.2H_2PO_4^-$ | Takusagawa and Koetzle, 1979 |
| Calcium di(hydrogen maleate) pentahydrate | $Ca^{2+}.2C_4H_3O_4^-.5H_2O$ | Hsu and Schlemper, 1980 |
| Imidazolium hydrogen maleate | $C_3H_5N_2^+.C_4H_3O_4^-$ | Hsu and Schlemper, 1980 |
| 2-(2-Chlorobenzoylimino)-1,3-thiazolidine | $C_{10}H_9N_2OSCl$ | Cohen-Addad *et al.*, 1981 |
| 9-Methyladenine[10] | $C_6H_7N_5$ | Craven and Benci, 1981 |
| 5,5-Diethylbarbituric acid | $C_8H_{12}N_2O_3$ | Craven *et al.*, 1982 |
| Imidazole | $C_3H_4N_2$ | Epstein *et al.*, 1982 |
| $\gamma$-Aminobutyric acid | $C_4H_9NO_2$ | Craven and Weber, 1983 |
| 1,1,2,2-ethanetetracarbonitrile | $C_6H_2N_4$ | Declercq *et al.*, 1983 |
| 3-benzoylimino-4-methyl-1,2,4-oxathiazane | $C_{11}H_{12}N_2O_2S$ | Cohen-Addad *et al.*, 1984 |
| Phosphorylethanolamine | $C_2H_8NO_4P$ | Swaminathan and Craven, 1984 |
| $\beta$-DL-Arabinose | $C_5H_{10}O_5$ | Longchambon *et al.*, 1985 |
| 4-methylpyridine | $C_6H_7N$ | Ohms *et al.*, 1985 |
| Alloxan | $C_4H_2N_2O_4$ | Swaminathan *et al.*, 1985 |
| 2,5-Diaza-1,6-dioxa-6a-thiapentalene | $C_3H_2N_2O_2S$ | Cohen-Addad *et al.*, 1988 |
| *meso*-2,3-Bis(dimethylamino)butanedinitrile | $C_8H_{14}N_4$ | Parfonry *et al.*, 1988 |
| Cytosine monohydrate[10] | $C_4H_5N_3O.H_2O$ | Weber and Craven, 1990 |
| 18-crown-6·2 cyanamide | $C_{12}H_{24}O_6.2H_2NCN$ | Koritsanszky *et al.*, 1991 |

TABLE 3. Continued.

| Structure | Molecular formula | Reference |
|---|---|---|
| 1-Methyluracil | $C_5H_6N_2O_2$ | Klooster et al., 1992 |
| Benzene | $C_6H_6$ | Su and Coppens, 1992 |
| Acetamide | $C_2H_5NO$ | Zobel et al., 1992 |
| Bullvalene | $C_{10}H_{10}$ | Koritsanszky et al., 1993 |
| 1,4-Dinitrobenzene | $C_6H_4N_2O_4$ | Tonogaki et al., 1993 |
| Methylammonium hydrogen succinate monohydrate | $CH_3NH_3^+.C_4H_5O_4^-.H_2O$ | Flensburg et al., 1995 |
| Benzamide | $C_7H_7NO$ | Ruble and Galvao, 1995 |
| L-arginine phosphate monohydrate | $C_6H_{15}N_4O_2^+.H_2PO_4^-.H_2O$ | Espinosa et al., 1996 |
| DMAN[‡] 1,2-dichloro hydrogen maleate salt | $C_{14}H_{19}N_2^+.C_4HCl_2O_4^-$ | Mallinson et al., 1997 |
| Creatine monohydrate | $C_4H_9N_3O_2.H_2O$ | Frampton, 1997 |
| 2-amino-4-pentynoic acid | $C_5H_7NO_2$ | Frampton, 1997 |
| Methylammonium hydrogen maleate | $CH_3NH_3^+.C_4H_3O_4^-$ | Madsen et al., 1998a |
| Benzoylacetone | $C_{10}H_{10}O_2$ | Madsen et al., 1998b |

[‡] 1,8-bis(dimethylamino)naphthalene
[1] Also Stewart, 1970
[2] Also Stevens, 1980; Stevens and Coppens, 1980; Coppens et al., 1981; Coppens et al., 1984
[3] Also Stevens and Hope, 1975; Terpstra et al., 1993; Kampermann et al., 1994
[4] Also Hansen and Coppens, 1978; Scheringer et al., 1978
[5] Also Kvick et al., 1979
[6] Also Harkema et al., 1977; Thomas, 1978
[7] Also Coppens and Lehmann, 1976; Hansen and Coppens, 1978
[8] Also Swaminathan et al., 1984
[9] Also He et al., 1988
[10] Also Eisenstein, 1988.

diffraction analyses with respect to non-H atom positional parameters. However, there is no disputing the fact that neutron diffraction is capable of determining the positions of H-atoms to higher accuracy and precision than X-ray diffraction. Thus, "Neutron diffraction data only were used to derive mean bond lengths involving hydrogen atoms" in the 1987 tabulation of average bond lengths in organic compounds (Allen et al., 1987). Furthermore, accurate analysis of anisotropic H-atom displacements is routine with neutron data, but not with X-ray diffraction data. These particular shortcomings in the X-ray method cannot be circumvented in high-order refinements due to the absence of core orbitals in the H-atom (Hirshfeld, 1992). Therefore, neutron diffraction has clear advantages over X-ray diffraction with respect to high precision refinement of H-atom positional and anisotropic displacement parameters.

For the reasons just outlined, structural parameters derived from neutron refinement play an important part in X-ray analyses of electron densities and chemical bonding in molecules containing H-atoms. The deformation density, for example, is defined as the difference between the total electron density and the density calculated with respect to a reference model based on unbiased positional and displacement parameters (Coppens, 1997). The role of neutron diffraction in deformation density studies is, therefore, to assist in the derivation of accurate positional and displacement parameters (particularly for H-atoms), unbiased by chemical bonding and lone pair effects (Coppens, 1974, 1977).

This assistance can be provided either by introducing structural parameters from a neutron refinement into the X-ray analysis of the electron density distribution (Coppens, 1967) or, alternatively, by refining structural and electron density parameters simultaneously using combined X-ray plus neutron diffraction data (Duckworth *et al.*, 1969; Coppens *et al.*, 1981; Cohen-Addad *et al.*, 1988). Either way, a problem arises in ensuring that data derived from two different diffraction experiments are compatible. It is common enough to find X-ray and neutron analyses in good agreement over positional parameters for non-H atoms, but differing significantly with respect to the atomic displacement parameters (Hirshfeld, 1992). This discrepancy is frequently greater than can be accounted for by inadequacies in the X-ray scattering model, and contributing factors can include unequal experimental temperatures, the effects of systematic errors in the diffraction measurements, or the effects of absorption, extinction, thermal diffuse scattering or multiple reflection (Scheringer *et al.*, 1978; Coppens *et al.*, 1981; Blessing, 1995). Whatever the source of the mismatch, appropriate correction factors may be applied to scale the neutron atomic displacement parameters to produce a satisfactory average match to the X-ray parameters. One approach is to calculate the correction factors from quantitative comparisons of the non-H atom anisotropic displacement parameters derived from the X-ray and neutron refinements (Blessing, 1995). The correction factors can then be used to scale the anisotropic displacement parameters of all atoms in a molecule, including the H-atoms (Blessing, 1995; Flensburg *et al.*, 1995).

Table 3 lists examples of X-ray analysis of electron density distribution in organic molecules where the analysis features a contribution from neutron diffraction data collected from the compound. In view of what has just been said above, it comes as no surprise to find extensive comparisons of X-ray and neutron structural parameters amongst the entries in Table 3. Electron density studies provide a powerful experimental means of investigating electronic properties and bonding interactions, and the contribution of neutron diffraction in this field has been summarised nicely by Coppens (1997):

> "The $X$-$N$ deformation densities[†] are important for the study of the charge density distribution in and around hydrogen atoms. Without the extra effort required for a neutron experiment, assumptions on the hydrogen atom location and vibrations must be made which introduce a considerable uncertainty in the results".

### 2.4.2 Single-Crystal Neutron Refinement Versus Ab Initio Molecular Orbital Calculation

> "Benzene is a compound of central importance in organic chemistry. It is fitting therefore that it should be repeatedly investigated by each scientific method whenever there is a significant advance with respect to the detail or accuracy that the method can offer" (Jeffrey *et al.*, 1987)

The study of benzene by Jeffrey *et al.* (1987) included a comparison of bond lengths derived from a high precision single-crystal neutron refinement with a high-level *ab initio* molecular orbital calculation. The neutron refinement (934 reflections, 54 positional plus

---

[†]The symbol $N$ indicates the use of structural parameters from neutron refinement, after appropriate scaling of the atomic displacement parameters, in the X-ray analysis of electron density distribution.

atomic displacement parameters) of deuterated benzene at 15 K gave mean observed bond lengths C–C = 1.3972(5) Å and C–$^2$H = 1.0864(7) Å. When corrected for molecular libration, the corresponding at-rest values were 1.3980 Å and 1.088 Å. This compared with 1.395 Å and 1.087 Å calculated *ab initio* at the MP-2 6-31G* level for the isolated molecule at rest.

This state-of-the-art comparison demonstrates that: (a) with the positions and displacement amplitudes of the nuclei determined to a sufficiently high degree of precision, and (b) appropriate thermal motion corrections applied to the bond lengths, the experimental refined geometry is in excellent agreement with the equilibrium nuclear positions (energy-optimized geometry) of the isolated benzene molecule.

One other interesting feature of the benzene geometry is the deviation from planarity. The refined structure shows atoms $^2$H(3) and $^2$H(3)' to be displaced from a plane through the other four deuterium atoms $^2$H(1,1',2,2') by ±0.0215(11) Å, producing a chair conformation (Jeffrey *et al.*, 1987; Jeffrey, 1992; primed and unprimed atoms related by a crystallographic centre of inversion). This is a result of the shear exerted on the molecule by the crystal environment. The carbon ring appears to be less deformable than the C–$^2$H bonds, and distortion of the carbon ring away from planarity to a chair conformation is barely significant.

### 2.4.3  Comparisons of Single-Crystal and Powder Neutron Refinements

The atomic positional and anisotropic displacement parameters determined for deuterated benzene and nitromethane (4.2 K) using time-of-flight neutron powder diffraction refinement compare very favourably with neutron single-crystal analyses of both compounds at 15 K (David *et al.*, 1992b; Jeffrey, 1992). The principal differences can be summarised as follows: (a) the lattice parameters are determined with much higher precision from the powder diffraction data; (b) single-crystal diffraction gives higher precision on all of the other structural parameters.

Essentially the same conclusions were reached in a study of the crystal structure of oxalic acid dihydrate at 100 K, refined using time-of-flight neutron powder data (*ca.* 22 h data collected from deuterated compound) and compared with neutron single-crystal diffraction data at the same temperature (Lehmann *et al.*, 1994).

In summary, high-resolution neutron powder diffraction is a credible rival to neutron single-crystal diffraction, at least for relatively simple structures. However, as the complexity of the structure increases, three-dimensional single-crystal data are much better able to maintain the observation-to-parameter ratio necessary to obtain high accuracy and precision.

### 2.4.4  Phase Transformations in Organic Molecular Crystals

Polymorphism amongst pharmaceuticals is particularly problematic where problems arise in the advanced stages of product development. The HIV protease inhibitor ritonavir is a case in point; Abbot International recently notified the European Agency for the Evaluation of Medicinal Products of manufacturing and batch release problems with the capsule formulation arising from the unexpected appearance of a new ritonavir polymorph with a much reduced solubility (SCRIP, 1998). There is, therefore, a strong incentive to identify potential problems early on, and a comprehensive polymorph screen will include a search for phase transformations in the solid state. It is the phase surveying capability of powder neutron diffraction in particular which we wish to highlight here.

Such surveys require: (a) accurate and precise lattice parameters determined from rapidly collected data; (b) sophisticated control of sample environment. Neutrons are highly penetrating and can pass through relatively thick-walled temperature and pressure cells (a 1Åbeam can pass through at least 13 cm of aluminium before it is attenuated by 10%; Biggin (1986)). Variable temperature and pressure experiments are therefore easier to perform on a neutron beamline than they are on an X-ray beamline, particularly with the fixed diffraction geometry utilised in time-of-flight methods. With the high-resolution neutron powder diffractometer HRPD at ISIS, temperatures in the range 1.2 K to 2300 K and pressures up to 3.5 kbar are readily accessible, making it relatively straightforward to survey $P$-$T$ space.

Some 50% of the citations given in Table 2 feature variable temperature powder neutron diffraction experiments. Alongside these examples, pyrene provides a powerful illustration of the phase surveying capability of powder neutron diffraction. The monoclinic unit cell at 4.2 K was refined against data collected from *ca.* 3 g of perdeuterated pyrene II on HRPD to give $a = 12.29995(8)$, $b = 9.98738(7)$, $c = 8.22012(7)$Å, $\beta = 96.4065(7)°$ (space group = $P2_1/a$). The measurement was repeated at 10 K, and then at 5 K intervals up to and including 185 K, collecting for *ca.* 20 minutes at each step, with 10 minutes for equilibration between steps (Love, 1997). The lattice parameters were refined at each temperature via the Pawley method (Figure 7).

Figure 8 summarises the evolution of the pyrene crystal structure with increasing temperature. The high precision of the lattice parameters arises from the high resolution ($\Delta d/d$) afforded by HRPD (essentially constant across the pattern), combined with the use of the full pattern to refine the lattice parameters. This makes HRPD ideally suited to following the evolution of structure, and the first order transition pyrene II→I is clearly evident from the 'jump' in volume of some 7 Å$^3$ at *ca.* 105 K in Figure 9.

The $P$-$T$ phase behaviour of pyrene has recently been mapped out using data from both HRPD and IRIS instruments at ISIS (Knight *et al.*, 1999). Other examples of powder neutron diffraction at non-ambient pressures include studies of pressure-induced structural changes in oxalic acid dihydrate (Putkonen *et al.*, 1985), cyclohexane (Wilding *et al.*, 1991) and benzoic acid (Brougham *et al.*, 1996).

It is important to maintain an awareness for the possibility of incorporating single-crystal diffraction data into a phase survey. For example, the single-crystal structure of pyrene II has recently been reported (Frampton *et al.*, 2000), and the refined X-ray structure at 93 K is in good agreement with the structure of pyrene II at 4.2 K refined using powder neutron diffraction data:

|  | Single-crystal X-ray, 93 K | Powder neutron, 4.2 K |
|---|---|---|
| $a$/Å | 12.358(6) | 12.30267(1) |
| $b$/Å | 10.020(4) | 9.98788(1) |
| $c$/Å | 8.260(4) | 8.22064(1) |
| $\beta$/° | 96.48(4) | 96.4046(1) |
| Space group | $P2_1/a$ | $P2_1/a$ |
| Reference | Frampton *et al.*, 2000 | Knight *et al.*, 1996 |

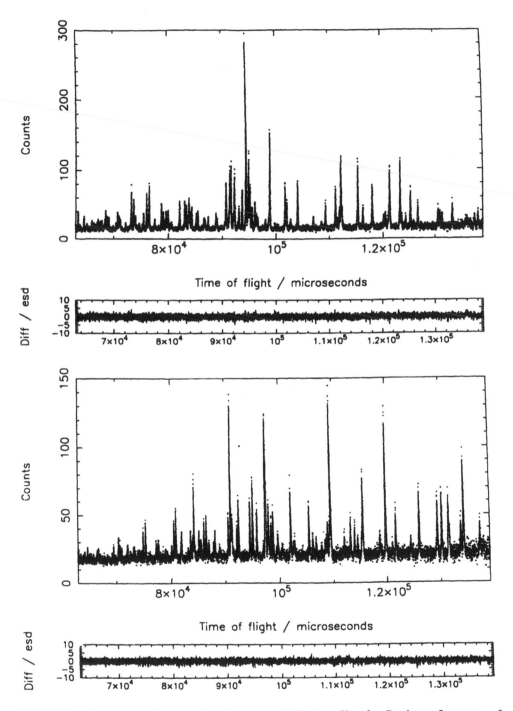

FIGURE 7. Final observed (points) and calculated (line) profiles for Pawley refinements of deuterated pyrene at 4.2 K (top) and 185 K (bottom). The difference plots are also shown.

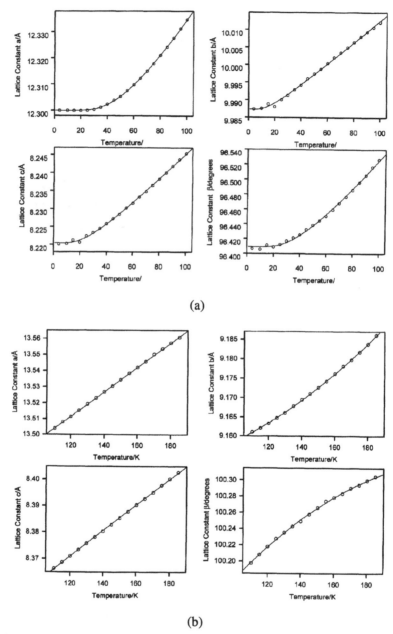

FIGURE 8. Calculated fits for the variation in lattice constants of (a) pyrene II; (b) pyrene I. For each lattice parameter, $l$, in pyrene II the solid line shows the fit to a quasi harmonic Einstein model for the thermal expansion: $l(T) = l_0 + k/[\exp(E/T) - 1]$. The linear fits to the $a$ and $c$ data of pyrene I are consistent with the fact that the temperature is greater than the Einstein temperature for pyrene ($\approx 80$ K, using the $T$-dependence of the unit cell volume of pyrene II). Both $b$ and $\beta$ deviate from linearity and were fitted empirically as a power series in $T$. Figures courtesy of K. Knight.

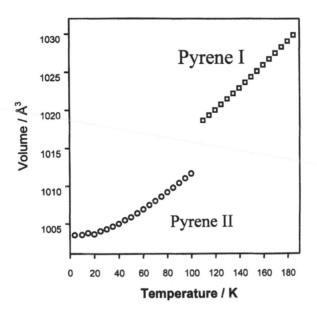

FIGURE 9.   Unit cell volume change on heating pyrene II from 4.2 K up to and through the transition to pyrene I. Figure courtesy of K. Knight.

However, the single-crystal approach is not always straightforward. In the case of hexamethylbenzene, for example, the compound undergoes a first order phase transition at *ca.* 116 K. Attempts to preserve a single crystal of hexamethylbenzene through the transition for diffraction analysis met with repeated failure (Hamilton *et al.*, 1969), with the crystal shattering during the transition. Powder diffraction comes into its own in such a situation, and the structure of hexamethylbenzene has since been determined at 20 K using powder neutron diffraction data (David, 1992).

A state-of-the-art demonstration of the amount of subtle structural detail that can be extracted from high resolution powder neutron diffraction data is given by David *et al.* (1993) in a case study of $C_{60}$. A plot of cubic lattice parameter versus temperature clearly identifies the first-order transition at 260 K (*via* a 'jump' of *ca.* 0.04 Å), and an orientational glass transition at 86 K. A detailed analysis of the structural parameters provided estimates of the energy difference between the major and minor orientations of the molecule, the different effective volumes of these two orientations and the reorientational activation energy. As the authors themselves point out, problems with a far greater number of structural variables than $C_{60}$ have been tackled with powder neutron data (some are listed in Section 2.3.2.). However, what is highly significant in the $C_{60}$ study is the amount, the detail and particularly the variety of structural information which emerges from it.

The complementary use of powder neutron diffraction alongside heat capacity and dielectric permittivity measurements is demonstrated in a study of the phase transition at 127 K in acetone (Ibberson *et al.*, 1995). A small isotopic effect is evident in the calorimetric data:

|                                              | $(CH_3)_2CO$       | $(C[^2H]_3)_2CO$   |
|----------------------------------------------|--------------------|--------------------|
| Molecular weight / g mol$^{-1}$              | 58.080             | 64.121             |
| Phase transition temperature / K             | 127                | 132                |
| Transition entropy / J mol$^{-1}$ K$^{-1}$   | 2.04               | 2.08               |
| Fusion temperature / K                       | $178.32 \pm 0.02$  | $178.78 \pm 0.02$  |
| Enthalpy of fusion / kJ mol$^{-1}$           | $5.612 \pm 0.002$  | $5.665 \pm 0.002$  |

Diffraction data from both orthorhombic phases of perdeuterated acetone (5 g) were also collected using HRPD. The phase transition was apparent from a discontinuity in the plots of lattice parameters *versus* temperature at around the expected temperature. However, in this instance, anisotropic line broadening in the diffraction profiles of both phases precluded full structure determinations.

The examples cited in this section are non-pharmaceutical, and one may ask why this is, given that the surveying capabilities of an instrument such as HRPD are well suited to pharmaceutical polymorphism problems. The answer lies partly in the need to access a central facility, and partly in the necessity to work with a deuterated sample. Both 'problems' are surmountable, given the appropriate motivation.

## 2.5 CONCLUDING REMARKS

It has been pointed out that "Chemistry was structural long before molecular geometries could be determined" (Wipff and Boudon, 1992). However, accurate molecular structures undoubtedly provide a greater insight into the workings of chemistry. The examples given in this chapter hopefully provide the reader with some good reasons to use neutron diffraction as means of investigating the structure and properties of organic molecular crystals.

### Acknowledgements

We are indebted to Bill David and Kevin Knight of the ISIS facility at Rutherford Appleton Laboratory, and Chris Frampton of Roche Drug Discovery, for their close collaboration.

### References

Albinati, A., Rouse, K.D. and Thomas, M.W. (1978a) Neutron powder diffraction analysis of hydrogen-bonded solids. I. Refinement of the structure of deuterated acetic acid at 4.2 and 12.5 K, *Acta Crystallogr., Sect. B*, **34**, 2184–2187.

Albinati, A., Rouse, K.D. and Thomas, M.W. (1978b) Neutron powder diffraction analysis of hydrogen-bonded solids. II. Structural study of formic acid at 4.5 K, *Acta Crystallogr., Sect. B*, **34**, 2188–2190.

Allen, F.H. (1986) A systematic pairwise comparison of geometric parameters obtained by X-ray and neutron diffraction, *Acta Crystallogr., Sect. B*, **42**, 515–522.

Allen, F.H., Kennard, O., Watson, D.G., Brammer, L, Orpen, A.G. and Taylor, R. (1987) Table of bond lengths determined by X-ray and neutron diffraction. Part 1. Bond lengths in organic compounds, *J. Chem. Soc. Perkin Trans II*, S1–S19.

Allen, F.H. (1992) Crystallographic databases: retrieval and analysis of precise structural information from the Cambridge Structural Database, in *Accurate Molecular Structures*, Domenicano, A. and Hargittai, I. (Eds), Ch. 15, 355–378, IUCr Monographs on Crystallography 1, Oxford University Press, New York.

Allen, F.H. and Kennard, O. (1993) 3D Search and research using the Cambridge Structural Database, *Chemical Design Automation News*, **8**, 1, 31–37.

Allen, F.H., Kennard, O. and Watson, D.G. (1994) Crystallographic databases: search and retrieval of information from the Cambridge Structural Database, in *Structure Correlation, Volume 1*, Bürgi, H.-B. and Dunitz, J.D. (Eds), Ch. 3, 71–110, VCH, Weinheim.

Almlöf, J., Kvick, Å. and Thomas, J.O. (1973) Hydrogen bond studies. 77. Electron density distribution in $\alpha$-glycine: $X$-$N$ difference Fourier synthesis vs *ab initio* calculations, *J. Chem. Phys.*, **59**, 3901–3906.

Antson, O.K., Tilli, K.J. and Andersen, N.H. (1987) Neutron powder diffraction study of deuterated $\beta$-acetonitrile, *Acta Crystallogr., Sect. B*, **43**, 296–301.

Arzi, E. and Sándor, E. (1975) Crystal structure and phase transition in deutero-methane, *Acta Crystallogr., Sect. A*, **31**, Supplement, Abstract 12.2–3.

Bacon, G.E. and Curry, N.A. (1956) A study of $\alpha$-resorcinol by neutron diffraction, *Proc. R. Soc. A*, **235**, 552–559.

Bacon, G.E. (1975a) *Neutron diffraction*, 3rd edition, Monographs on the physics and chemistry of materials, Oxford University Press, London.

Bacon, G.E. (1975b) Neutron diffractometers, in *Neutron diffraction*, 3rd edition, Ch. 1.4, 7–18, Monographs on the physics and chemistry of materials, Oxford University Press, London.

Bacon, G.E. (1975c) Primary and secondary extinction, in *Neutron diffraction*, 3rd edition, Ch. 3.1, 65–71, Monographs on the physics and chemistry of materials, Oxford University Press, London.

Bacon, G.E. (1975d) The use of powdered crystals, in *Neutron diffraction*, 3rd edition, Ch. 3.9, 92–93, Monographs on the physics and chemistry of materials, Oxford University Press, London.

Bacon, G.E. (1975e) Single-crystal measurements, in *Neutron diffraction*, 3rd edition, Ch. 4.6, 124–129, Monographs on the physics and chemistry of materials, Oxford University Press, London.

Bats, J.W., Coppens, P. and Koetzle, T.F. (1977) The experimental charge density in sulfur-containing molecules: a study of the deformation electron density in sulfamic acid at 78 K by X-ray and neutron diffraction, *Acta Crystallogr., Sect. B*, **33**, 37–45.

Becker, P., Coppens, P. and Ross, F.K. (1973) Valence electron distribution in cubic tetracyanoethylene by the combined use of X-ray and neutron diffraction, *J. Am. Chem. Soc.*, **95**, 7604–7609.

Biggin, S. (Ed.) (1986) *Neutron beams: their use and potential in scientific research*, Science and Engineering Research Council.

Blessing, R.H. (1995) On the differences between X-ray and neutron thermal vibration parameters, *Acta Crystallogr., Sect. B*, **51**, 816–823.

Brougham, D.F., Horsewill, A.J., Ikram, A., Ibberson, R.M., McDonald, P.J. and Pinter-Krainer, M. (1996) The correlation between hydrogen bond tunneling dynamics and the structure of benzoic acid dimers, *J. Chem. Phys.*, **105**, 979–982.

Bruno, I.J., Cole, J.C., Lommerse, J.P.M., Rowland, R.S., Taylor, R. and Verdonk, M.L. (1997) IsoStar: a library of information about nonbonded interactions, *J. Comp.-Aided Mol. Design*, **11**, 525–537.

Carlile, C.J., Ibberson, R.M., Fillaux, F. and Willis, B.T.M. (1990) The crystal structure of 4-methyl pyridine at 4.5 K, *Z. Kristallogr.*, **193**, 243–250.

Clarke, S.J., Cockcroft, J.K. and Fitch, A.N. (1993) The structure of solid CF$_3$I, *Z. Kristallogr.*, **206**, 87–95.

Cockcroft, J.K. and Fitch, A.N. (1991) The structure of solid dichlorodifluoromethane CF$_2$Cl$_2$ by powder neutron diffraction, *Z. Kristallogr.*, **197**, 121–130.

Cockcroft, J.K. and Fitch, A.N. (1994) Structure of solid trichlorofluoromethane, CFCl$_3$, by powder neutron diffraction, *Z. Kristallogr.*, **209**, 488–490.

Cohen-Addad, C., Savariault, J.-M. and Lehmann, M.S. (1981) 2-(2-Chlorobenzoylimino)-1,3-thiazolidine: structure refinement from neutron diffraction data at 113 K and charge density deformation maps, *Acta Crystallogr., Sect. B*, **37**, 1703–1706.

Cohen-Addad, C., Lehmann, M.S., Becker, P., Párkányi, L. and Kálmán, A. (1984) Nature of S• • •O interaction in short X–S• • ••O contacts: charge density experimental studies and theoretical interpretation, *J. Chem. Soc. Perkin Trans. II*, 191–196.

Cohen-Addad, C., Lehmann, M.S., Becker, P. and Davy, H. (1988) Structure of 1,6-dioxa-6a-thiapentalene, C$_5$H$_4$O$_2$S, and comparison with a new structure refinement of 2,5-diaza-1,6-dioxa-6a-thiapentalene, C$_3$H$_2$N$_2$O$_2$S, from X-ray and neutron data at 122 K. Preliminary charge-density study, *Acta Crystallogr., Sect. B*, **44**, 522–527.

Cole, J.C., Taylor, R. and Verdonk, M.L. (1998) Directional preferences of intermolecular contacts to hydrophobic groups, *Acta Crystallogr., Sect. D*, **54**, 1183–1193.

Coppens, P. (1967) Comparative X-ray and neutron diffraction study of bonding effects in s-Triazine, *Science*, **158**, 1577–1579.

Coppens, P., Sabine, T.M., Delaplane, R.G. and Ibers, J.A. (1969) An experimental determination of the asphericity of the atomic charge distribution in oxalic acid dihydrate, *Acta Crystallogr., Sect. B*, **25**, 2451–2458.

Coppens, P. and Vos, A. (1971) Electron density distribution in cyanuric acid. II. Neutron diffraction study at liquid nitrogen temperature and comparison of X-ray and neutron diffraction results, *Acta Crystallogr., Sect. B*, **27**, 146–158.

Coppens, P. (1974) Some implications of combined X-ray and neutron diffraction studies, *Acta Crystallogr., Sect. B*, **30**, 255–261.

Coppens, P. and Lehmann, M.S. (1976) Charge density studies below liquid nitrogen temperature. II. Neutron analysis of *p*-nitropyridine *N*-oxide at 30 K and comparison with X-ray results, *Acta Crystallogr., Sect. B*, **32**, 1777–1784.

Coppens, P. (1977) Experimental electron densities and chemical bonding, *Angew. Chem. Int. Ed. Engl.*, **16**, 32–40.

Coppens, P., Boehme, R., Price, P.F. and Stevens, E.D. (1981) Joint X-ray and neutron data refinement of structural and charge density parameters, *Acta Crystallogr., Sect. A*, **37**, 857–863.

Coppens, P., Dam, J., Harkema, S., Feil, D., Feld, R., Lehmann, M.S., Goddard, R., Krüger, C., Hellner, E., Johansen, H., Larsen, F.K., Koetzle, T.F., McMullan, R.K., Maslen, E.N. and Stevens, E.D. (1984) International Union of Crystallography, Commission on Charge, Spin and Momentum Densities, Project on comparison of structural parameters and electron density maps of oxalic acid dihydrate, *Acta Crystallogr., Sect. A*, **40**, 184–195.

Coppens, P. (1997) Deformation densities, in *X-ray charge densities and chemical bonding*, Ch. 5.2, 94–115, IUCr Texts on Crystallography 4, Oxford University Press, New York.

Craven, B.M. and McMullan, R.K. (1979) Charge density in parabanic acid from X-ray and neutron diffraction, *Acta Crystallogr., Sect. B*, **35**, 934–945.

Craven, B.M. and Benci, P. (1981) The charge density and hydrogen bonding in 9-methyl-adenine at 126 K, *Acta Crystallogr., Sect. B*, **37**, 1584–1591.

Craven, B.M., Fox Jr, R.O. and Weber, H.-P. (1982) The charge density in polymorph II of 5,5-diethylbarbituric acid (barbital) at 198 K, *Acta Crystallogr., Sect. B*, **38**, 1942–1952.

Craven, B.M. and Weber, H.-P. (1983) Charge density in the crystal structure of $\gamma$-aminobutyric acid at 122 K — an intramolecular methylene H bridge, *Acta Crystallogr., Sect. B*, **39**, 743–748.

David, W.I.F. (1992) Transformations in neutron powder diffraction, *Physica B*, 180 & 181, 567–574.

David, W.I.F., Ibberson, R.M. and Matthewman, J.C. (1992a) Profile analysis of neutron powder diffraction data at ISIS, *Rutherford Appleton Laboratory Report*, RAL-92-032.

David, W.I.F., Ibberson, R.M., Jeffrey, G.A. and Ruble, J.R. (1992b) The crystal structure analysis of deuterated benzene and deuterated nitromethane by pulsed-neutron powder diffraction: a comparison with single crystal neutron diffraction analysis, *Physica B*, **180 & 181**, 597–600.

David, W.I.F. and Ibberson, R.M. (1992) A reinvestigation of the structure of tetrahydrofuran by high-resolution neutron powder diffraction, *Acta Crystallogr., Sect. C*, **48**, 301–303.

David, W.I.F., Ibberson, R.M. and Matsuo, T. (1993) High resolution neutron powder diffraction: a case study of the structure of $C_{60}$, *Proc. R. Soc. Lond. A*, **442**, 129–146.

Declercq, J.P., Tinant, B., Parfonry, A., Van Meerssche, M., Legrand, E. and Lehmann, M.S. (1983) Combined X-ray and neutron study of 1,1,2,2-ethanetetracarbonitrile, $C_6H_2N_4$, at 158 K, *Acta Crystallogr., Sect. C*, **39**, 1401–1405.

Delaplane, R.G. and Ibers, J.A. (1969) An X-ray study of $\alpha$-oxalic acid dihydrate, $(COOH)_2.2H_2O$, and of its deuterium analogue, $(COOD)_2.2D_2O$: isotope effect in hydrogen bonding and anisotropic extinction effects, *Acta Crystallogr., Sect. B*, **25**, 2423–2437.

Ding, J., Steiner, T., Zabel, V., Hingerty, B.E., Mason, S.A. and Saenger, W. (1991) Topography of cyclodextrin inclusion complexes. 28. Neutron diffraction study of the hydrogen bonding in partially deuterated $\gamma$-cyclodextrin.$15.7D_2O$ at $T = 110$ K, *J. Am. Chem. Soc.*, **113**, 8081–8089.

Dougherty, D.A. (1996) Cation-$\pi$ interactions in chemistry and biology: a new view of benzene, Phe, Tyr and Trp, *Science*, **271**, 163–168.

Duckworth, J.A.K., Willis, B.T.M. and Pawley, G.S. (1969) Joint refinement of neutron and X-ray diffraction data, *Acta Crystallogr., Sect. A*, **25**, 482–484.

Duckworth, J.A.K., Willis, B.T.M. and Pawley, G.S. (1970) Neutron diffraction study of the atomic and molecular motion in hexamethylenetetramine, *Acta Crystallogr., Sect. A*, **26**, 263–271.

Eisenstein, M. (1988) Static deformation densities for cytosine and adenine, *Acta Crystallogr., Sect. B*, **44**, 412–426.

Epstein, J., Ruble, J.R. and Craven, B.M. (1982) The charge density in imidazole by X-ray diffraction at 103 and 293 K, *Acta Crystallogr., Sect. B*, **38**, 140–149.

Espinosa, E., Lecomte, C., Molins, E., Veintemillas, S., Cousson, A. and Paulus, W. (1996) Electron density study of a new non-linear optical material: L-arginine phosphate monohydrate (LAP). Comparison between $X - X$ and $X - (X + N)$ refinements, *Acta Crystallogr., Sect. B*, **52**, 519–534.

Etter, M.C., Jahn, D.A., Donahue, B.S., Johnson, R.B. and Ojala, C. (1986) Growth and characterization of small molecule organic crystals, *J. Crystal Growth*, **76**, 645–655.

Filhol, A., Bravic, G., Rey-Lafon, M. and Thomas, M. (1980) X-ray and neutron studies of a displacive phase transition in N,N-dimethylnitramine (DMN), *Acta Crystallogr., Sect. B*, **36**, 575–586.

Finney, J.L. (1995) The complementary use of X-ray and neutron diffraction in the study of crystals, *Acta Crystallogr., Sect. B*, **51**, 447–467.

Fitch, A.N. and Cockcroft, J.K. (1992) The structure of solid tribromofluoromethane $CFBr_3$ by powder neutron diffraction, *Z. Kristallogr.*, **202**, 243–250.

Fitch, A.N. and Cockcroft, J.K. (1993) The structure of solid carbon tetrafluoride, *Z. Kristallogr.*, **203**, 29–39.

Flensburg, C., Larsen, S. and Stewart, R.F. (1995) Experimental charge density study of methylammonium hydrogen succinate monohydrate. A salt with a very short O–H–O hydrogen bond, *J. Phys. Chem.*, **99**, 10130–10141.

Frampton, C.S., Wilson, C.C., Shankland, N. and Florence, A.J. (1997) Single-crystal neutron refinement of creatine monohydrate at 20 K and 123 K, *J. Chem. Soc., Faraday Trans.*, **93**, 1875–1879.

Frampton, C.S. (1997) personal communication.

Frampton, C.S., Knight, K.S., Shankland, N. and Shankland, K. (2000) Single-crystal X-ray diffraction analysis of pyrene II at 93 K, *J. Mol. Struct.*, **520**, 29–32.

Gerlach, P.N., Torrie, B.H. and Powell, B.M. (1986) The crystal structures and phase transition of methyl bromide, *Mol. Phys.*, **57**, 919–930.

Glusker, J.P., Lewis, M. and Rossi, M. (1994a) Modifications of the diffraction pattern, in *Crystal Structure Analysis for Chemists and Biologists*, Ch. 6.6, 209–219, VCH Publishers, New York.

Glusker, J.P., Lewis, M. and Rossi, M. (1994b) Recognition and receptors, in *Crystal Structure Analysis for Chemists and Biologists*, Ch. 17, 731–782, VCH Publishers, New York.

Glusker, J.P., Lewis, M. and Rossi, M. (1994c) Structure-activity results, in *Crystal Structure Analysis for Chemists and Biologists*, Ch. 18, 783–821, VCH Publishers, New York.

Griffin, J.F. and Coppens, P. (1975) Valence electron distribution in perdeuterio-$\alpha$-glycylglycine. A high-resolution study of the peptide bond, *J. Am. Chem. Soc.*, **97**, 3496–3505.

Hamilton, W.C. (1969) Comparison of X-ray and neutron diffraction structural results: a study in methods of error analysis, *Acta Crystallogr., Sect. A*, **25**, 194–206.

Hamilton, W.C., Edmonds, J.W., Tippe, A. and Rush, J.J. (1969) Methyl group rotation and the low temperature transition in hexamethylbenzene. A neutron diffraction study, *Discuss. Faraday Soc.*, **48**, 192–204.

Hansen, N.K. and Coppens, P. (1978) Testing aspherical atom refinements on small-molecule data sets, *Acta Crystallogr., Sect. A*, **34**, 909–921.

Hanson, J.C., Sieker, L.C. and Jensen, L.H. (1973) Sucrose: X-ray refinement and comparison with neutron refinement, *Acta Crystallogr., Sect. B*, **29**, 797–808.

Harkema, S., De With, G. and Keute, J.C. (1977) Lithium formate monohydrate: a comparison of the $X$-$N$ difference density and structural parameters derived from different experiments, *Acta Crystallogr., Sect. B*, **33**, 3971–3973.

He, X.M., Swaminathan, S., Craven, B.M. and McMullan, R.K. (1988) Thermal vibrations and electrostatic properties of parabanic acid at 123 and 298 K, *Acta Crystallogr., Sect. B*, **44**, 271–281.

Hewat, A.W. (1986) High-resolution neutron and synchrotron powder diffraction, *Chemica Scripta*, **26A**, 119–130.

Hirshfeld, F.L. (1992) The role of electron density in X-ray crystallography, in *Accurate Molecular Structures*, Domenicano, A. and Hargittai, I. (Eds), Ch. 10, 237–269, IUCr Monographs on Crystallography 1., Oxford University Press, New York.

Hsu, B. and Schlemper, E.O. (1980) $X-N$ deformation density studies of the hydrogen maleate ion and the imidazolium ion, *Acta Crystallogr., Sect. B*, **36**, 3017–3023.

Ibberson R.M., David, W.I.F. and Knight, K.S. (1992a) The high resolution powder diffractometer (HRPD) at ISIS — a user guide, *Rutherford Appleton Laboratory Report*, RAL-92-031.

Ibberson, R.M., David, W.I.F. and Prager, M. (1992b) Accurate determination of hydrogen atom positions in $\alpha$-toluene by neutron powder diffraction, *J. Chem. Soc., Chem. Commun.*, 1438–1439.

Ibberson, R.M., David, W.I.F., Yamamuro, O., Miyoshi, Y., Matsuo, T. and Suga, H. (1995) Calorimetric, dielectric, and neutron diffraction studies on phase transitions in ordinary and deuterated acetone crystals, *J. Phys. Chem.*, **99**, 14167–14173.

Ibberson, R.M. and Prager, M. (1995) The *ab initio* crystal-structure determination of perdeuterodimethylacetylene by high-resolution neutron powder diffraction, *Acta Crystallogr., Sect. B*, **51**, 71–76.

Ibberson, R.M. and Prager, M. (1996) The *ab initio* crystal structure determination of vapour-deposited methyl fluoride by high-resolution neutron powder diffraction, *Acta Crystallogr., Sect. B*, **52**, 892–895.

Jeffrey, G.A., Ruble, J.R., McMullan, R.K. and Pople, J.A. (1987) The crystal structure of deuterated benzene, *Proc. R. Soc. Lond. A*, **414**, 47–57.

Jeffrey, G.A. and Saenger, W. (1991a) Significance of small molecule crystal structural studies, in *Hydrogen Bonding in Biological Structures*, Ch. 1.4, 11–14, Springer-Verlag, Berlin.

Jeffrey, G.A. and Saenger, W. (1991b) X-ray and neutron crystal structure analysis, in *Hydrogen Bonding in Biological Structures*, Ch. 3.2, 52–58, Springer-Verlag, Berlin.

Jeffrey, G.A. and Saenger, W. (1991c) O–H• • •O hydrogen bonding in crystal structures of cyclic and linear oligoamyloses: cyclodextrins, maltotriose, and maltohexaose, in *Hydrogen Bonding in Biological Structures*, Ch. 18, 309–350, Springer-Verlag, Berlin.

Jeffrey, G.A. (1992) Accurate crystal structure analysis by neutron diffraction, in *Accurate Molecular Structures*, Domenicano, A. and Hargittai, I. (Eds), Ch. 11, 270–298, IUCr Monographs on Crystallography 1, Oxford University Press, New York.

Johnson, C.K., Gabe, E.J., Taylor, M.R. and Rose, I.A. (1965) Determination by neutron and X-ray diffraction of the absolute configuration of an enzymatically formed $\alpha$-monodeuterioglycolate, *J. Am. Chem. Soc.*, **87**, 1802–1804.

Johnson, C.K. (1971) ORTEPII, *Oak Ridge National Laboratory Report 3794*, revised.

Jouanneaux, A., Fitch, A.N. and Cockcroft, J.K. (1992) The crystal structure of $CBrF_3$ by high-resolution powder neutron diffraction, *Mol. Phys.*, **71**, 45–50.

Kampermann, S.P., Ruble, J.R. and Craven, B.M. (1994) The charge-density distribution in hexamethylenetetramine at 120 K, *Acta Crystallogr., Sect. B*, **50**, 737–741.

Keen, D.A. and Wilson, C.C. (1996) Single crystal diffraction at ISIS. User guide for the SXD instrument, *Rutherford Appleton Laboratory Technical Report*, RAL-TR-96-083.

Klooster, W.T., Swaminathan, S., Nanni, R. and Craven, B.M. (1992) Electrostatic properties of 1-methyluracil from diffraction data, *Acta Crystallogr., Sect. B*, **48**, 217–227.

Knight, K.S., Shankland, K., David, W.I.F., Shankland, N. and Love, S.W. (1996) The crystal structure of perdeuterated pyrene II at 4.2 K, *Chem. Phys. Letters*, **258**, 490–494.

Knight, K.S., Shankland, K. and Shankland, N. (1999) Solid state polymorphic transformations induced by variations in temperature and pressure, *Acta Crystallogr., Sect. A*, **55**, Supplement, Abstract M07.EE.004.

Koetzle, T.F., Takusagawa, F. and Kvick, Å. (1978) Structure and charge-density analysis of pyridine-2,3-dicarboxylic acid: a material with a short, asymmetric intramolecular O• • •O hydrogen bond, *Acta Crystallogr., Sect. A*, **34**, Supplement, Abstract 02.3–8.

Koritsanszky, T., Buschmann, J., Denner, L., Luger, P., Knöchel, A., Haarich, M. and Patz, M. (1991) Low-temperature X-ray and neutron diffraction studies on 18-crown-6•2 cyanamide including electron-density determination, *J. Am. Chem. Soc.*, **113**, 8388–8398.

Koritsanszky, T., Buschmann, J. and Luger, P. (1993) Charge-density study on bullvalene ($C_{10}H_{10}$), *Z. Naturforschung., Sect. A*, **48**, 55–57.

Koski, H.K. and Sándor, E. (1975) Neutron powder diffraction study of the low-temperature phase of solid acetylene-$d_2$, *Acta Crystallogr., Sect. B*, **31**, 350–353.

Kvick, Å., Thomas, R. and Koetzle, T.F. (1976) Hydrogen bond studies. CI. A neutron diffraction study of 2-amino-5-chloropyridine, *Acta Crystallogr., Sect. B*, **32**, 224–231.

Kvick, Å., Koetzle, T.F. and Stevens, E.D. (1979) Deformation electron density of $\alpha$-glycylglycine at 82 K. II. The X-ray diffraction study. *J. Chem. Phys.*, **71**, 173–179.

Larsen, A.C. and Von Dreele, R.B. (1994) GSAS — General structure analysis system, *Los Alamos National Laboratory Report*, LAUR 86-748.

Lehmann, A., Luger, P., Lehmann, C.W. and Ibberson, R.M. (1994) Oxalic acid dihydrate — an accurate low-temperature structural study using high-resolution neutron powder diffraction, *Acta Crystallogr. Sect. B.*, **50**, 344–348.

Longchambon, F., Gillier-Pandraud, H., Wiest, R., Rees, B., Mitschler, A., Feld, R., Lehmann, M. and Becker, P. (1985) Etude structurale et densité de déformation electronique $X - N$ à 75 K dans la région anomère du $\beta$-DL-arabinose, *Acta Crystallogr., Sect. B*, **41**, 47–56.

Love, S.W. (1997) Refinement of organic molecular crystal structures from powder diffraction data, Thesis, University of Strathclyde, Glasgow, UK.

Lundgren, J.-O. and Tellgren, R. (1974) Hydrogen bond studies. LXXXVI. An asymmetric non-centred $H_5O_2^+$ ion: neutron diffraction study of picrylsulphonic acid tetrahydrate, $[H_5O_2]^+$ $[C_6H_2(NO_2)_3SO_3]^-.2H_2O$, *Acta Crystallogr., Sect. B*, **30**, 1937–1947.

Madsen, D., Flensburg, C. and Larsen, S. (1998a) Properties of the experimental crystal charge density of methylammonium hydrogen maleate. A salt with a very short intramolecular O–H–O hydrogen bond, *J. Phys. Chem. A*, **102**, 2177–2188.

Madsen, G.K.H., Iversen, B.B., Larsen, F.K., Kapon, M., Reisner, G.M. and Herbstein, F.H. (1998b) Topological analysis of the charge density in short intramolecular O–H● ● ●O hydrogen bonds. Very low temperature X-ray and neutron diffraction study of benzoylacetone, *J. Am. Chem. Soc.*, **120**, 10040–10045.

Mallinson, P.R., Wozniak, K., Smith, G.T. and McCormack, K.L. (1997) A charge density analysis of cationic and anionic hydrogen bonds in a "proton sponge" complex, *J. Am. Chem. Soc.*, **119**, 11502–11509.

Matthews, D.A. and Stucky, G.D. (1971) Bonding and valence electron distributions in molecules. Experimental determination of aspherical electron charge density in tetracyanoethylene oxide, *J. Am. Chem. Soc.*, **93**, 5954–5959.

Myers, R., Torrie, B.H. and Powell, B.M. (1983) Crystal structures of solid bromoform, *J. Chem. Phys.*, **79**, 1495–1504.

Ohms, U., Guth, H., Treutmann, W., Dannöhl, H., Schweig, A. and Heger, G. (1985) Crystal structure and charge density of 4-methylpyridine ($C_6H_7N$) at 120 K, *J. Chem. Phys.*, **83**, 273–279.

Parfonry, A., Declercq, J.-P., Tinant, B., Van Meerssche, M. and Schweiss, P. (1988) Electron-density analysis at 155 K by X-ray and neutron diffraction of *meso*-2,3-bis(dimethylamino)butanedinitrile, $C_8H_{14}N_4$, *Acta Crystallogr., Sect. B*, **44**, 435–440.

Pawley, G.S. (1981) Unit cell refinement from powder diffraction scans, *J. Appl. Crystallogr.*, **14**, 357–361.

Pawley, G.S. and Hewat, A.W. (1985) The crystal structure of chlorotrifluoromethane, $CF_3Cl$; neutron powder diffraction and constrained refinement, *Acta Crystallogr., Sect. B*, **41**, 136–139.

Pawley, G.S. and Whitley, E. (1988) Structure of sym-$C_2F_4Br_2$, *Acta Crystallogr., Sect. C*, **44**, 1249–1251.

Pollock, J. McC., Ubbelohde, A.R. and Woodward, I. (1956) Hydrogen bonds in crystals X. The isotope effect and thermal expansion of non-cooperative hydrogen bonds in furoic acid, *Proc. R. Soc. A*, **235**, 149–158.

Prager, M., David, W.I.F. and Ibberson, R.M. (1991) Methyl rotational excitations in *p*-xylene: a test of pair interaction potentials, *J. Chem. Phys.*, **95**, 2473–2480.

Press, W. and Eckert, J. (1976) Structure of solid ethylene-d$_4$, *J. Chem. Phys.*, **65**, 4362–4364.

Prystupa, D.A., Torrie, B.H., Powell, B.M. and Gerlach, P.N. (1989) Crystal structures of methylene bromide and methylene iodide, *Mol. Phys.*, **68**, 835–851.

Putkonen, M.-L., Feld, R., Vettier, C. and Lehmann, M.S. (1985) Powder neutron diffraction analysis of the hydrogen bonding in deutero-oxalic acid dihydrate at high pressures, *Acta Crystallogr., Sect. B*, **41**, 77–79.

Refson, K. and Pawley, G.S. (1986) The structure and orientational disorder in solid $n$-butane by neutron powder diffraction, *Acta Crystallogr., Sect. B*, **42**, 402–410.

Rietveld, H.M. (1969) A profile refinement method for nuclear and magnetic structures, *J. Appl. Crystallogr.*, **2**, 65–71.

Rosenfield Jr., R.E. and Murray-Rust, P. (1982) Analysis of the atomic environment of quaternary ammonium groups in crystal structures, using computerized data retrieval and interactive graphics: modeling acetylcholine-receptor interactions, *J. Am. Chem. Soc.*, **104**, 5427–5430.

Ruble, J.R. and Galvao, A. (1995) Electrostatic potentials from charge-density studies of benzamide at 123 K, *Acta Crystallogr., Sect. B*, **51**, 835–838.

Sabine, T.M., Cox, G.W. and Craven, B.M. (1969) A neutron diffraction study of $\alpha$-oxalic acid dihydrate, *Acta Crystallogr., Sect. B*, **25**, 2437–2441.

Scheringer, C., Kutoglu, A. and Mullen, D. (1978) $X - N$ maps from room-temperature data — a warning, *Acta Crystallogr., Sect. A*, **34**, 481–483.

SCRIP (1998) No. 2370, 13, PJB Publications Ltd.

Sears, V.F. (1992) Neutron scattering lengths and cross sections, *Neutron News*, **3**, 29–37, Gordon and Breach.

Seiler, P., Martinoni, B. and Dunitz, J.D. (1984) Can X-ray diffraction distinguish between protium and deuterium atoms? *Nature*, **309**, 435–438.

Shankland, N., Love, S.W., Watson, D.G., Knight, K.S., Shankland, K. and David, W.I.F. (1996) Constrained Rietveld refinement of $[\beta\text{-}^1\mathrm{H}_1]$decadeuteriodopamine deuteriobromide using powder neutron diffraction data, *J. Chem. Soc., Faraday Trans.*, **92**, 4555–4559.

Shankland, N., Florence, A.J., Cox, P.J., Wilson, C.C. and Shankland, K. (1998a) Conformational analysis of Ibuprofen by crystallographic database searching and potential energy calculation, *Int. J. Pharm.*, **165**, 107–116.

Shankland, N., Florence, A.J and Cannell, R.J.P. (1998b) Crystallization and final stages of purification in *Natural Products Isolation*, Cannell, R.J.P. (Ed.), Ch. 9, 261–278, Methods in Biotechnology 4, Humana Press, Totowa, New Jersey.

Steiner, T., Mason, S.A. and Saenger, W. (1991) Topography of cyclodextrin inclusion complexes. 27. Disordered guest and water molecules. Three-center and flip-flop O–H• • •O hydrogen bonds in crystalline $\beta$-cyclodextrin ethanol octahydrate at $T$ = 295 K: a neutron and X-ray diffraction study, *J. Am. Chem. Soc.*, **113**, 5676–5687.

Steiner, T. and Saenger, W. (1991) H• • •H van der Waals distance in cooperative O–H• • •O–H• • •O hydrogen bonds determined from neutron diffraction data, *Acta Crystallogr., Sect. B*, **47**, 1022–1023.

Steiner, T. and Saenger, W. (1992a) Covalent bond lengthening in hydroxyl groups involved in three-center and in cooperative hydrogen bonds. Analysis of low-temperature neutron diffraction data, *J. Am. Chem. Soc.*, **114**, 7123–7126.

Steiner, T. and Saenger, W. (1992b) Geometric analysis of non-ionic O–H• • •O hydrogen bonds and non-bonding arrangements in neutron diffraction studies of carbohydrates, *Acta Crystallogr., Sect. B*, **48**, 819–827.

Steiner, T. and Saenger, W. (1992c) Geometry of C–H• • •O hydrogen bonds in carbohydrate crystal structures. Analysis of neutron diffraction data, *J. Am. Chem. Soc.*, **114**, 10146–10154.

Steiner, T. and Saenger, W. (1993a) Role of C–H• • •O hydrogen bonds in the coordination of water molecules. Analysis of neutron diffraction data, *J. Am. Chem. Soc.*, **115**, 4540–4547.

Steiner, T. and Saenger, W. (1993b) Distribution of observed C–H bond lengths in neutron crystal structures and temperature dependence of the mean values, *Acta Crystallogr., Sect. A*, **49**, 379–384.

Steiner, T. and Saenger, W. (1994) Lengthening of the covalent O–H bond in O–H• • •O hydrogen bonds re-examined from low-temperature neutron diffraction data of organic compounds, *Acta Crystallogr., Sect. B*, **50**, 348–357.

Stevens, E.D. and Hope, H. (1975) Accurate positional and thermal parameters of hexamethylenetetramine from $K$-shell X-ray diffraction data, *Acta Crystallogr., Sect. A*, **31**, 494–498.

Stevens, E.D., Lehmann, M.S. and Coppens, P. (1977) Experimental electron density distribution of sodium hydrogen diacetate. Evidence for covalency in a short hydrogen bond, *J. Am. Chem. Soc.*, **99**, 2829–2831.

Stevens, E.D. (1980) Comparison of theoretical and experimental electron density distributions of oxalic acid dihydrate, *Acta Crystallogr., Sect. B*, **36**, 1876–1886.

Stevens, E.D. and Coppens, P. (1980) Experimental electron density distributions of hydrogen bonds. High-resolution study of $\alpha$-oxalic acid dihydrate at 100 K, *Acta Crystallogr., Sect. B*, **36**, 1864–1876.

Stewart, R.F. (1970) Valence structure from X-ray diffraction data: an $L$-shell projection method, *J. Chem. Phys.*, **53**, 205–213.

Su, Z. and Coppens, P. (1992) On the mapping of electrostatic properties from the multipole description of the charge density, *Acta Crystallogr., Sect. A*, **48**, 188–197.

Swaminathan, S. and Craven, B.M. (1984) Electrostatic properties of phosphorylethanolamine at 123 K from crystal diffraction, *Acta Crystallogr., Sect. B*, **40**, 511–518.

Swaminathan, S., Craven, B.M., Spackman, M.A. and Stewart, R.F. (1984) Theoretical and experimental studies of the charge density in urea, *Acta Crystallogr., Sect. B*, **40**, 398–404.

Swaminathan, S., Craven, B.M. and McMullan, R.K. (1985) Alloxan — electrostatic properties of an unusual structure from X-ray and neutron diffraction, *Acta Crystallogr., Sect. B*, **41**, 113–122.

Takusagawa, F. and Koetzle, T.F. (1979) A study of the charge density in putrescine diphosphate at 85 K, *Acta Crystallogr., Sect. B*, **35**, 867–877.

Taylor, J.C. and Sabine, T.M. (1972) Isotope and bonding effects in ammonium oxalate monohydrate, determined by the combined use of neutron and X-ray diffraction analyses, *Acta Crystallogr., Sect. B*, **28**, 3340–3351.

Tellgren, R. and Olovsson, I. (1971) Hydrogen bond studies. XXXXVI. The crystal structures of normal and deuterated sodium hydrogen oxalate monohydrate $NaHC_2O_4.H_2O$ and $NaDC_2O_4.D_2O$, *J. Chem. Phys.*, **54**, 127–134.

Tellgren, R., Thomas, J.O. and Olovsson, I. (1977) Hydrogen bond studies. CX. A neutron diffraction and deformation electron density study of sodium hydrogen oxalate monohydrate, $NaHC_2O_4.H_2O$, *Acta Crystallogr., Sect. B*, **33**, 3500–3504.

Terpstra, M., Craven, B.M. and Stewart, R.F. (1993) Hexamethylenetetramine at 298 K: new refinements, *Acta Crystallogr., Sect. A*, **49**, 685–692.

Thomas, J.O. (1973) Hydrogen-bond studies. LXXV. An X-ray diffraction study of normal and deuterated hydrazinium hydrogen oxalate, $N_2H_5HC_2O_4$ and $N_2D_5DC_2O_4$, *Acta Crystallogr., Sect. B*, **29**, 1767–1776.

Thomas, J.O., Tellgren, R. and Almlöf, J. (1975) Hydrogen bond studies. XCVI. $X - N$ maps and *ab initio* MO-LCAO-SCF calculations of the difference electron density in non-centrosymmetric lithium formate monohydrate, $LiHCOO.H_2O$, *Acta Crystallogr., Sect. B*, **31**, 1946–1955.

Thomas, J.O. (1977) Hydrogen bond studies. CXXII. A neutron diffraction and $X - N$ deformation-electron-density study of dimethylammonium hydrogen oxalate, $(CH_3)_2NH_2HC_2O_4$, at 298 K, *Acta Crystallogr., Sect. B*, **33**, 2867–2876.

Thomas, J.O. (1978) $X - N$ and multipole deformation electron density maps for the non-centrosymmetric structure lithium formate monohydrate, $LiHCOO.H_2O$, *Acta Crystallogr., Sect. A*, **34**, 819–823.

Thomas, J.O. and Liminga, R. (1978) The electron and proton distribution in the short O–H–O bond in hydrazinium hydrogen oxalate, $N_2H_5HC_2O_4$, *Acta Crystallogr., Sect. B*, **34**, 3686–3690.

Tonogaki, M., Kawata, T., Ohba, S., Iwata, Y. and Shibuya, I. (1993) Electron-density distribution in crystals of *p*-nitrobenzene derivatives, *Acta Crystallogr., Sect. B*, **49**, 1031–1039.

Torrie, B.H., Weng, S.-X. and Powell, B.M. (1989) Structure of the $\alpha$-phase of solid methanol, *Mol. Phys.*, **67**, 575–581.

Torrie, B.H., Binbrek, O.S. and Von Dreele, R. (1993) Crystal structure of chloroiodomethane, *Mol. Phys.*, **79**, 869–874.

Trevino, S.F., Prince, E. and Hubbard, C.R. (1980) Refinement of the structure of solid nitromethane, *J. Chem. Phys.*, **73**, 2996–3000.

Trueblood, K.N. and Dunitz, J.D. (1983) Internal molecular motions in crystals. The estimation of force constants, frequencies and barriers from diffraction data. A feasibility study, *Acta Crystallogr., Sect. B*, **39**, 120–133.

Wang, Y., Blessing, R.H., Ross, F.K. and Coppens, P. (1976) Charge density studies below liquid nitrogen temperature: X-ray analysis of *p*-nitropyridine *N*-oxide at 30 K, *Acta Crystallogr., Sect. B*, **32**, 572–578.

Watkin, D.J. (1972) A simple small-scale recrystallization apparatus, *J. Appl. Crystallogr.*, **5**, 250.

Weber, H.-P. and Craven, B.M. (1990) Electrostatic properties of cytosine monohydrate from diffraction data, *Acta Crystallogr., Sect. B*, **46**, 532–538.

Wilding, N.B., Hatton, P.D. and Pawley, G.S. (1991) High-pressure phases of cyclohexane-$d_{12}$, *Acta Crystallogr., Sect. B*, **47**, 797–806.

Wilding, N.B., Crain, J., Hatton, P.D. and Bushnell-Wye, G. (1993) Structural studies of cyclohexane IV, *Acta Crystallogr., Sect. B*, **49**, 320–328.

Wilson, C.C. (1990) The data analysis of reciprocal space volumes, in *Neutron scattering data analysis 1990*, IOP Series 107, Johnson, M.W. (Ed.), Ch. 2, 145–163, Adam Hilger, Bristol.

Wilson, C.C., Shankland, N. and Florence, A.J. (1996) A single-crystal neutron diffraction study of the temperature dependence of hydrogen-atom disorder in benzoic acid dimers, *J. Chem. Soc., Faraday Trans.*, **92**, 5051–5057.

Wilson, C.C. (1997a) Neutron diffraction of *p*-hydroxyacetanilide (Paracetamol): libration or disorder of the methyl group at 100 K, *J. Mol. Struct.*, **405**, 207–217.

Wilson, C.C. (1997b) Zero point motion of the librating methyl group in *p*-hydroxyacetanilide, *Chem. Phys. Letters*, **280**, 531–534.

Wipff, G. and Boudon, S. (1992) The importance of accurate structure determination in organic chemistry, in *Accurate Molecular Structures*, Domenicano, A. and Hargittai, I. (Eds), Ch. 16, 379–411, IUCr Monographs on Crystallography 1., Oxford University Press, New York.

Yamamuro, O., Matsuo, T., Suga, H., David, W.I.F., Ibberson, R.M. and Leadbetter, A.J. (1992) Neutron diffraction and calorimetric studies of methylammonium iodide, *Acta Crystallogr., Sect. B*, **48**, 329–336.

Young, R.A. (1995) Introduction to the Rietveld Method, in *The Rietveld Method*, Young, R.A. (Ed.), Ch. 1, 1–38, IUCr Monographs on Crystallography 5, Oxford University Press, Oxford.

Zobel, D., Luger, P., Dreissig, W. and Koritsanszky, T. (1992) Charge density studies on small organic molecules around 20 K: oxalic acid dihydrate at 15 K and acetamide at 23 K, *Acta Crystallogr., Sect. B*, **48**, 837–848.

# 3. The Liquid Structure of Halocarbons

ANDREW N. BURGESS[1], KATHLEEN A. JOHNSON[2], KATHARINE A. MORT[3], and
W. SPENCER HOWELLS[4]

[1]*ICI Technology, Runcorn;* [2]*University of Liverpool;* [3]*University of Keele;* [4]*Rutherford
Appleton Laboratory, Oxon, UK*

## 3.1 INTRODUCTION

The development of insights into structure-property relationships holds the promise that in
some cases, one might be able to apply the relationship in reverse. That is, to use a desired
property as the starting point and work out what structure would be required to produce
such a property.

The halocarbons comprise an industrially important and chemically interesting set of
molecules. Consider, for instance, the methane derivatives: depending on the degree of
substitution, the molecules can be quasi-spherical such as $CF_4$ or symmetric tops such as
$CClF_3$. They can have a large dipole moment, for example, $CHF_3$, or none. The halogens
also display a wide range of polarizability. These features are expected to play a role in the
nature of the liquid structure: is the liquid structure of a particular halomethane driven by
Coulombic or steric effects? Does the polarizability of the halogen effect the structure? Does
the known anisotropy associated with chlorine containing molecules have a significant effect
when modelling these compounds? Understanding the changes in structure and properties
resulting from the substitution of anything from one to all four of the hydrogens of methane
by any combination of fluorine, chlorine or bromine represents an interesting and perhaps
achievable challenge. To tackle the structures of all the possible members of this set of
molecules is not, however, an attractive prospect; nor hopefully necessary. If the goal of
developing transferable potentials to predict the properties and structure of these relatively

simple liquid systems can be achieved, this approach could be used with more confidence elsewhere.

In the case where the material of interest is a liquid and the properties are thermodynamic in nature, the relationship can be expressed via the Clausius Virial Theorem:

$$pV_m = RT - \frac{N_A}{6} \int_0^\infty \frac{dv(r)}{dr} \rho(r) 4\pi r^3 dr \qquad (1)$$

Thus, the pressure $p$ of a liquid at a given temperature $T$ is related to the potential energy $v(r)$ arising from the interatomic separations in the liquid of radial density, $\rho(r)$, and it is necessary therefore to determine not only the structure of the liquid but also the internal potential energy associated with that structure.

One approach to this problem is to use a combination of diffraction and molecular simulation techniques. Liquid diffraction yields the total radial distribution function which can be used to refine the potential parameters employed in a molecular dynamics or Monte Carlo simulation of the liquid. Careful examination of the simulation so fitted to the diffraction data then provides a mechanism for examining the individual pairwise contributions to the total radial distribution function. This provides a useful method of unravelling the details of liquid structure in systems where isotopic substitution may not be an option, either because the isotopes are too expensive or do not exist.

A number of studies comparing liquid structures derived from simulation studies with the results of neutron diffraction have been made over recent years. These include $CH_3Cl$ (Bohm *et al.*, 1985), $CH_2Cl_2$ (Kneller and Gieger, 1989; Bohm *et al.*, 1985), $CHCl_3$ (Evans, 1983), $CBrF_3$ (Burgess *et al.*, 1995; Mort *et al.*, 1998), $CClF_3$ (Mort *et al.*, 1998), $CCl_2F_2$ (Hall *et al.*, 1992; Mort *et al.*, 1998) and $CHF_3$ (Mort *et al.*, 1997). There has been particular interest and discussion regarding the liquid structure of $CCl_4$ (Egelstaff *et al.*, 1971; Gubbins *et al.*, 1973; Lowden & Chandler, 1974; Bermejo *et al.*, 1988; MacDonald *et al.*, 1982).

In recent years, we have determined the liquid structure of some of the more accessible halocarbons, in particular, the methane derivatives and the insights we have gained to date are presented here.

### 3.1.1. Liquid Diffraction

Liquid diffraction experiments can be performed using either X-ray or neutron scattering techniques. However, many of the halomethanes are gases under ambient conditions and it is convenient to be able to use a sample cell capable of withstanding modest operating pressures while not itself contributing much to the scattering. This may be readily achieved in neutron scattering by the use of a cell constructed from the null scattering alloy, TiZr. Zirconium has a positive scattering length, while that of titanium is negative. Hence, rather analogous to the case of null scattering $H_2O:D_2O$ mixtures, a TiZr alloy of the right atomic ratio results in a zero coherent scattering length density.

In the case of neutrons, the total differential scattering cross-section per molecule $\left(\frac{\delta\sigma}{\delta\Omega}\right)_{total}$ is the sum of a coherent and incoherent contribution. The coherent component comprises scattering arising from correlations between different nuclei (the distinct part) and the same nuclei (the self part). The distinct part in turn arises from correlations arising within the same molecule (intramolecular) and between different molecules (intermolecular). Thus, we obtain,

$$\left(\frac{\delta\sigma}{\delta\Omega}\right)_{total} = \left(\frac{\delta\sigma}{\delta\Omega}\right)_{intra}^{dist} + \left(\frac{\delta\sigma}{\delta\Omega}\right)_{inter}^{dist} + \left(\frac{\delta\sigma}{\delta\Omega}\right)^{self} + \left(\frac{\delta\sigma}{\delta\Omega}\right)^{incoh} \tag{2}$$

The distinct terms may be more simply expressed in terms of the scattering function $S(Q)$,

$$S(Q) = F(Q) + D_m(Q) \tag{3}$$

where $F(Q)$, the intramolecular form factor, is given by,

$$F(Q) = \frac{1}{\left(\sum b_i\right)^2} \sum_{ij} b_i b_j \frac{\sin(Qr_{ij})}{Qr_{ij}} \exp\left(-\frac{1}{2}\langle u_{ij}^2\rangle Q^2\right) \tag{4}$$

where $b_i$ is the scattering length of the $i$th nucleus, $r_{ij}$ is the distance between nuclei $i$ and $j$ and the exponential term is the Debye-Waller factor where $u_{ij}$ is the amplitude of vibration of atom $i$ relative to atom $j$.

The term $D_m(Q)$ can also be expressed in terms of the scattering-length weighted sum of the partial pairwise intermolecular structure factors

$$D_m(Q) = \sum_{ij} b_i b_j S_{ij}(Q) \tag{5}$$

$S(Q)$ is related to $G(r)$ by,

$$4\pi r\rho[g(r) - 1] = \frac{2}{\pi}\int_0^\infty QS(Q)\sin(Qr)dQ \tag{6}$$

So that $G(r)$ is related to the individual pair distribution functions $g_{ij}(r)$ via

$$G(r) = \sum_{ij} b_i b_j g_{ij}(r) \tag{7}$$

However, straightforward Fourier transformation of the scattering data can give rise to anomalies arising from truncation errors. If equation (6) which relates $g(r)$ and $S(Q)$ is reversed one obtains,

$$S(Q) = \frac{4\pi\rho}{Q}\int_0^\infty r[g(r) - 1]\sin(Qr)dr \tag{8}$$

Equation (8) provides a way of using a Monte Carlo approach to the generation of a $g(r)$ which can then be fitted to the experimentally derived $S(Q)$. In our work we use this approach and employ the Minimum Information Method of Soper (Soper, 1990) to perform this fitting procedure. The basic principle behind the Minimum Information method is to produce the smoothest pair distribution function consistent with the experimental data. Rather than transform the $S(Q)$ into a pair distribution function, a pair distribution function is produced and then converted into an $S(Q)$ via a Fourier transform and compared to the experimental data. This avoids the problem of cut-offs in the Fourier transform, and hence ringing. Peaks and troughs are introduced into the pair distribution function at random and the moves are accepted based on two criteria:

1. The change in the pair distribution function increases the agreement to the experimental data.
2. The change in the pair distribution function reduces the amount of 'noise' in the pair distribution function.

Here the definition of noise is the number of turning points in the pair distribution function. If the criteria are balanced correctly then the pair distribution function produced will have the minimum number of peaks that are consistent with the data. Thus spurious peaks in the pair distribution function that have no physical meaning will be avoided.

For details regarding a fuller account of the data reduction procedure such as the necessary corrections for inelastic scattering, the reader is referred to other more specialized sources, e.g. (Soper *et al.*, 1989).

### 3.1.2  Computer Simulation of Fluids

There are three principal ways of simulating the structure of relatively complex fluids like halocarbons at the atomic level: Molecular Dynamics (MD), Metropolis Monte Carlo (MMC) and Reverse Monte Carlo (RMC). In the MMC and RMC methods, alterations in the molecular centres of mass and rotations are made on a random basis within constraints defined by the user. In MMC, the energy change associated with this random alteration to the liquid structure is calculated based on the intermolecular or interatomic potentials being used. If the energy is found to decrease as a result of the move, it is accepted. If the energy is found to increase, the move is accepted with a Boltzmann probability (Metropolis *et al.*, 1953).

In RMC, no potential is used and the random changes in the structure are assessed by calculating the resulting $S(Q)$ and comparing it with experimental scattering data. Acceptance of the move results if the comparison improves but 'bad' moves can also be accepted with a probability defined by the statistical errors in the data (McGreevy *et al.*, 1990; Wicks, 1993).

MD simulations are intrinsically different to both MMC and RMC. Potential energy expressions are used to calculate the intermolecular forces resulting from a given separation and Newton's equations of motion are applied to calculate the resulting accelerations. The new atomic positions so derived after some short time interval (typically femtoseconds) are used to generate a new set of forces and so the cycle is continued for thousands of time steps (picoseconds to nanoseconds of simulation time). After an initial equilibration period, the time-averaged results are used to accumulate $g(r)$ and the internal energy of the system. Excellent texts on the details of the technique and its application exist for the interested reader e.g. (Allen & Tildesley, 1987).

In our work, we use the molecular dynamics program DLPOLY (Forester & Smith) with an effective pair potential derived from the Lennard-Jones site-site potential energy function with the addition of electrostatic interactions via simple fractional charges on the atoms, derived from *ab initio* molecular orbital calculations with Mulliken population analysis. The form of the potential is given by

$$\Phi_{ab}(r_{ij}) = 4\varepsilon_{ab}\left[\left(\frac{\sigma_{ab}}{r_{ij}}\right)^{12} + \left(\frac{\sigma_{ab}}{r_{ij}}\right)^{6}\right] + \frac{q_a q_b}{4\pi\varepsilon_0 r_{ij}} \tag{9}$$

where $a$ and $b$ denote two sites on different molecules $i$ and $j$, $q_{\text{site}}$ is the fractional charge on the site, $r_{ij}$ is the distance between site $a$ on molecule $i$ and site $b$ on $j$, $\varepsilon$ and $\sigma$ are the LJ interaction parameters, and $\varepsilon_0$ is the permittivity of free space.

The parameters for the unlike interactions are derived using Lorentz-Berthelot mixing rules, i.e.,

$$\sigma_{ab} = \frac{1}{2}(\sigma_{aa} + \sigma_{bb}); \; \varepsilon_{ab} = (\varepsilon_{aa}\varepsilon_{bb})^{\frac{1}{2}} \tag{10}$$

The Lennard-Jones parameters are adjusted in order to achieve agreement with both the experimental thermodynamic data and the neutron weighted $g(r)$. If it is found that no satisfactory agreement can be obtained, the calculated charges and the Lorentz-Berthelot derived unlike interactions are also adjusted.

## 3.2 RESULTS

### 3.2.1 Neutron Diffraction

All of the halogenated methane derivatives we have studied have been gases at room temperature and so have been contained in pressure vessels for the neutron measurements as described earlier. We aimed to study the structure of the coexisting liquid at the temperature of the measurements.

It is helpful to divide the halomethanes into two broad categories: those which contain hydrogen and those which do not. From the former category, we have studied $CHF_3$ and $CH_2F_2$. Measurements have also been made on $CClF_2H$, but a clear interpretation of the liquid structure of this fluid has so far not proved to be possible. In the latter category, we have made measurements on $CClF_3$, $CBrF_3$ and $CCl_2F_2$. Standard data analysis techniques were used to obtain the total atomic structure factors $S(Q)$ and the total pair distribution functions $g(r)$ for all five of these molecules (Hall *et al.*, 1991; Burgess *et al.*, 1995; Mort *et al.*, 1996; Mort *et al.*, 1998). The hydrogen containing molecules were more difficult to analyse and considerable care was taken with the data corrections in obtaining the structure factors reproduced in Figure 1 (Mort *et al.*, 1997). The minimum information method was used to transform the structure factors into the neutron weighted total pair distribution functions illustrated in Figure 2. These total pair distribution functions contain several features that demonstrate the power of neutron diffraction in obtaining liquid structures.

Firstly, each of the pair distribution functions show sharp peaks at low $r$. These are due to intramolecular interatomic separations. The position and shape of these peaks can be fitted to obtain these separations and the corresponding Debye Waller factors for the molecule in question. For some of the molecules reported here, there were significant differences between the liquid phase intramolecular structure obtained from the neutron measurements

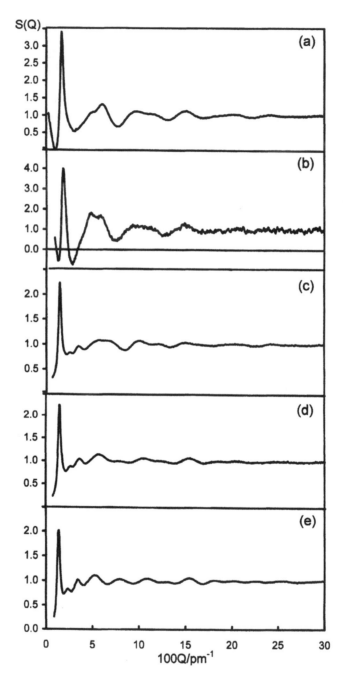

FIGURE 1.  Atomic structure factor $S(Q)$ for liquid halomethanes: (a) trifluoromethane; (b) difluoromethane; (c) bromotrifluoromethane; (d) chlorotrifluoromethane; (e) dichlorodifluoromethane.

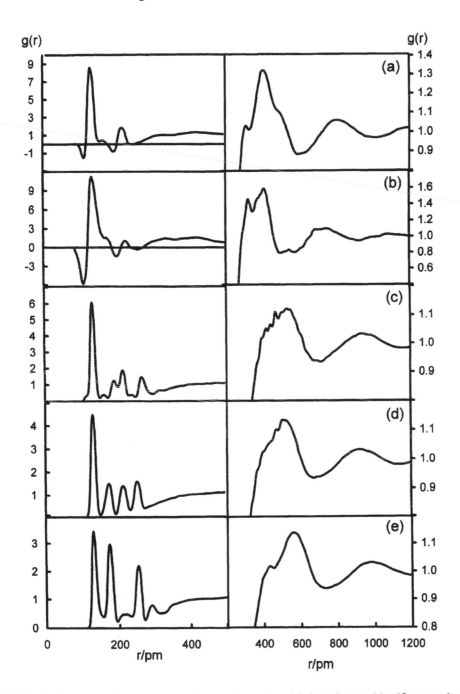

FIGURE 2. Total pair distribution functions $g(r)$ for liquid halomethanes: (a) trifluoromethane; (b) difluoromethane; (c) bromotrifluoromethane; (d) chlorotrifluoromethane; (e) dichlorodifluoromethane.

and that obtained by conventional means for isolated gas-phase molecules. For example it appears that the CH bond length in $CH_2F_2$ is significantly longer in the liquid phase (approximately 115 pm) than in the gas phase (approximately 109 pm).

Secondly, for the hydrogen containing species, it will be observed that some of the sharp peaks are negative; this is due to the negative scattering length of hydrogen. This negative scattering can be an advantage if measurements can also be made on the deuterated (and hence positively scattering) sample as the contrast allows clear assignment of the peaks. However, if isotopically substituted measurements are not available then the presence of positive and negative peaks very close to one another can make the determination of intramolecular separations more difficult.

Finally, the total pair distribution functions of all of the species show clear intermolecular structure in the first shell. If the molecules were arranged randomly, the first shell would appear as a simple maximum with no additional structure. This is not the case for any of the halomethanes. In addition, for many of the species, the existence of more shells is apparent. Indeed, for the highly structured halomethanes such as trifluoromethane it is possible to observe five co-ordination shells at low temperature although at higher temperatures this reduces to three.

### 3.2.2 Molecular Dynamics Simulations

In order to compare a radial distribution function resulting from a molecular dynamics simulation with that obtained by neutron diffraction, the MD result must be neutron weighted. That is to say, the individual pairwise contributions to the total $g(r)$ must be weighted according to the product of the scattering lengths of the atoms in the pair. For example, in order to calculate the neutron weighted $g(r)$ for $CCl_2F_2$, the appropriately weighted sum is:

$$g(r) = 0.0321 g_{CC} + 0.1849 g_{CCl} + 0.1092 g_{CF} + 0.2664 g_{ClCl}$$
$$+ 0.0929 g_{FF} + 0.3146 g_{ClF} \tag{11}$$

This example illustrates the point that some pairs can contribute much more to the sum than others and therefore their pair distribution functions may ultimately be rather better determined.

Extensive work is required to match the shape of the first co-ordination shell of the total pair distribution functions obtained from MD simulation and diffraction experiments of even these relatively simple molecules. However, an additional constraint which must be observed is that the energy of the simulation agrees with experimental measurements of the enthalpy of vaporization $\Delta H_V$:

$$\Delta H_V = (PE + KE + p.V_m)_g - (PE + KE + p.V_m)_l \tag{12}$$

where $PE$ and $KE$ are the potential and kinetic energies, $p$ is the pressure and $V_m$ the molar volumes of the gas and liquid states. For a liquid well below its boiling point, one may make the approximation that $p.V_{m_l}$ and $PE_g$ are both close to zero and the following ideal approximation holds fairly well:

$$\Delta H_V \approx RT - PE_1 \tag{13}$$

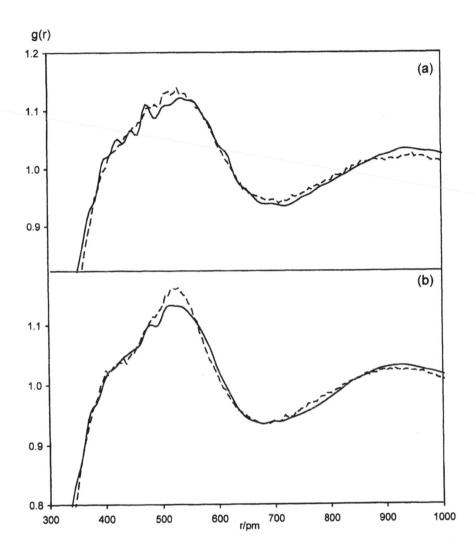

FIGURE 3. Total pair distribution functions $g(r)$ obtained from the neutron scattering (solid line) and from the computer simulation (dashed line) for (a) $CBrF_3$ and (b) $CClF_3$ at 153 K.

thus allowing a ready comparison between the experimental enthalpy of vaporization and the energy of the liquid simulation.

Good agreement has been found for non-hydrogen containing molecules such as $CClF_3$ and $CBrF_3$: the best simulation results are compared with the shape of the first co-ordination shell in Figure 3. For the hydrogen containing molecules, matching the shapes of the curves is much more difficult. For example, Figure 4a shows the best agreement we have obtained for $CHF_3$. This difficulty is partly due to the negative scattering length of hydrogen which

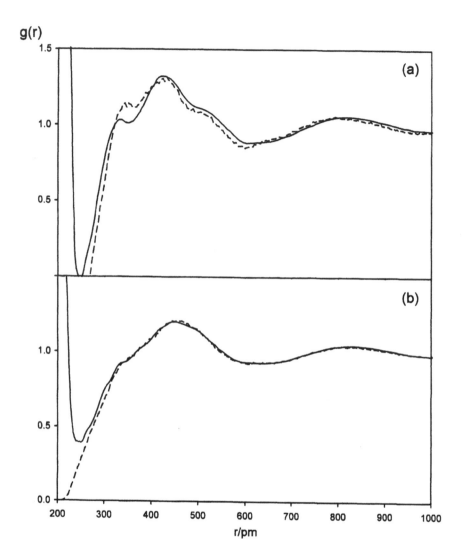

FIGURE 4. Total pair distribution functions $g(r)$ obtained from the neutron scattering (solid line) and from the computer simulation (dashed line) for (a) $CHF_3$ and (b) $CDF_3$ at 153 K.

means that small changes in the partial pair distribution functions leads to massive changes in the total pair distribution function. This is illustrated by our results for $CDF_3$. The same simulation, with the partial structure factors weighted for $CDF_3$ instead of $CHF_3$, gives very good agreement with the scattering result (see Figure 4b). Another factor which may explain the greater difficulty in fitting the radial distribution functions of hydrogenous species is the much higher incoherent scattering component can make the data analysis more prone to error due to the larger inelastic scattering corrections.

TABLE 1. Comparison between the experimental potential energies for the molecules studied with those obtained from molecular dynamics simulations.

| | Temperature (K) | $PE_{exp}$ (kJ mol$^{-1}$) | $P_{sim}$ (kJ mol$^{-1}$) |
|---|---|---|---|
| $CBrF_3$ | 153 | −19.2 | −19.6 |
| $CClF_3$ | 153 | −15.9 | −15.8 |
| $CCl_2F_2$ | 153 | −23.1 | −23.6 |
| $CHF_3$ | 153 | −17.6 | −17.6 |
| | 250 | −10.3 | −9.74 |

Table 1 indicates the extent of agreement achieved with respect to the energy of the system for each of these cases.

### 3.2.3. Structural Interpretation

A liquid has a dynamic structure wherein the molecules are continuously moving and reorienting themselves. It is more helpful therefore, to think about the time averaged or spatially averaged structure; hence the convention of referring to the pair distribution function. However, it can be useful to consider detailed aspects of this average structure; for instance, to think about possible orientations of a pair of molecules, and to analyse how each of these orientations may contribute to the total pair distribution functions. A pair of tetrahedral molecules can be regarded as being able to arrange in two differing ways which we refer to as the rocket and the straddle arrangements (Hall *et al.*, 1992). Other researchers have used the terms corner-to-face and corner-to-corner to describe the same arrangements. In a highly symmetrical molecule such as $CCl_4$, a more or less complete picture of the structure of the liquid can be visualised by specifying the relative populations of these two possible orientations. If the molecules have less symmetry, a collection of different rocket and straddle arrangements becomes possible, and a description of the liquid structure involves consideration of all these different arrangements. For illustration, all of the possible rocket and straddle arrangements of two $CBrF_3$ molecules are shown in Figure 5. In a well-ordered liquid, such as $CHF_3$, it may be necessary to describe the structure of the liquid in terms of groups of more than two molecules. This makes a simple description of the structure more difficult.

Once a simulation reproduces the shape of the first co-ordination shell and gives the correct internal energy, it is possible to look at the forms of the partial pair distribution functions to establish the arrangement of the molecules in the liquid state. A knowledge of the positions and the areas of the peaks in the partial pair distribution functions leads to information about the separation between two atoms of a particular type and of their co-ordination number.

To demonstrate the interpretation of the structures it is best to consider two different systems. The first, and simplest is $CBrF_3$. Figure 6 shows all the different pair distribution functions that must be added together to obtain the total pair distribution function of the fluid. By measuring the relative heights and the positions of the peaks, it is possible to conclude that the straddle arrangement with the bromine atoms facing each other

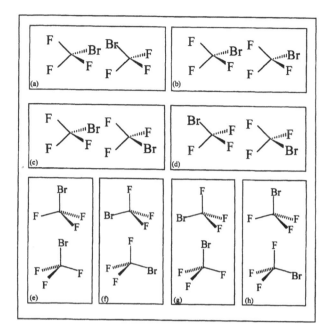

FIGURE 5. The possible arrangements for two bromotrifluoromethane molecules. Structures (a) to (d) are "straddle" arrangements and (e) to (h) are "rocket" arrangements.

(Figure 5(a)) is the most likely arrangement of two molecules in the liquid state. However, the rocket arrangement with the bromine atom interacting with the three fluorine atoms in a symmetrical manner (5(e)) and the more symmetrical straddle (5(b)) also occur (Burgess *et al.*, 1995). In addition it is possible to establish that the other straddle arrangements (5(c) and (d)) and the rocket arrangements (5(f), (g) and (h)) are relatively much less likely to occur in the liquid state at 153 K. The interpretation of the structure of $CClF_3$ is very similar, indicating that the driving forces for the liquid structure of both of these fluids is the same: namely, the steric effect of the large halogen atom (Mort *et al.*, 1998).

As a second example, $CHF_3$ provides an interesting contrast. The number of partial pair distribution functions is the same as for $CBrF_3$ but, as can be seen in Figure 7, their shapes are very different. Interpretation of the structure in terms of the different separations of atoms that can occur when two $CHF_3$ molecules align was impossible. Instead it is necessary to consider the separations that occur when a group of five molecules align; the main problem is in assigning the short HF separations. Figure 8 illustrates the final structure. This arrangement satisfies all of the observed atom separations and the relevant co-ordination numbers. This suggests a rigidity in the structure of $CHF_3$ that is absent in $CBrF_3$ and $CClF_3$. However it is important to realise that, when visualising the liquid structure, this is not a static arrangement but is rather the juxtaposition of molecules that arises most often.

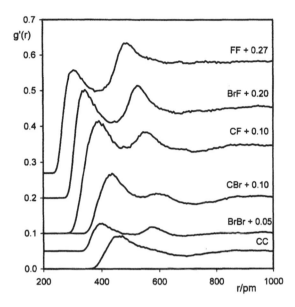

FIGURE 6. Individual pair distribution functions $g'(r)$ as obtained from the computer simulation for CBrF$_3$ at 153 K. In the figure each pair distribution function has been neutron weighted. The total pair distribution function is obtained by a simple summation of these pair distribution functions. For clarity each curve has been displaced vertically by the amount indicated on the diagram.

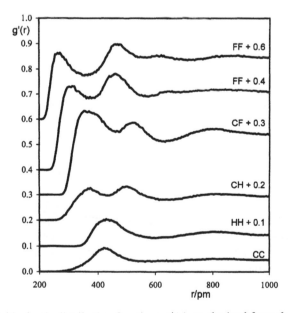

FIGURE 7. Individual pair distribution functions $g'(r)$ as obtained from the computer simulation for CHF$_3$ at 153 K.

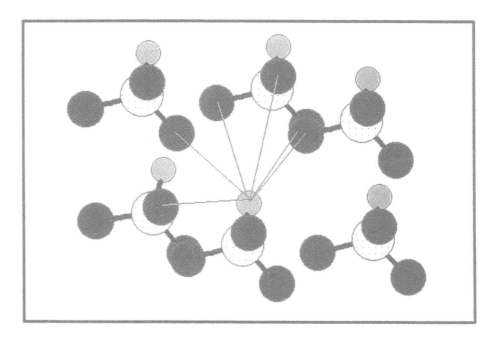

FIGURE 8. A possible liquid-state structure for trifluoromethane. The low ($r < 600$ pm) atomic separations this structure are consistent with those obtained in the simulation. Those distances indicated by fine lines drawn from fluorine atoms on different molecules to one of the hydrogen atoms are all less than 300 pm.


### 3.2.4. *Transferability of Potentials*

Table 2 shows the Lennard Jones potential parameters for $CBrF_3$, $CClF_3$ and $CCl_2F_2$. The other parameters in the potential, namely the fractional charges on the atoms, are also included for completeness.

Two things will be observed: firstly, in many cases, the Lorentz-Berthelot mixing rule values require modification and secondly, the transferability of parameters between molecules is rather poor.

One possible explanation for this is that while a potential description based on Lennard-Jones plus point charges can be made to account for the structure and energy of these systems quite well, such a model is nevertheless an approximation of the true situation. In other words, for any given system we are essentially looking at the way in which the model potential maps onto the true potential. This mapping is subtly different for each molecule and so the interatomic potential model parameters can assume different values for different molecules.

Thus, the goal of complete transferability is likely to depend on the development and use of more advanced potential descriptions.

TABLE 2. Parameters used in DLPOLY for the successful simulation of the halogenated methanes: $q_i$ is the partial charge on atom $i$. The subscripts $ij$ refer to the Lennard Jones parameters used to describe the intermolecular potential between atom i on one molecule and atom $j$ on a different molecule. $X$ is either Br or Cl as appropriate. The quantities in brackets are the values obtained from Lorentz-Berthelot mixing rules.

| | $CBrF_3$ | $CClF_3$ | $CCl_2F_2$ |
|---|---|---|---|
| $q_C/e$ | 0.1464 | 0.3562 | 0.1843 |
| $q_F/e$ | −0.0252 | −0.1018 | −0.0835 |
| $q_X/e$ | −0.0708 | −0.0508 | −0.0086 |
| $\varepsilon_{CC}/J\ mol^{-1}$ | 395 | 387 | 613 |
| $\sigma_{CC}/pm$ | 335 | 320 | 340 |
| $\varepsilon_{CF}/J\ mol^{-1}$ | 350 (350.5) | 332 (331.5) | 336 (461) |
| $\sigma_{CF}/pm$ | 308.5 (308.5) | 305 (302.5) | 312 (305) |
| $\varepsilon_{FF}/J\ mol^{-1}$ | 311 | 284 | 347 |
| $\sigma_{FF}/pm$ | 282 | 285 | 270 |
| $\varepsilon_{XF}/J\ mol^{-1}$ | 571 (571.5) | 531 (530.5) | 400 (627) |
| $\sigma_{XF}/pm$ | 320 (322) | 295 (315) | 320 (302.5) |
| $\varepsilon_{XC}/J\ mol^{-1}$ | 644 (644) | 620 (619) | 592 (832) |
| $\sigma_{XC}/pm$ | 349 (348.5) | 325 (332.5 ) | 365 (337.5) |
| $\varepsilon_{XX}/J\ mol^{-1}$ | 1050 | 991 | 1131 |
| $\sigma_{XX}/pm$ | 362 | 345 | 335 |

# References

Allen, M.P. and Tildesley, D.J. (1987) *Computer Simulation of Liquids*, Oxford University Press.

Bermejo, F.J., Enciso, E., Alonso, J., Garcia, N. and Howells, W.S. (1988) How well do we know the structure of simple molecular liquids — CCl₄ revisited, *Mol. Phys.* **64**, 1169

Bohm, H.J., Meissner, C. and Ahlrichs, R. Molecular dynamics simulation of liquid CH₂Cl₂ and CHCl₃ with new pair potentials, *Mol. Phys.* **54**, 1261.

Burgess, A.N., Mort, K.A., Johnson, K.A., Cooper, D.L., Rogers, S.C. and Howells, W.S. (1995) Neutron diffraction plus molecular dynamics — a powerful approach for understanding liquid structure. *Nuclear Instruments and Methods in Phys. Res. A* **354**, 81.

Egelstaff, P.A., Page, D.I. and Powells, J.G. (1971) Orientational correlations in molecular liquids by neutron scattering. Carbon tetrachloride and germanium tetrabromide. *Mol. Phys.* **20**, 881.

Evans, M.W. (1983) A review and computer simulation of the structure and dynamics of liquid chloroform. *J. Mol. Liquid* **25**, 211.

Forester, T.R. and Smith, W. CCP5 Program Library, Daresbury Laboratory.

Gubbins, K.E., Gray, C.G., Egelstaff, P.A. and Ananth, M.S. (1973) Angular Correlation effects in neutron diffraction from molecular fluids. *Mol. Phys.* **25**, 1353.

Hall, C.D., Johnson, K.A., Burgess, A.N., Winterton, N. and Howells, W.S. (1991) The structure of fluid fluoroform, chlorodifluoromethane, and dichlorodifluoromethane by neutron diffraction. *Mol. Phys.* **74**, 27.

Hall, C.D., Johnson, K.A., Burgess, A.N., Winterton, N. and Howells, W.S. (1992) The structure of fluid dichlorodifluoromethane — a comparison between molecular-dynamics simulation and neutron-diffraction results. *Mol. Phys.* **76**, 1061

Kneller, G.R. and Gieger, A. (1989) Molecular dynamics studies and neutron scattering experiments on methylene chloride. I Structure. *Mol. Phys.* **68**, 487.

Lowden, L.J. and Chandler, D. (1974) Theory of intermolecular pair correlations for molecular liquids. Applications to the liquids carbon tetrachloride, carbon disulfide, carbon diselenide and benzene. *J. Chem. Phys.* **61**, 5228.

McDonald, I.R., Bounds, D.G. and Klein, M.L. (1982) Molecular dynamics calculations for the liquid and cubic plastic crystal phases of carbon tetrachloride. *Mol. Phys.* **45**, 521.

McGreevy, R.L., Howe, M.A., Keen, D.A. and Clausen, K.N. (1990) Reverse monte-carlo (RMC) simulation — modeling structural disorder in crystals, glasses and liquids from diffraction data. *Neutron Scattering Data Analysis*, (Ed) M.W Johnson, IoP Conference Series **107**, 165.

Metropolis, N., Rosenbluth, A.W., Rosenbluth, M.N., Teller, A.H. and Teller, E.J. (1953) Equation of state calculations by fast computing machines. *J. Phys. Chem.* **21**, 1087.

Mort, K.A., Johnson, K.A., Cooper, D.L., Burgess, A.N. and Howells, W.S. (1997) The liquid structure of trifluoromethane *Mol. Phys.***90**, 415.

Mort, K.A., Johnson, K.A., Cooper, D.L., Burgess, A.N. and Howells, W.S. (1998) Liquid structure of halomethanes. *J Chem. Soc., Fraday Trans.*, **94**, 765.

Soper, A.K., Howells, W.S. and Hannon, A.C. (1989) ATLAS — Analysis of time-of-flight diffraction data from liquids and amorphous samples. *RAL–89–046*, Rutherford Appleton Laboratory.

Soper, A.K. (1990) An amateur guide to the pitfalls of maximum-entropy. *Neutron Scattering Data Analysis*, (Ed) M.W Johnson, IoP Conference Series **107**, 57.

Wicks, J.D. (1993) *D. Phil Thesis* Lincoln College, Oxford University.

# 4. Using SANS to Study Adsorbed Layers in Colloidal Dispersions

STEPHEN M. KING[1,*], PETER C. GRIFFITHS[2] and TERRY COSGROVE[3]

[1]*Large-Scale Structures Group, ISIS Facility, Rutherford Appleton Laboratory, Chilton, Didcot, Oxon. OX11 0QX, UK*
[2]*Department of Chemistry, University of Wales Cardiff, P.O. Box 912, Cardiff CF10 3TB, UK*
[3]*School of Chemistry, University of Bristol, Cantock's Close, Bristol BS8 1TS, UK*

## 4.1 INTRODUCTION

Adsorbed layers of polymer or surfactant molecules, or combinations of the two, are important because they are ubiquitous in everyday products. For example, they may be used to modify the behaviour of a product during processing or to stabilize that product against flocculation. As such polymer adsorption is found in applications as diverse as cosmetic and pharmaceutical preparations, paint and agrochemical formulations and even foodstuffs. Polymer adsorption is also important in the biological arena where, for example, it mediates in cell adhesion processes. Control of the structure of an adsorbed layer is therefore of paramount importance to the colloid scientist. Equally important however, is the ability to probe the arrangement of the molecules in the adsorbed layer and to extract quantitative information from what might well be a complex, multi-component system. In these respects small-angle neutron scattering (SANS) is certainly a successful technique.

SANS, together with small-angle X-ray (SAXS) and light (LS) scattering, is but one of a family of related transmission diffraction techniques. Each of these techniques is capable of providing information about the size, shape and orientation of some component

---

*Correspondence.

of the sample, but on much larger length scales than the atomic resolution of typical crystallographic experiments. The actual length scales probed, and consequently the type of sample that can be studied, and indeed the sample environment, all depend on the nature of the radiation employed. For example, LS obviously cannot be used to study optically opaque samples, whilst the shorter wavelengths employed in SANS (and SAXS) probe smaller length scales than LS. Thus to a large extent these techniques are complementary. They also share several similarities. Perhaps the most important of these is the fact that, with minor adjustments (to account for the different types of radiation), each technique is based on the same fundamental equations. Hence the Guinier and Zimm plots, traditionally methods of analysis employed by the light and X-ray communities, can just as easily be used to interpret SANS data.

This chapter gives only a brief overview of the technique of SANS in order to concentrate on its application to the study of adsorbed layers. Those readers requiring more detailed information are therefore referred to the many excellent texts available on LS (Guinier & Fournet, 1955; van de Hulst, 1957; Kerker, 1969; Brown, 1993), SAXS (Glatter & Kratky, 1982; Lindner & Zemb, 1991), and neutron scattering in general (Marshall & Lovesey, 1971; Willis, 1973; Kostorz, 1982; Lovesey, 1984; Skold & Price, 1986; Feigin & Svergun, 1987; Newport *et al.*, 1988). There are also several texts and papers dealing with the wider application of SANS in colloid and polymer science (Ottewill, 1982 & 1991; Hunter, 1989; Fleer *et al.*, 1993; Higgins & Stein, 1978; Maconnachie & Richards, 1978; Richards, 1989; Arrighi *et al.*, 1993; Higgins & Benoit, 1994; Perkins, 1994; King, 1999).

## 4.2  SMALL-ANGLE NEUTRON SCATTERING

In any SANS experiment a collimated, though not necessarily monochromatic, neutron beam with wavelengths typically between 0.1 and 3 nm, and having a diameter of perhaps 10 mm, is directed at a sample illuminating a small volume, $V$. Since the pathlength of a typical sample might be 1–5 mm, the volume of sample needed for an experiment is usually between 0.5 and 2 cm$^3$.

Some of the neutrons are transmitted by the sample, some are absorbed and some are scattered (mostly elastically). A detector, or detector element, of dimensions $dx$ by $dy$ positioned at some distance, $L_{sd}$, and scattering angle, $\theta$, from the sample then records the flux of neutrons scattered into a solid angle element, $\Delta\Omega (= dx\, dy/L_{sd}^2)$. For a given wavelength, $\lambda$, this flux, $I(\lambda, \theta)$, may be expressed in general terms in the following way (Newport *et al.*, 1988)

$$I(\lambda, \theta) = I_0(\lambda)\Delta\Omega\eta(\lambda)T(\lambda)V\frac{\partial\Sigma}{\partial\Omega}(Q) \tag{1}$$

where $I_0(\lambda)$ is the incident flux, $\eta(\lambda)$ is the detector efficiency, $T(\lambda)$ is the sample transmission and $(d\Sigma/d\Omega)(Q)$ is a function known as the *differential scattering cross-section*. Although this function is specific to neutron scattering (see the introductory chapter of this book for more details), analogous functions exist for X-rays and light. The first three terms of Equation 1 are clearly instrument-specific whilst the last three terms are sample-dependent. From an experimental viewpoint it is the last term which is the most important and so is considered in more detail below.

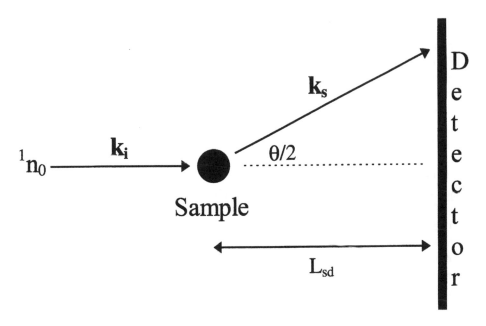

FIGURE 1. A diagrammatic representation of the geometry of a SANS experiment. See main text for an explanation of the symbols.

### 4.2.1 The Scattering Vector, Q

In neutron scattering the quantity $Q$ is known as the *scattering vector*. It is the modulus of the resultant wavevector between the incident, $k_i$, and scattered, $k_s$, wavevectors (see Figure 1) and its value is given by

$$Q = /\mathbf{Q}/ = /\mathbf{k}_s - \mathbf{k}_i/ = \frac{4\pi n}{\lambda} \sin(\theta/2) \qquad (2)$$

$Q$ has dimensions of $(\text{length})^{-1}$ and is normally quoted in $\text{nm}^{-1}$ or $\text{Å}^{-1}$. To a very good approximation, the neutron refractive index, $n$, may be taken as unity. In fact, and in contrast to optical refractive indices, $n$ is actually very slightly less than unity. This allows neutrons to be totally externally reflected from a surface (provided that the angle of incidence is less than some small critical angle), a property that is exploited in the rapidly expanding field of neutron reflectometery (NR) (Penfold & Thomas, 1990).

Substituting Equation 2 into Bragg's Law of Diffraction, viz.

$$\lambda = 2d \sin(\theta/2) \qquad (3)$$

then yields the very useful expression

$$d = \frac{2\pi}{Q} \qquad (4)$$

Here $d$ is a molecular-level length scale; in the original Bragg derivation $d$ was the separation between atomic planes in a crystal. Equations 2 and 4 are central to SANS experiments because through their combined use it is possible to both quickly and rapidly "size" the scattering centres in a sample from the position of any diffraction peak in $Q$-space and to configure an instrument (i.e., ensure that its "$Q$-range" will cover the length scales of interest). As an example, the SANS instrument at the ISIS Spallation Neutron Source, called $LOQ$, uses neutrons with wavelengths between 0.2 and 1.0 nm to provide a $Q$-range of approximately 0.08 to 14 nm$^{-1}$, thereby allowing it to probe length scales of between 0.4 and 80 nm, though longer length scales can often be accessed in systems exhibiting one-dimensional order (e.g., liquid crystals, shear-aligned micelles and layered clays).

Equation 2 also highlights the different approaches to SANS measurements at the different types of neutron source. At a steady-state source (e.g., a reactor) it is usual to vary $Q$ by effectively scanning $\theta$ at a pre-selected value of $\lambda$. Different ranges of $\theta$ are obtained by physically moving the detector. On a pulsed (e.g., accelerator-based) source it is more common to employ a "fixed instrument geometry" (effectively a constant $\theta$ mode) and to obtain a range of $Q$ values by time-sorting the arrival of neutrons with different wavelengths (from a polychromatic incident beam) at the detector. One of the consequences of this approach is that pulsed source SANS instruments typically have a greater dynamic range in $Q(Q_{\mathrm{max}}/Q_{\mathrm{min}})$ available in one instrument configuration than their steady-state source counterparts, even though the full $Q$-range actually accessible by both types of instrument is generally quite similar. Fixed-geometry instruments are therefore ideal for studying systems where a range of length scales are present, or possibly, where the length scales are uncertain. On the other hand, steady-state source instruments sometimes have the advantage that if the $Q$-range of interest is known then it may be possible to maximize the count rate by utilizing a wavelength close to the peak of the neutron flux distribution. However, it should be noted that the effective wavelength resolution of the two types of instrument is rather different and it is the wavelength resolution which largely determines the $Q$-resolution of an instrument. On a pulsed source instrument the wavelength resolution is principally (but not entirely) determined by the resolution with which the different neutron times-of-flight (typically a few milliseconds) can be measured. Consequently the wavelength resolution of such instruments is generally good and a value of $\Delta\lambda/\lambda \leq 5\%$ would not be uncommon. On the other hand a steady-state source SANS instrument never operates in a truly monochromatic manner. Instead a bandpass wavelength selector is employed and this also serves to enhance flux. For such instruments $\Delta\lambda/\lambda$ may be as high as 10 to 20% and mathematical deconvolution procedures may be necessary to recover any fine structure in the scattering pattern.

### 4.2.2  The Differential Scattering Cross-Section, $(d\Sigma/d\Omega)(Q)$

The first objective of any SANS experiment is to determine the differential scattering cross-section, $(d\Sigma/d\Omega)(Q)$, since this quantity contains all the information on the shape, size, and interactions between the scattering centres in the sample. (*We shall use the term "scattering centre" as a generic description for that part of a system scattering neutrons; a scattering centre need not be a complete molecule*). In this context the differential cross-section is given by

$$\frac{\partial \Sigma}{\partial \Omega}(Q) = N_p V_p^2 (\Delta \rho)^2 P(Q) S(Q) + B \tag{5}$$

where $N_p$ is the number concentration of scattering centres (given the subscript "p" for "particles"), $V_p$ is the volume of one scattering centre, $(\Delta \rho)^2$ is the *contrast*, $P(Q)$ is a function known as the *form or shape factor*, $S(Q)$ is the *interparticle structure factor* and $B$ is the background signal. The principle terms of Equation 5 are discussed below. $(d\Sigma/d\Omega)(Q)$ has dimensions of (length)$^{-1}$ and is normally expressed in units of cm$^{-1}$.

*The contrast term, $(\Delta \rho)^2$*

The contrast, $(\Delta \rho)^2$, is simply the square of the difference in neutron scattering length density between that part of the sample of interest, $\rho_p$, and the surrounding medium or matrix, $\rho_m$; i.e., $(\Delta \rho)^2 = (\rho_p - \rho_m)^2$.

The value of $\rho$ for a molecule of $i$ atoms may be readily calculated from the simple expresssion

$$\rho = \sum_i b_i \frac{\delta N_A}{m} = N \cdot \sum_i b_i \tag{6}$$

where $\delta$ is the bulk density of the molecule, $m$ is its relative molar mass, $N$ is the number density of scattering centres and $b_i$ is the *coherent neutron scattering length* of nucleus $i$. A table of selected scattering lengths is given in Chapter 1. For a more comprehensive listing the reader should consult one of the many compilations that are available (see for example Sears, 1992). For polymers it is only necessary to calculate $\rho$ for one repeat unit since this is what constitutes a scattering centre. $\rho$ has dimensions of (length)$^{-2}$ and is normally expressed in units of $10^{10}$ cm$^{-2}$ or $10^{-6}$ Å$^{-2}$. Note that whilst $\rho$ can be negative, $(\Delta \rho)^2$ cannot.

Clearly, if $(\Delta \rho)^2$ is zero then Equations 5 and 1 are also zero and there is no small-angle scattering. When this condition is met the scattering centres are said to be at '*contrast match*'. Since the scattering from a multi-component sample is essentially a contrast-weighted summation of the scattering from each individual component (plus a contribution from interference terms), the technique of 'contrast matching' can dramatically simplify scattering patterns. Usually the key to this simplification is deuterium-for-hydrogen substitution, a consequence of their markedly different scattering lengths (see Chapter 1). In the study of adsorbed layers it is common practice to contrast match the substrate and the dispersion medium, typically by mixing hydrogenous and deuterated forms of the medium in an appropriate ratio so that the only scattering that is observed arises solely from the adsorbed layer. Conversely, a hydrogenous polymer dissolved in a deuterated solvent will, so far as the neutrons are concerned, be highly visible. A selection of neutron scattering length densities for the fully hydrogenated and perdeuterated forms of common solvents and polymers are shown in Table 1, whilst the scattering length densities of some common substrates are shown in Table 2.

TABLE 1.  Neutron scattering length densities for the hydrogenated (h) and perdeuterated (d) forms of some common solvents and polymers at 25°C.

| Solvent | $\rho$ (h-form) ($10^{10}$ cm$^{-2}$) | $\rho$ (d-form) ($10^{10}$ cm$^{-2}$) | Polymer | $\rho$ (h-form) ($10^{10}$ cm$^{-2}$) | $\rho$ (d-form) ($10^{10}$ cm$^{-2}$) |
|---|---|---|---|---|---|
| Water | −0.56 | +6.38 | PB | −0.47 | +6.82 |
| Octane | −0.53 | +6.43 | PE | −0.33 | +8.24 |
| Cyclohexane | −0.28 | +6.70 | PS | +1.42 | +6.42 |
| Toluene | +0.94 | +5.66 | PEO | +0.64 | +6.46 |
| Chloroform | +2.39 | +3.16 | PDMS | +0.06 | +4.66 |
| Carbon Tet. | +2.81 | | PMMA | +1.10 | +7.22 |

TABLE 2.  Neutron scattering length densities for some common inorganic substrates.

| Substrate | $\rho$ ($10^{10}$ cm$^{-2}$) | Substrate | $\rho$ ($10^{10}$ cm$^{-2}$) |
|---|---|---|---|
| Silicon | +2.07 | SiO$_2$ | +3.15 |
| Quartz | +3.47 | TiO$_2$ | +2.57 |

As an illustration, consider an aqueous dispersion of silica particles sterically stabilized by an adsorbed layer of *deuterated* poly(ethylene oxide) (PEO). When the dispersion medium is heavy water, $D_2O$, the polymer will effectively be invisible and the scattering will come only from the silica particles. However, if the dispersion medium were to be replaced by a mixture 53.5% $D_2O$ and 46.5% $H_2O$ the silica particles would be at contrast match and the scattering would arise from the adsorbed polymer alone.

The scattering length density of a molecule is remarkably sensitive to the value of the bulk density $\delta$ used in its calculation and careful consideration should always be given to this aspect. Of course, there may be occasions when the bulk density either cannot be measured conventionally, or its meaning is in doubt (perhaps because of a non-uniform mass distribution or porosity, for example). In these instances a contrast matching experiment can be used to determine the scattering length density of the material in question and therefore its *apparent* bulk density.

*The form factor, $P(Q)$*

The form factor, $P(Q)$, is a function that describes how $(d\Sigma/d\Omega)(Q)$ is modulated by interference effects between radiation scattered by *different parts of the same scattering centre*. Consequently it is sensitive to the shape of the scattering centre. The general form of $P(Q)$ is given by the following expression (van de Hulst, 1957)

$$P(Q) = \frac{1}{V_p^2} \left| \int_0^{V_p} \exp[i\ f(Q\alpha)]dV_p \right|^2 \qquad (7)$$

where $\alpha$ is a "shape parameter" that might, for example, represent a length or a radius of gyration. Fortunately, analytical expressions exist for most common shapes. Expressions

TABLE 3. Selected scattering form factors.

| | |
|---|---|
| Sphere of radius $R_p$ | $P(Q) = \left[ \frac{3(\sin(QR_p) - QR_p \cos(QR_p))}{(QR_p)^3} \right]^2$ |
| Disc of negligible thickness and radius $R_p$ ($J_1$ is a first-order Bessel function of the first kind) | $P(Q) = \frac{2}{(QR_p)^2} \left[ 1 - \frac{J_1(2QR_p)}{QR_p} \right]$ |
| Rod of negligible cross-section and length $L$ ($S_i$ is the Sine integral function) | $P(Q) = \frac{2S_i(QL)}{QL} - \frac{\sin^2(QL/2)}{(QL/2)}$ |
| Gaussian random coil with radius of gyration $R_g$, polydispersity $(Y+1)$, and $U = \frac{(QR_g)^2}{(1+2Y)}$ | $P(Q) = \frac{2[(1+UY)^{-1/Y} + U - 1]}{(1+Y)U^2}$ |
| Concentrated polymer solution with screening length $\xi = R_g \left( \frac{\phi}{\phi^*} \right)^z$ and excluded volume exponent $v$, where $z = \frac{v}{(1-3v)}$ | $P(Q) = P(0) \left[ \frac{1}{1+(Q\xi)^2} \right]$ |

for more complex topologies, such as concentric cylinders (Livsey, 1987) or "hinged rods" (Muroga, 1988) can usually be based on functions for simpler shapes. A selection of form factors can be found in Table 3.

Although these expressions are derived for scattering centres of a definite size ($R_p$, $R_g$ and $L$), in reality a range of dimensions is likely to be present in the system under study. This obviously affects the scattering. It is necessary, therefore, to convolute the form factor with a suitable *particle size distribution function* which simply states how some dimension of the scattering centre varies about the mean value of that dimension. Two of the most commonly encountered particle size distributions, the Log-Normal and the Schultz distributions, are given in the Appendix.

*The structure factor, $S(Q)$*

The interparticle structure factor, $S(Q)$, given by

$$S(Q) = 1 + \frac{4\pi N_p}{QV} \int_0^\infty [g(r) - 1]r \sin(Qr)dr \tag{8}$$

is a function describing how $(d\Sigma/d\Omega)(Q)$ is modulated by interference effects between radiation scattered by *different scattering centres*. Consequently it is dependent on the degree of local order in the sample, for example as might arise in a dispersion of charged particles interacting via an electrostatic potential. The corollary of this, is that SANS can be used to gain information about the relative positions of the scattering centres, usually through the *radial distribution function*, or r.d.f.

$$r.d.f. = \frac{4\pi N_p r^2}{V} g(r) \tag{9}$$

where $r$ is a radial distance outward from the centre of any scattering centre in the sample, and the density distribution $g(r)$ is obtained from Equation 8 by Fourier inversion using suitable extrapolation procedures. Note that $S(Q)$ tends to unity at high-$Q$. Also as $N_p$ tends to zero (i.e., as the concentration of scattering centres becomes more dilute) the region where $S(Q) = 1$ moves to lower values of $Q$. So long as $P(Q)$ is invariant with concentration this allows $S(Q)$ to be obtained from Equation 5 by measuring $(d\Sigma/d\Omega)(Q)$ at two rather different particle concentrations (see for example Ottewill, 1982). The r.d.f. is typically a damped, oscillating density distribution function whose maxima correspond to the distance of each nearest-neighbour coordination shell.

The function $\ln g(r)$ is directly related to the potential energy function governing the interactions between the scattering centres. This means that calculated scattering data can be model-fitted to the observed $(d\Sigma/d\Omega)(Q)$ data using one of the many approximate forms of $S(Q)$ that have been developed to describe particular types of system (see for example Ashcroft & Lekner, 1966; Hayter & Penfold, 1981; Hansen & Hayter, 1982). These approximations generally differ in the type of interaction potential that they incorporate. Farsaci (Farsaci, 1989) has developed a more empirical function which does not incorporate a potential.

## 4.3 SANS FROM ADSORBED LAYERS

A properly designed and executed SANS experiment may provide values for the surface area per unit volume of the substrate, estimates of the thickness of the adsorbed layer, measures of how much polymer or surfactant is present within this layer, and a description of how the polymer or surfactant is arranged in the interfacial region — *the volume fraction profile*. Some of this information can be obtained by other techniques (e.g., nmr, ellipsometery, neutron reflectometery, Rutherford backscattering and evanescent wave induced fluorescence) but SANS is the only technique to have provided detailed information on the form of the volume fraction profile (Fleer *et al.*, 1993).

Unfortunately, meaningful interpretation of the SANS data is generally only possible when the scattering centres (or whatever else may be of interest, such as pores for example) conform to some simple geometry. The practical consequence of this is that SANS experiments designed to investigate the structure of adsorbed layers normally utilize flat surfaces or dispersions of approximately spherical polymer latices, inorganic oxide particles or emulsion droplets. To minimize inter-layer interactions, excluded volume effects, and possible multiple scattering effects from the substrate particles, the dispersion concentration is normally kept <5% by weight.

There are essentially four distinct ways of analysing the scattering from adsorbed layers and we shall consider each in turn. As will become apparent, these different approaches vary considerably in their complexity, but this is more than compensated for by the information that they provide. They can be classified as follows:

1. Those which assume a simple geometrical model for the scattering centre and only provide size information (e.g., the use of limiting laws such as the Guinier approximation)

2.  Those which assume a simple geometrical model for the scattering centre and a simple functional form for the volume fraction profile (e.g., model-fitting using the "core-shell" model)
3.  Those which do not assume a form for the volume fraction profile and which may neglect concentration fluctuations in the adsorbed layer (e.g., the mathematical transformation — or "inversion" — approaches of Crowley, 1984 and of Auvray *et al.*, 1992)
4.  Those which assume a functional form for the volume fraction profile but treat concentration fluctuations in the adsorbed layer (e.g., the "scaling" methodologies of Auvray *et al.* and of Cosgrove *et al.*)

### 4.3.1  The Guinier Approximation

This approach (Guinier, 1955) relies on the fact that the form factor for a sphere of radius $R_p$ (see Table 3) can be approximated by a series expansion of $\cos(QR_p)$ when $QR_p$ is small, giving

$$P(Q) \approx \left(1 - \frac{(QR_p)^2}{10} + \cdots\right) \tag{10}$$

A binomial expansion of Equation 10 then yields the rather more useful result

$$P(Q) \approx \exp\left(-\frac{(QR_p)^2}{5}\right) \tag{11}$$

Hence a graph of $[\ln(d\Sigma/d\Omega)(Q)]$ versus $Q^2$ should be linear at small values of $Q^2$ with a slope of $(-R_p^2/5)$. An example of a "Guinier plot" is shown in Figure 2.

Equation 11 can be reformulated in terms of the *radius-of-gyration*, $R_g$, of the sphere using the relationship

$$\frac{R_p^2}{5} = \frac{R_g^2}{3} \tag{12}$$

For quasi-spherical scattering centres the Guinier approximation is generally valid for $QR_p < (\pi/2)$. However, neglecting the higher-order terms in the binomial expansion means that it rapidly becomes less reliable as the scattering centres become less spherical.

The Guinier approximation provides a straightforward means of particle sizing which could be used, in principle, to determine the thickness of an adsorbed layer from the difference in size between particles or droplets with and without adsorbed polymer (cf. the method of determining hydrodynamic thicknesses by Photon Correlation Spectroscopy, PCS). The problem with this approach is that, unlike PCS, SANS is relatively insensitive to the few, well-solvated, highly-extended polymer "tails" at the periphery of the adsorbed layer. Consequently without some *a priori* knowledge of the polymer volume fraction profile, it is difficult to correlate any adsorbed layer thickness measured in this way with any physical length scale in the real system.

### 4.3.2  The "Core-Shell" Model

This approach assumes that a particle and its adsorbed layer can be modelled as two homogeneous, concentric spheres, see Figure 3.

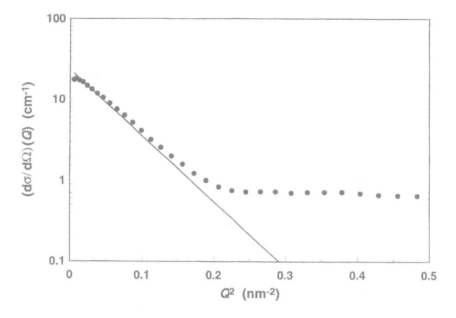

FIGURE 2. A Guinier plot of the scattering from a dispersion of relatively monodisperse calcium carbonate particles. The continuous line is the least-squares fit to the initial linear region and used to extract the particle size in accordance with Equation 11. In this instance $R_p = 9.7$ nm.

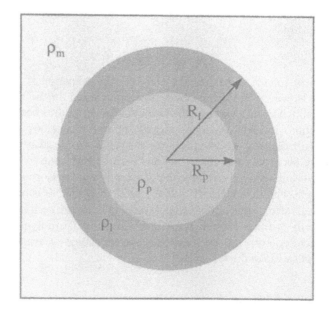

FIGURE 3. A diagrammatic representation of the "core-shell" model. See main text for an explanation of the symbols.

The overall scattering from such a "composite particle" is then the scattering from a sphere of radius $R_l$ and scattering length density $\rho_l$, *minus* the scattering from a sphere of radius $R_p$ and scattering length density $\rho_l$, *plus* the scattering from a sphere of radius $R_p$ and scattering length density $\rho_p$. Equation 5 then becomes

$$\frac{\partial \Sigma}{\partial \Omega}(Q) = \frac{16\pi^2}{9} N_p P(Q) S(Q) + B \tag{13}$$

where

$$P(Q) = [(\rho_l - \rho_m)^2 F(Q, R_l)^2] + [2(\rho_l - \rho_m)(\rho_p - \rho_l) F(Q, R_p) F(Q, R_l)]$$
$$+ [(\rho_p - \rho_l)^2 F(Q, R_p)^2] \tag{14}$$

and

$$F(Q, R_l)^2 = R_l^6 P(Q, R_l)$$
$$F(Q, R_p)^2 = R_p^6 P(Q, R_p)$$

$P(Q, R_l)$ and $P(Q, R_p)$ are the form factors for spheres of radii $R_l$ and $R_p$, respectively (see Table 3). Clearly, when $\rho_l = \rho_m$, that is when the adsorbed layer is contrast matched to the dispersion medium, only the scattering from the core of the particle is observed. A more useful scenario though, occurs when $\rho_p = \rho_m$, i.e. when the core is contrast matched to the dispersion medium. Then

$$P_l(Q) = [(\rho_l - \rho_m) F(Q, R_l) - (\rho_l - \rho_m) F(Q, R_p)]^2$$
$$= [(\rho_l - \rho_m) F(Q, R_l) + (\rho_p - \rho_l) F(Q, R_p)]^2 \tag{15}$$

Equation 15 is the form factor describing the scattering from a spherical shell of thickness $(R_l - R_p)$ and scattering length density $\rho_l$.

Equations 13–15 can be fitted to the scattering from a particle or droplet with an adsorbed layer using a suitable least-squares algorithm. Typically $N_p$, $R_p$, $\rho_m$ and $\rho_p$ will be constrained to their known values, and $R_l$ and $\rho_l$ will be the variable parameters. Comparing the value of $\rho_l$ obtained from such a fitting procedure with its calculated value may give some indication of the degree of solvent penetration into, or homogeneity of, the adsorbed layer.

The drawback with this approach is that it assumes the scattering length density within the adsorbed layer is uniform and so it assumes a step function for the shape of the polymers volume fraction profile; what is more commonly referred to as a block profile, see Figure 4.

As will be shown later, neither of these assumptions are very satisfactory. However, the core-shell model does work reasonably well for adsorbed layers of short chain surfactants since these tend to be rather more uniform and homogeneous than adsorbed polymer layers, see Figure 5. Two excellent examples of the application of this approach to such systems are the work of Ottewill and coworkers (Markovic *et al.*, 1984, 1986, 1987; Ottewill *et al.*, 1992) and Cummins (Cummins *et al.*, 1990).

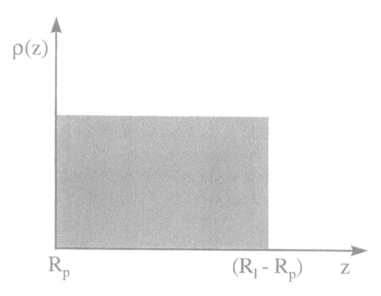

FIGURE 4. A diagrammatic representation of the neutron scattering length density distribution (or the adsorbed layer volume fraction profile) for the "core-shell" model depicted in Figure 3.

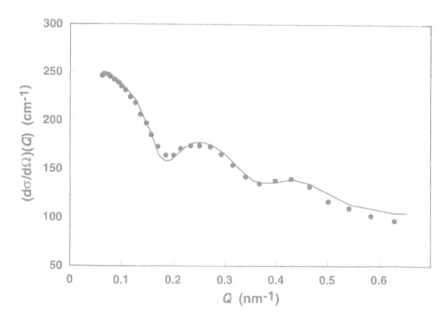

FIGURE 5. The scattering from an aqueous dispersion of colloidal silica particles stabilised by an adsorbed layer of the surfactant hexaethylene glycol monododecyl ether ($C_{12}E_6$). The continuous line is a non-linear least-squares fit of Equations 13 and 14 to the data. The oscillations in the scattering are a manifestation of the underlying spherical form factor and show that these data are from a relatively monodisperse system.

### 4.3.3  The Inversion Technique

The great advantage of this approach is that it does not presuppose a form for the volume fraction profile. In fact, this technique is actually capable of determining it (Cosgrove, 1990).

The volume fraction profile, $\Phi(z)$, and the segment density distribution, $\rho(z)$, of an adsorbed polymer are inter-related through

$$\Phi(z) = \frac{\rho(z)\Gamma}{\delta} \qquad (16)$$

where $\Gamma$ is the amount of polymer adsorbed (with dimensions of mass per unit area) and $\delta$ is once again the bulk density. $\Phi(z)$ provides the most detailed description of the structure of an adsorbed layer that can be obtained experimentally or calculated theoretically. In addition to showing how the average density of polymer segments varies with distance, $z$, in the direction perpendicular to the interface, it also provides several measures of the layer thickness, the adsorbed amount (from the integral under the curve) and the average fraction of bound segments (from the segment density within, say, the range $0 \leq z < 1$ nm) (Fleer et al., 1993).

The inversion technique has essentially evolved in two forms; one due to Crowley (Crowley, 1984) and coworkers from studies of both grafted (Cosgrove et al., 1981; Barnett, 1982) and, in particular, physically adsorbed polymers (Barnett et al., 1981; Cosgrove et al., 1990; Fleer et al., 1993; Cosgrove et al., 1999), and the other due to Auvray (Auvray, 1986) and coworkers, principally from studies of grafted polymer layers (Auvray & de Gennes, 1986; Auvray & Cotton, 1987; Auvray et al., 1992; Auroy et al., 1990, 1991, 1992; Auroy & Auvray, 1992; Caucheteux et al., 1989). The principle difference between the two forms is that Auvray and coworkers specifically consider the contribution to the scattering that arises from segment density fluctuations in the adsorbed layer, whereas Crowley and coworkers generally ignore this contribution. The latter argue that such fluctuations are generally unimportant except in a few special cases, such as in systems with dense grafted layers (Auvray & de Gennes, 1986; Cosgrove et al., 1987, 1994; and also Forsman & Latshaw, 1990). Generally however, both approaches are valid since the scattering from the fluctuation term is indeed very weak for polymer layers that are neither very dense nor comprise high molecular weight polymers.

The starting point for both derivations is a variation of Equation 5, viz.

$$\frac{\partial \Sigma}{\partial \Omega}(Q) = N_p P(Q) S(Q) + B$$

where (cf. Equations 13–15)

$$P(Q) = [(\rho_p - \rho_m) F_p(Q) + (\rho_l - \rho_m) F_l(Q)]^2 \qquad (17)$$

$F_p(Q)$ represents the intraparticle form factor for the core particle and $F_l(Q)$ the intralayer form factor for the adsorbed polymer. The inversion technique therefore properly considers the separate contributions to the scattering from both the particle and the adsorbed layer.

If the curvature of the surface is small compared to $Q$, that is, if $QR_p \gg 1$, then Equation 17 may be expanded to give

$$P(Q) = [(\rho_p - \rho_m)^2 F_p(Q)^2] \tag{18a}$$

$$+ [2(\rho_p - \rho_m)(\rho_l - \rho_m) F_p(Q) F_l(Q)] \tag{18b}$$

$$+ [(\rho_l - \rho_m)^2 F_l(Q)^2] \tag{18c}$$

from which it can be seen that Equation 18a (called the $I_2(Q)$ term by Crowley *et al.* or the $I_{gg}(Q)$ term by Auvray *et al.*; the subscript "g" is for "grain") describes the contribution to $P(Q)$ arising from the core particle, Equation 18c (called the $I_0(Q)$ by Crowley *et al.* or the $I_{pp}(Q)$ term by Auvray *et al.*; the subscript "p" here is for "polymer") describes the contribution from the adsorbed layer, and Equation 18b (called the $I_1(Q)$ term by Crowley *et al.* or the $I_{pg}(Q)$ term by Auvray *et al.*) is a particle-surface interference term.

Using an independent notation, where the subscript "p" signifies the "particle" and the subscript "l" signifies the (adsorbed) layer, these three terms may be written explicitly thus:

$$I_{pp}(Q) = (\delta_p - \delta_m)^2 \frac{2\pi A_P}{Q^4} \left[ 1 + \frac{1}{Q^2 R_p^2} \right] \tag{19}$$

$$I_{pl}(Q) = (\delta_p - \delta_m)(\delta_l - \delta_m) \frac{4\pi A_p}{Q^4} \left[ \int_0^t \Phi(z) \cos(Qz) dz - QR_p \int_0^t \Phi(z) \sin(Qz) dz \right] \tag{20}$$

$$I_{ll}(Q) = (\delta_l - \delta_m)^2 2\pi A_p \left[ \frac{1}{Q^2} \left| \int_0^t \Phi(z) \exp(iQz) dz \right|^2 + \bar{I}_{ll} \right] \tag{21}$$

where $A_p$ is the surface area per unit volume of a particle (usually expressed as $S_p/V_p$ in the Auvray derivation, where $S_p = 4\pi R_p^2$ for spherical particles), $t$ is the maximum extent of the adsorbed layer and $\bar{I}_{ll}$ is related to the density-density correlation function that describes fluctuations in the adsorbed layer. It is the $\bar{I}_{ll}$ term in Equation 21 that the Crowley derivation does not treat implicitly.

### The $I_{pp}(Q)$ term

Equation 19 is simply a statement of the well known Porod law (Porod, 1951). The $Q^{-4}$ dependence of the particle scattering is a general result for a surface fractal. When $\rho_l = \rho_m$ this term can be used to determine the surface area-to-volume ratio of the substrate.

### The $I_{pl}(Q)$ term

If $QR_p \gg 1$, Equation 20 reduces to the sine Fourier transform of $\Phi(z)$, that is

$$I_{pl}(Q) \approx -(\rho_p - \rho_m)(\rho_l - \rho_m) \frac{4\pi A_p}{Q^3} \int_0^t \Phi(z) \sin(Qz) dz \tag{22}$$

Thus, by subtracting the scattering obtained at contrast match for the particles ($\rho_p = \rho_m$) from the scattering obtained "off-contrast" ($\rho_l \neq \rho_p \neq \rho_m$), multiplying the result by $Q^3$ and taking the Fourier transform it is possible to obtain $\Phi(z)$. In fact, this was the method used to derive the very first volume fraction profile from SANS data (Cosgrove et al., 1981; Barnett, 1982). This route to $\Phi(z)$ has both advantages and disadvantages. In its favour is the fact that inversion only requires a Fourier transform and the procedure is not affected by the form of $\Phi(z)$. This aspect is discussed further in the next subsection. A further advantage is that it is independent of the fluctuation term $\bar{I}_{ll}$. The downside, however, is that when the particles are not at contrast match the $I_{pp}(Q)$ term is the dominant contribution to the observed scattering. This much stronger signal can be subtracted (either by using the scattering obtained when the adsorbed layer is at contrast match, i.e. when $\rho_l = \rho_m$, or by using Porod's law), but the resulting $I_{pl}(Q)$ scattering can suffer from a poor signal-to-noise ratio. One way to overcome this drawback is to maximize $A_p$. For example, Auvray and coworkers have demonstrated that $I_{pl}(Q)$ can be obtained with good statistics when the polymers are adsorbed in dispersions of macroporous particles (Auvray & Cotton, 1987; Auroy et al., 1991).

*The $I_{ll}(Q)$ term*

The more straightforward route for obtaining $\Phi(z)$ is from the scattering when the particles are at contrast match ($\rho_p = \rho_m$). Under this condition $P(Q)$ in Equations 18a–18c reduces to Equation 21. Unfortunately, because it is the modulus of the integral that is measured experimentally, the transformation of the first term in the brackets is complicated by the need to introduce a phase factor, $\exp(i\varphi Q)$, where $\varphi$ is unknown. Fortunately the particle surface can be used as a phase reference point since $\Phi(z) = 0$ for $z < 0$. Equation 21 is thus what mathematicians refer to as a causal function. This allows $\varphi$ to be determined using a dispersion integral relationship between $\partial\varphi/\partial Q$ and $\partial Ln|\int \Phi(z)\exp(iQz)dz|/\partial Q$ (Barnett, 1982; Crowley, 1984).

With a knowledge of $\varphi$, $\Phi(z)$ can then obtained by Hilbert transformation of the $I_{ll}(Q)$ scattering data. This normally requires that the data be extrapolated to both low and high $Q$ to prevent Gibbs oscillations from being introduced through premature truncation. The limiting behaviour of Equation 21 is $I_{ll}(Q) \propto Q^{-2}$ as $Q$ tends to zero, and $I_{ll}(Q) \propto Q^{-4}$ as $Q$ tends to infinity.

This procedure always generates a possible $\Phi(z)$, provided that $\Phi(z)$ does not contain any zeroes in the upper complex plane. What this means in practice is that whilst some rapidly decaying functional forms for $\Phi(z)$ can be problematical, for the most part very few functions are limited by this constraint since $\Phi(z)$ will still approximate the real volume fraction profile, except at large $z$.

In principle any physically realistic $\Phi(z)$ can be back-transformed into "calculated" scattering data and compared with actual experimental scattering data using a Maximum Entropy minimization (Finch, 1988). Unfortunately, the Hilbert transformation is not guaranteed to produce a *unique* profile and so back-transformation must be treated with caution.

Figure 6 shows the volume fraction profile for a block copolymer adsorbed at the liquid-liquid interface in an emulsion (Washington et al., 1996) and was obtained by

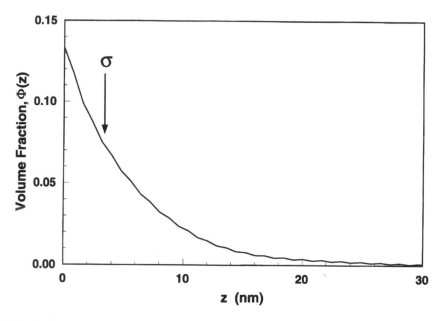

FIGURE 6. The volume fraction profile for the PEO-PPO-PEO block copolymer Pluronic F127 (Poloxamer 407) adsorbed at the interface in a perfluorodecalin-in-water mesoscopic emulsion system (Washington *et al.*, 1996). Also shown is the position of the second moment of the profile.

inversion of the $I_{ll}(Q)$ scattering. The shape of this profile, whilst typical of a physically adsorbing block copolymer, is clearly very different to the block profile assumed by the core-shell model discussed previously and illustrated schematically in Figure 4. It shows that whilst the bulk of each adsorbed polymer molecule resides within 3–4 nm of the interface, there are polymer segments in "tails" extending some 20–30 nm. Figure 6 convincingly demonstrates the advantage of the inversion techniques.

When the polymer chains in the adsorbed layer are "grafted" (terminally-attached) to the interface the volume fraction profile often displays a maximum. This is discussed in more detail below.

Provided $R_p \gg t$, the exponential term in Equation 21 can be expanded to yield

$$I_{ll}(Q) = (\rho_l - \rho_m)^2 2\pi A_p \left[ \frac{1}{Q^2} \left| M \left[ 1 + iQz - \frac{(Qz)^2}{2} + \cdots \right] \right|^2 + \bar{I}_{ll} \right] \quad (23)$$

where the normalisation constant, $M$, (Auvray uses the symbol $\gamma$) is given by

$$M = \int_0^t \Phi(z)dz = \frac{\Gamma}{\delta} \quad (24)$$

Neglecting fluctuations, Equation 23 then approximates to

$$I_{ll}(Q) \approx (\rho_l - \rho_m)^2 \frac{2\pi A_p M^2}{Q^2} \exp(-Q^2\sigma^2) \quad (25)$$

for $Q\sigma < 1$, where $\sigma$, *the second moment about the mean*, or *standard deviation*, of the volume fraction profile is defined as

$$\sigma^2 = \langle z^2 \rangle - \langle z \rangle^2 \tag{26}$$

and the *root-mean-square thickness*, $t_{rms}$, is defined as

$$t_{rms} = \langle z^2 \rangle^{1/2} \tag{27}$$

where

$$\langle z^n \rangle = M^{-1} \int_0^t \Phi(z) z^n dz \tag{28}$$

Physically, $\sigma$ provides an estimate of the average distance of the centre-of-mass of an adsorbed polymer from the interface. For a block profile (cf. the "core-shell" model) of thickness $t$, $\sigma^2 = t^2/12$.

Substituting expressions for $A_p$ and $M$ (Equation 24) into Equation 25, and noting that the latter expression is per unit volume, then leads to the very useful result

$$\frac{\partial \Sigma}{\partial \Omega}(Q) \approx (\rho_l - \rho_m)^2 \left[ \frac{6\pi\phi_p}{Q^2} \frac{\Gamma^2}{\delta^2 R_p} \exp(-Q^2\sigma^2) \right] + B \tag{29}$$

Equation 29 describes the scattering from the adsorbed polymer layer in a dilute, non-interacting, system where the substrate is at contrast match, and is the surface analogue of the Guinier approximation discussed earlier. This expression can be model-fitted to the observed scattering to obtain values for $\Gamma$ and $\sigma$, and an example of such a fit is shown in Figure 7. From it, values of $\Gamma = 1.9$ mg m$^{-2}$ and $\sigma = 2.3$ nm were obtained.

An additional parameter that can be determined with a knowledge of the volume fraction profile is the average bound fraction, $\langle p \rangle$, given by the expression

$$\langle p \rangle = M^{-1} \int_0^l \Phi(z) dz \tag{30}$$

This is the average fraction of segments in adsorbed polymer molecules that are bound to the interface or, in other words, the fraction of segments in "trains". Although intuitively it may seem sensible to choose $l$ to be the length of one segment, say, $\leq 0.5$ nm, in practice it is better to choose a larger value in order to offset the effects of any interfacial inhomogeneities. By scaling experimental data from simple systems to simulated data from contemporary mean-field theories it has been suggested that one lattice layer in a simulation corresponds to a distance of 1.3 nm (Cosgrove *et al.*, 1990). On this basis a more pragmatic approach is to use $l \approx 1$ nm. For the polymer volume fraction profile depicted in Figure 6 it is found that $\langle p \rangle \approx 14\%$.

Equations 23, 25 and 29 work well if the adsorbed layer has a reasonable extent but the limiting behaviour of Equation 21 causes problems if the adsorbed layer is relatively thin, as is often the case with short-chain surfactants. Acknowledging this limitation, Crowley has recently formulated an extension to the method outlined above (Gladman *et al.*, 1995). The principles of this treatment are outlined later in this chapter.

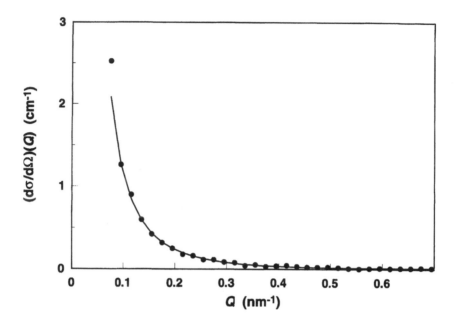

FIGURE 7.  The scattering data from which the volume fraction profile in Figure 6 was derived. The continuous line is a non-linear least-squares fit of the data to Equation 29.

*The $\bar{I}_{ll}$ Term*

Auvray *et al.* have shown that it is possible to draw analogies between de Gennes scaling description of semi-dilute polymer solutions (de Gennes, 1979) and the structure of adsorbed layers (Auroy *et al.*, 1991). They show that any concentration fluctuations in the adsorbed layer manifest themselves as a contribution to the scattering (cf. Equations 21 and 23) of the form

$$\bar{I}_{ll} \propto \frac{1}{(aQ)^n} \tag{31}$$

where $a$ is the size of a polymer segment. A slightly modified functional form is necessary with dense adsorbed layers, as is discussed below.

Three different categories of fluctuation contribution have been identified. Each is characterized by its own scaling exponent $n$, but only two have actually been verified experimentally. In all cases $n$ is positive, and so $\tilde{I}_{ll}$ decays as $Q$ increases. $\tilde{I}_{ll}$ therefore makes less of a contribution to the scattering in the high-$Q$ limit but unfortunately, in this regime $(d\Sigma/d\Omega)(Q)$ is normally of the same order as $B$. This means that the data are frequently subject to large statistical errors which may render any precise measurement and meaningful interpretation of the fluctuation contribution difficult.

### (a) Physically adsorbed layers where $R_p \gg t$

When the thickness of the adsorbed layer is relatively small compared to the radius of the substrate particle, $\Phi(z)$ is predicted to decay as $z^{-4/3}$ over the range $a < z < Q^{-1}$. Under such conditions, $n = 4/3$. However, experimental evidence suggests that this behaviour is not universal (Fleer et al., 1993).

### (b) Physically adsorbed layers where $R_p \leq t$

When the thickness of the adsorbed layer is of the same order as the radius of the substrate particle, $n = {}^5/_3$. However, the scaling of $\Phi(z)$ has not yet been established.

### (c) Grafted adsorbed layers

When the adsorbed polymer is grafted (terminally-attached) to the surface of the substrate, or possibly where a block copolymer is adsorbed at high coverage, a "brush-like" adsorbed layer develops. This type of adsorbed layer structure can be characterized by parabolic or Gaussian volume fraction profiles ($\Phi(z) \propto z^{-2}$) – as the majority of polymer segments are forced to occupy the region just above the interface – and some length scale, $\xi$. $\xi$, which is similar in concept to the correlation length of a polymer gel or network, is related to the distance between anchor points. In this category of adsorbed layer the fluctuation contribution varies as

$$\tilde{I}_{ll} \propto \frac{1}{1 + (\xi Q)^2} \tag{32}$$

Clearly, in the low-$Q$ limit $\tilde{I}_{ll}$ tends to unity and so it can be ignored providing that the other contributions to $I_{ll}$ are not of the same order. In the high-$Q$ limit, $\tilde{I}_{ll} \propto Q^{-2}$.

An example of the volume fraction profile from a grafted layer is shown in Figure 8. To obtain this profile scattering data were generated using different functional forms for the volume fraction profile and then model-fitted to the experimental scattering data. In this example the most realistic functional form for the volume fraction profile was a Gaussian.

### 4.3.4 The Scaling Methodologies

By combining Equations 21 and 31 a very generic description of the scattering from an adsorbed layer can be formulated, viz.

$$\frac{\partial \Sigma}{\partial \Omega}(Q) = \frac{U}{Q^2} + \frac{V}{Q^n} + B \tag{33}$$

where $U$ and $V$ are constants and ${}^4/_3 \leq n \leq 2$. Multiplying Equation 33 through by $Q^n$, and neglecting the term in $Q^{n-2}$ on the grounds that it contributes less than the other two (except when $n = 2$), then gives

$$Q^n \frac{\partial \Sigma}{\partial \Omega}(Q) \approx W + B Q^n \tag{34}$$

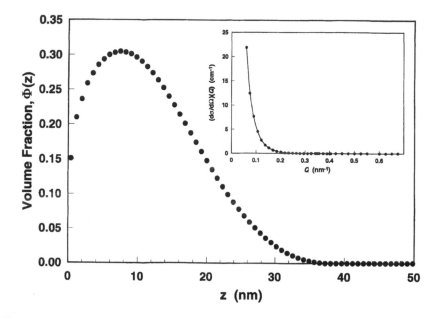

FIGURE 8. The Gaussian volume fraction profile that best describes the structure of the "brush"-like adsorbed layer formed by perdeuterated PS ($M_w = 24000$) terminally-attached to the surface of colloidal silica dispersed in dimethylformamide. The grafting density was $\Gamma = 6.5$ mg$^{-2}$. The corresponding experimental (•) and calculated (—) scattering data are compared in the inset.

Hence a graph of $Q^n (d\Sigma/d\Omega)(Q)$ versus $Q^n$ should be linear with a gradient equal to the background signal, and an intercept, $W$, proportional to the magnitude of the fluctuation contribution.

### 4.3.5  SANS From Thin Adsorbed Layers

For the case of thin adsorbed layers on large particles there is a narrow range of $Q$ over which the "sheet-like" scattering from the adsorbed layer dominates other contributions (specifically, the particle and particle-layer interference scattering when $Q \leq \sigma^{-1}$ (where $\sigma$ is the standard deviation of the volume fraction profile that has already been met previously), and the specular Porod's law scattering when $Q \geq \pi/R_p$) (Gladman *et al.*, 1995). The principle drawback of this new formalism is that it cannot provide the volume fraction profile for the adsorbed layer.

The derivation involves combining the principal terms of Equations 19 and 23 with Equation 22. On multiplying through by $Q^4$ and taking the low-$Q$ limit it may be shown that

$$Q^4 \frac{\partial \Sigma}{\partial \Omega}(Q) \approx 2\pi N_p A_p [M_{\text{eff}}^2 Q^2 + (\rho_p - \rho_m)^2] \qquad (35)$$

where $M_{\text{eff}}^2$, the magnitude of the "sheet" scattering contribution (called $\Gamma_{\text{eff}}^2$, by Crowley), is given by

$$M_{\text{eff}}^2 = M[M(\rho_l - \rho_m)^2 - 2\langle z \rangle (\rho_l - \rho_m)(\rho_p - \rho_m)] \tag{36}$$

where $M$ was defined in Equation 24. Crowley et al. go on to define the *contrast ratio*

$$\Delta = \frac{(\rho_p - \rho_m)}{(\rho_l - \rho_p)} \tag{37}$$

and the *mean excess volume fraction* within the layer

$$\Phi_{xs} = \frac{M}{2\langle z \rangle} \tag{38}$$

Using the notation of Crowley, it then follows that

$$\frac{M_{\text{eff}}^2}{(1 + \Delta)} = m(Q)\Delta + g(Q) \tag{39}$$

where

$$m(Q) = -2M\langle z \rangle (\rho_l - \rho_p)^2 \tag{40a}$$

$$g(Q) = M^2(\rho_l - \rho_m)(\rho_l - \rho_p) \tag{41a}$$

If the particles are at contrast match, $\rho_m = \rho_p$ and $\Delta = 0$, thus

$$m(Q) = -M^2 \Phi_{xs}^{-1}(\rho_l - \rho_p)^2 \tag{40b}$$

$$g(Q) = M_{\text{eff}}^2 = M^2(\delta_l - \delta_p)^2 \tag{41b}$$

Hence, by performing scattering measurements *as a function of contrast* it is possible to obtain $M_{\text{eff}}^2$ from a graph of $Q^4(d\Sigma/d\Omega)(Q)$ versus $Q^2$; and $M$ (and therefore $\Gamma$), $\langle z \rangle$ and $\Phi_{xs}$ from a graph of $[M_{\text{eff}}^2/(1 + \Delta)]$ versus $\Delta$ extrapolated to $\Delta = 0$.

An example of scattering data plotted in the form of Equation 35 is shown in Figure 9. Each of the four sets of data represents a slightly different $\Delta$. In this example a low molecular weight ($m = 1600$) polymeric surfactant was adsorbed on titania particles having a mean diameter of several microns. Using the method of analysis described above, the average distance of surfactant monomers from the particle surface, $\langle z \rangle$, was determined to be 1.4 nm, whilst the volume fraction of surfactant in the adsorbed layer was found to be $\Phi_{xs} \approx 40\%$.

### 4.3.6 SANS From Adsorbed Layers In Ultra-Concentrated Dispersions

Crowley and coworkers have also developed a description of the scattering from film-forming polymer latex dispersions (Sanderson et al., 1992). As the dispersion medium evaporates the particle volume fraction increases until eventually capillary forces drive the polymer latex particles together. Locally-flat bilayers develop where the particle surfaces meet and, in some instances, the initially spherical particles may deform into rounded polyhedra. This has implications for any adsorbed stabiliser. Given the considerable number of technological applications for latex films, for example as surface coatings and adhesives, there is obvious interest in techniques that can investigate the bilayer region and determine the fate of the stabiliser. SANS is the only technique that can study all stages of the film formation; from dispersion to dry film.

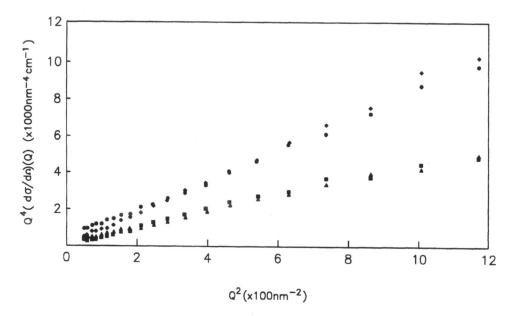

FIGURE 9. The scattering from surfactant stabilised titania particles plotted according to Equation 35. Each dataset represents a different contrast ratio. Reprinted with permission from Gladman *et al.*, 1995. Copyright 1995 The Royal Swedish Academy of Sciences.

Uniquely, this description allows the radius of curvature of the particles, the bilayer thickness and the disjoining pressure or osmotic pressure to be determined. It relies on the fact that when $QR_p > \pi$ only specular scattering from the bilayer region is observed. This scattering is insensitive to larger-scale effects such as the packing of the particles and is the sum of two contributions. The first of these is simply the self-scattering from the individual particle surfaces forming the bilayer and was introduced earlier as $I_{pp}(Q)$ (Equation 19). The second contribution, $\Delta I_{pp}(Q)$ is due to interference between adjacent particle surfaces in the bilayer. This contribution is the important one when $\phi_p > 0.74$. Thus one may write

$$P(Q) = I_{pp}(Q) + \Delta I_{pp}(Q) \tag{42}$$

where it can be shown that

$$\Delta I_{pp}(Q) = -(\rho_p - \rho_m)^2 \frac{4\pi A_B}{Q^4} \cos(Q\Delta_0) \tag{43}$$

The quantity $\Delta_0$ (called $\delta_0$ by Crowley) is the "bilayer thickness", the distance between two opposing particle surfaces.

(Note that Crowley *et al.* refer to the $P(Q)$, $I_{pp}(Q)$ and $\Delta I_{pp}(Q)$ terms as $I_B(Q)$, $I_0(Q)$ and $\Delta I_B(Q)$ respectively, where the subscript "B" stands for "bilayer". What they term $I_0(Q)$ should not be confused with that in this chapter.)

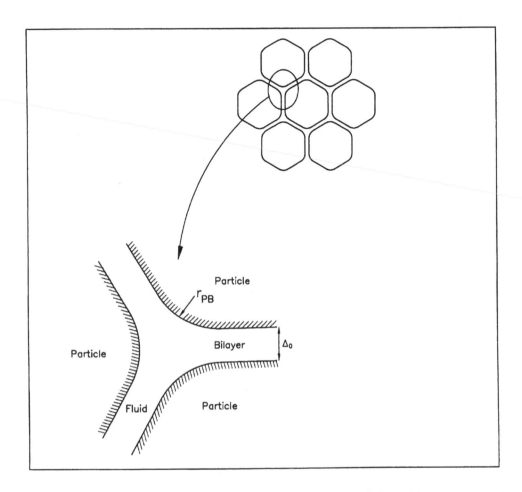

FIGURE 10. A diagrammatic representation of the "Plateau borders" formed by a concentrated film-forming polymer latex on drying. Reprinted with permission from Sanderson *et al.*, 1992. Copyright 1992 American Chemical Society.

The surface area of a bilayer, $A_B$, is given by

$$A_B = \frac{A_C}{2} = \frac{A_p^*}{2} - a_B L_p r_{PB} \tag{44}$$

where $A_C$ is the contact area per particle, $A_p^*$ is the surface area of a polyhedral particle (Crowley *et al.* use a rounded rhombic dodecahedron in their model), $L_p$ is the length of the edges of a polyhedral particle, $r_{PB}$ is the radius of curvature of a particle (the subscript "PB" refers to the "Plateau borders" formed by three contacting particles, see Figure 10) and $a_B$ is a geometrical normalization factor equal to $1/\sqrt{3}$. (Note that although Crowley *et al.* use the same symbols as in Equation 44, theirs are multiplied by $N_p$ and so represent *total* values for the whole system). It can then be shown that

$$r_{PB} \approx \frac{\gamma}{\Pi_D} \tag{45}$$

where $\gamma$ is the surface tension and $\Pi_D$ is the disjoining pressure. When the deformation of the latex particles is great, $\Pi_D$ may be replaced by the osmotic pressure $\Pi$.

Neglecting the scattering from the stabiliser, substituting Equation 43 into Equation 42, expanding and multiplying through by $Q^4$, it may then be shown that

$$Q^4 \frac{\partial \Sigma}{\partial \Omega}(Q) \approx 2\pi N_p (\rho_p - \rho_m)^2 [A_p - 2A_B \cos(Q\Delta_0)] \tag{46}$$

Hence a graph of $Q^4 (d\Sigma/d\Omega)(Q)$ versus $Q$ oscillates about a mean value proportional to $A_p$ with an amplitude proportional to $A_B$ and a period of $2\pi/\Delta_0$. The first maximum occurs at $Q = \pi/\Delta_0$. This is illustrated in Figure 11 which shows how the scattering from a dispersion of block copolymer latex particles ($R_p = 100$ nm) changes during the film forming process. As the dispersion medium evaporates the peak moves to higher $Q$ and broadens considerably, indicating that $\Delta_0$ is decreasing.

When $Q\Delta_0 < \pi$, Equation 46 may be further approximated to

$$Q^4 \frac{\partial \Sigma}{\partial \Omega}(Q) \approx 2\pi N_p (\rho_p - \rho_m)^2 [A_F + A_B(Q\Delta_0)^2] \tag{47}$$

where $A_F$ is the "free" surface area of a particle; i.e. those regions of the particle surfaces which are not in contact, such as the Plateau borders (Plateau, 1861). Thus when $Q\Delta_0$ is small the scattering is dominated by that from the free surface. Crowley *et al.* refer to Equation 47 as the *free-surface/bilayer (FSB) approximation*. When applying Equation 47 the background scattering should first be removed as (constant $\times$ $Q$) and $Q^4(d\Sigma/d\Omega)(Q)$ should be scaled so that it tends to unity at high $Q$ values.

When $r_{PB} > \Delta_0$, a much better approximation, called the *contact region (CR) approximation* by Crowley *et al.*, may be obtained by evaluating the interference contributions to the scattering from opposing surfaces in the regions where the particle surfaces curve away from one another.

This approach is quite flexible in describing the Porod's law scattering from particles in contact, for example the CR approximation does not depend on any detailed model for the particles or their contact geometry. However, its derivation is not particularly straightforward and so will not be discussed here. The interested reader requiring further information is directed to the original paper.


## 4.4 SUMMARY

Adsorbed layers of polymer or surfactant molecules are technologically important because they are present in many different everyday products. Small-angle neutron scattering (SANS) is one of the most successful experimental techniques for probing the number and arrangement of molecules in an adsorbed layer and can yield quantitative information that can be directly compared with other measurements.

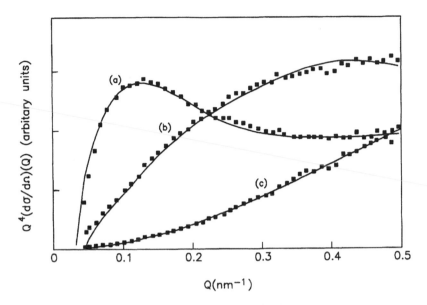

FIGURE 11. The scattering from film-forming block copolymer latex particles plotted according to
Equation 46. Each dataset represents a different drying time. The continuous lines are non-linear least
squares fits of Equations 44–46 to the data assuming values of: (a) $r_{PB} = 40$ nm, $\Delta_0 = 17.5$ nm; (b)
$r_{PB} = 34$ nm, $\Delta_0 = 6.5$ nm; and (c) $r_{PB} = 10$ nm, $\Delta_0 = 3$ nm. Reprinted with permission from
Sanderson *et al.*, 1992. Copyright 1992 American Chemical Society.

In this chapter we have given a brief overview of the technique of SANS and
a comprehensive discussion of its application in the study of adsorbed polymer and
surfactant layers. In doing so we have brought together, for the first time, several different
methodologies in one consistent, evolutionary treatment. We hope that this work will be of
use to both existing and potential users of SANS.

## 4.5 APPENDIX

The two most commonly encountered particle size distribution functions are the *zeroth-order
logarithmic distribution* (also known as the *log-normal distribution*) and the *Schultz
distribution*.

### 4.5.1 The Log-Normal Distribution

The fraction of radii, $n(R)$, having a radius, $R$, is given by

$$n(R) = \frac{\exp\left[-\frac{(\ln R - \ln R_m)^2}{2\beta_0^2}\right]}{(2\pi)^{\frac{1}{2}} \beta_0 R_m \exp(\beta_0^2/2)}$$

where $R_m$ is the modal value of $R$, and $\beta_0$ is a measure of the width and skewness, of the distribution. The mean (or average, $\bar{R}$) and modal values of $R$ are inter-related by

$$\ln \bar{R} = \ln R_m + 1.5\beta_0^2$$

The standard deviation, $\beta$, of the distribution is given by

$$\beta = R_m[\exp(4\beta_0^2) - \exp(3\beta_0^2)]^{\frac{1}{2}} \approx R_m\beta_0 \quad (\text{for } \beta_0 \ll 1)$$

### 4.5.2 *The Schultz Distribution*

The fraction of radii, $n(R)$, having a radius, $R$, is given by

$$n(R) = \frac{\left(\frac{\mu+1}{\bar{R}}\right)^{\mu+1} R^\mu \exp\left[-\left(\frac{\mu+1}{\bar{R}}\right) R\right]}{\Gamma(\mu+1)}$$

where $\bar{R}$ is the mean value of $R$, $\Gamma(x)$ is the Gamma function, $\mu$ is a measure of the width of the distribution and is given by the expression

$$\mu = \frac{\left(1 - \frac{\beta^2}{\bar{R}^2}\right)}{\frac{\beta^2}{\bar{R}^2}}$$

For further information on the Schultz distribution, see; M. Kotlarchyk, S.-H. Chen, *J. Chem. Phys.*, (1983), **79**, 2461.

## REFERENCES

Arrighi, V., Fernandez, M.L. and Higgins, J.S. (1993/94) 'Looking' at polymers with neutrons, *Sci. Prog.* **77**, 71 (A short introductory article on how neutron scattering can be used to study the structure and dynamics of polymers in solution and in the melt).

Ashcroft, N.W. and Lekner, J. (1966) Structure and resistivity of liquid metals, *Phys. Rev.*, **145**, 83.

Auroy, P., Auvray, L. and Léger, L. (1990) The study of grafted polymer layers by neutron scattering, *J. Phys., Cond. Matt.*, **2**, SA317.

Auroy, P., Auvray, L. and Léger, L. (1991) Structures of end-grafted polymer layers: A small-angle neutron scattering study, *Macromol.*, **24**, 2523.

Auroy, P., Auvray, L. and Léger, L. (1991) Scattering by grafted polymers, *Physica A*, **172**, 269.

Auroy, P. and Auvray, L. (1992) Collapse-stretching transition for polymer brushes. Preferential solvation, *Macromol.*, **25**, 4134.

Auroy, P., Mir, Y. and Auvray, L. (1992) Local structure and density profile of polymer brushes, *Phys. Rev. Lett.*, **69**, 93.

Auvray, L. (1986) On the intensity scattered at small angles by a polymer layer adsorbed on a wall, *C.R. Acad. Sci. (Paris), Ser. 2*, **302**, 859.

Auvray, L. and De Gennes, P.G. (1986) Neutron scattering by adsorbed polymer layers, *Europhys. Lett.*, **2**, 647.

Auvray, L. and Cotton, J.P. (1987) Self-similar structure of an adsorbed polymer layer: Comparison between theory and scattering experiments, *Macromol.*, **20**, 202.

Auvray, L., Auroy, P. and Cruz, M. (1992) Structure of polymer layers adsorbed from concentrated solutions, *J. Phys. I, France*, **2**, 943.

Barnett, K.G., Cosgrove, T., Vincent, B., Burgess, A.N., Crowley, T.L., King, T., Turner, J.D. and Tadros, Th.F., Neutron scattering, nuclear magnetic resonance and photon correlation studies of polymers adsorbed at the solid-solution interface, *Polymer*, **22**, 283.

Barnett, K.G. (1982) Ph.D. Thesis, University of Bristol, UK.

Brown, W. (editor) (1993) *Dynamic Light Scattering: The Method and some Applications*, Clarendon Press, Oxford.

Caucheteux, I., Hervet, H., Rondelez, F., Auvray, L. and Cotton, J.P. (1989) Polymer adsorption at the solid-liquid interface: The interfacial concentration profile, *Proc. Int. Symp. Trends Phys. Phys. Chem. Polym.*, **63**.

Cosgrove, T. (1990) Volume fraction profiles of adsorbed polymers, *J. Chem. Soc. Faraday Trans.*, **86**, 1323.

Cosgrove, T., Crowley, T.L., Vincent, B., Barnett, K.G. and Tadros, Th.F. (1981) The configuration of adsorbed polymers at the solid-solution interface, *Faraday Symposia of the Chemical Society*, No. 16, Royal Society of Chemistry, London.

Cosgrove, T., Heath, T.G., Ryan, K. and Crowley, T.L., Neutron scattering from adsorbed polymer layers, *Macromol.*, **20**, 2879.

Cosgrove, T., Crowley, T.L., Ryan, K. and Webster, J.R.P. (1990) The effects of solvency on the structure of an adsorbed polymer layer and dispersion stability, *Coll. Surf.*, **51**, 255.

Cosgrove, T. and Ryan, K. (1990) NMR and neutron-scattering studies on poly(ethylene oxide) terminally attached at the polystyrene/water interface, *Langmuir*, **6**, 136.

Cosgrove, T., Heath, T.G. and Ryan, K. (1994) Terminally attached polystyrene chains on modified silicas, *Langmuir*, **10**, 3500.

Cosgrove, T., King, S.M. and Griffiths, P.C. (1999) Small-angle neutron methods in polymer adsorption studies, in *Colloid-Polymer Interactions: From Fundamentals to Practice*, Farinato, R.S., Dubin, P.L. (editors), John Wiley, 1999.

Crowley, T.L. (1984) D.Phil. Thesis, University of Oxford, UK.

Cummins, P.G., Staples, E. and Penfold, J. (1990) Study of surfactant adsorption on colloidal particles, *J. Phys. Chem.*, **94**, 3740.

De Gennes, P.G. (1979) *Scaling Concepts in Polymer Physics*, Cornell University Press, Ithaca, New York.

Farsaci, F. (1989) Dynamical behaviour of structured macromolecular solutions, *Phys. Chem. Liq.* **20**, 205.

Feigin, L.A. and Svergun, D.I. (1987) in *Structure Analysis by Small Angle X-ray and Neutron Scattering*, Taylor, G.W. (editor), Plenum.

Finch, N.A. (1988) Ph.D. Thesis, University of Bristol, UK.

Fleer, G.J., Cohen Stuart, M.A., Scheutjens, J.M.H.M., Cosgrove, T. and Vincent, B. (1993) *Polymers at Interfaces*, Chapman & Hall, London.

Forsman, W.C. and Latshaw, B.E. (1990) Polymer molecules at interfaces: Studies by small-angle neutron scattering, *Mat. Res. Soc. Symp. Proc.*, **171**, 355.

Gladman, J.M., Crowley, T.L., Schofield, J.D. and Eaglesham, A. (1995) The structure of surface-adsorbed surfactants as determined by small-angle neutron scattering, *Phys. Script.*, **T57**, 146.

Glatter, O. and Kratky, O. (editors) (1982) *Small Angle X-ray Scattering*, Academic Press. (Divided into methodology and applications. Many chapters and references in each section.)

Guinier, A. and Fournet, G. (1955) *Small Angle Scattering of X-rays*, John Wiley (A very comprehensive text with 506 references).

Hansen, J.P. and Hayter, J.B. (1982) A rescaled MSA structure factor for dilute charged colloidal dispersions, *Mol. Phys.*, **46**, 651.

Hayter, J.B. and Penfold, J. (1981) An analytic structure factor for macroion solutions, *Mol. Phys.*, **42**, 109.

Higgins, J.S. and Stein, R.S. (1978) Recent developments in polymer applications of small-angle neutron, X-ray and light scattering, *J. Appl. Cryst.*, **11**, 346 (A little dated now, but still a useful review with 272 references).

Higgins, J.S. and Benoit, H.C. (1994) Polymers and Neutron Scattering, Oxford Series on Neutron Scattering in Condensed Matter, Volume 8, Clarendon Press. (A comprehensive, yet readable, text covering all aspects of neutron scattering from polymers.)

Hunter, R.J. (1989) *Foundations in Colloid Science*, Volume 2, Clarendon Press.

Kerker, M. (1969) *The Scattering of Light and other Electromagnetic Radiation*, Academic Press. (Probably still the definitive text on light scattering.)

King, S.M. (1999) Small-Angle Neutron Scattering, in *Modern Techniques for Polymer Characterisation*, Pethrick, R.A. and Dawkins, J.V. (editors), John Wiley. (A practical but readable guide to SANS, covering all aspects from choosing an instrument to analysing the data. Includes 115 references.)

Kostorz, G. (1982) Neutron Scattering, in *Treatise on Materials Science and Technology*, Volume 15, Kostorz, G. (editor), Academic Press. (This text was why Glatter & Kratky did not expand their book to encompass SANS.)

Lindner, P. and Zemb, Th. (editors) (1991) *Neutron, X-ray and Light Scattering*, Holland. (Proceedings of the European Workshop on Neutron, X-ray and Light Scattering as an Investigative Tool in Colloidal and Polymeric Systems, held in Bombannes, France, 1990.)

Livsey, I. (1987) Neutron scattering from concentric cylinders, *J. Chem. Soc., Far. Trans. 2*, **83**, 1445.

Lovesey, S.W. (1984) *The Theory of Neutron Scattering from Condensed Matter*, Volumes 1 & 2, Oxford University Press. (The definitive texts on the theory of neutron scattering.)

Maconnachie, A. and Richards, R.W. (1978) Neutron scattering and amorphous polymers, *Polymer*, **19**, 739. (A little dated now, but nonetheless a useful review of how different neutron scattering techniques can be used to investigate structure and dynamics in bulk polymers. Approximately half of the article is devoted to scattering theory, instrumentation and experimental techniques. 116 references.)

Markovic, I., Ottewill, R.H., Cebula, D.J., Field, I. and Marsh, J.F. (1984) Small-angle neutron scattering studies on non-aqueous dispersions of calcium carbonate. Part I. The Guinier approach, *Coll. Polym. Sci.*, **262**, 648.

Markovic, I. and Ottewill, R.H. (1986) Small-angle neutron scattering studies on non-aqueous dispersions of calcium carbonate. Part II. Determination of the form factor for concentric spheres, *Coll. Polym. Sci.*, **264**, 65.

Markovic, I. and Ottewill, R.H. (1986) Small-angle neutron scattering studies on non-aqueous dispersions of calcium carbonate. Part III. Concentrated dispersions, *Coll. Polym. Sci.*, **264**, 454.

Markovic, I. and Ottewill, R.H. (1987) Structural and dynamic features of concentrated non-aqueous dispersions, *Coll. Surf.*, **24**, 69.

Marshall, W. and Lovesey, S.W. (1971) *Theory of Thermal Neutron Scattering*, Clarendon Press, Oxford.

Muroga, Y. (1988) Conformational analysis of broken rodlike chains. 1. Scattering function of rods joined together by flexible coils, *Macromol.*, **21**, 2751.

Newport, R.J.; Rainford, B.D. and Cywinski, R. (editors) (1986) *Neutron Scattering at a Pulsed Source*, Adam Hilger. (The lectures from The Summer School on Neutron Scattering at a Pulsed Source, held at Mansfield College, Oxford and ISIS, UK, 1986.)

Ottewill, R.H. (1982) Small Angle Neutron Scattering, in *Colloidal Dispersions*, Goodwin, J.W. (editor), Special Publication No. 43, Royal Society of Chemistry. (A short, eminently readable, introductory account of how SANS may be used to study sterically-stabilised colloidal dispersions. Also see the chapter on "Concentrated Dispersions" by the same author.)

Ottewill, R.H. (1991) Small angle neutron scattering from colloidal dispersions, *J. Appl. Cryst.*, **24**, 436.

Ottewill, R.H., Sinagra, E., MacDonald, I.P., Marsh, J.F. and Heenan, R.K. (1992) Small-angle neutron scattering studies on non-aqueous dispersions. Part V. Magnesium carbonate dispersions in hydrocarbon media, *Coll. Polym. Sci.*, **270**, 602.

Penfold, J. and Thomas, R.K. (1990) The application of the specular reflection of neutrons to the study of surfaces and interfaces, *J. Phys. Cond. Matter*, **2**, 1369. (A review article with 102 references.)

Perkins, S.J. (1994) High Flux X-ray and Neutron Solution Scattering, in *Methods in Molecular Biology*, Volume 22, Jones, C., Mulloy, B. and Thomas, A.H. (editors), Humana Press. (A short introductory article on what SAXS and SANS can do for molecular biologists.)

Plateau, J. (1861) *Mem. Acad. Roy. Soc. Belg.*, **33**, Sixth series.

Porod, G. (1951) *Kolloid Z.*, **124**, 83.

Richards, R.W. (1989) Small-angle neutron scattering studies of polymers: Selected aspects, *J. Macromol. Sci. Chem.*, **A26**, 787. (A readable overview of the application of SANS to the study of multiphase polymer systems. Some typographical errors.)

Sanderson, A.R., Crowley, T.L., Morrison, J.D., Barry, M.D., Morton-Jones, A.J. and Rennie, A.R., The formation of bilayers and Plateau borders during the drying of film-forming latices as investigated by small-angle neutron scattering, *Langmuir*, **8**, 2110.

Sears, V.F. (1992) Neutron scattering lengths and cross-sections, *Neutron News*, **3**, 26.

Skold, K. and Price, D.L. (editors) (1986) Neutron Scattering, Parts A, B & C, in *Methods in Experimental Physics*, Celotta, R. (editor-in-chief), Academic Press.

Washington, C., King, S.M. and Heenan, R.K., Structure of block copolymers adsorbed to perfluorocarbon emulsions, *J. Phys. Chem.*, **100**, 7603.

Willis, B.J.M. (1973) *Chemical Applications of Neutron Scattering*, Oxford University Press, Oxford.

Van de Hulst, H.C. (1957) *Light Scattering by Small Particles*, John Wiley.

# 5. Crystalline and Amorphous Polymers

KEISUKE KAJI

*Institute for Chemical Research, Kyoto University, Uji, Kyoto-FU 611, Japan*

## 5.1 INTRODUCTION

Light hydrogen and deuterium atoms possess very different neutron scattering amplitudes, termed *scattering lengths*: $b_H = -0.374 \times 10^{-12}$ cm and $b_D = 0.667 \times 10^{-12}$ cm, respectively, though they have almost the same chemical properties. Such a large difference in scattering lengths gives rise to three distinct small-angle neutron scattering (SANS) methods, namely a *labelling method* (LM), *a contrast enhancement method* (CEM) and a *contrast variation method* (CVM). The LM enables the determination of chain conformations in the bulk state since a dilute solid solution of labeled molecules can be realized by dispersing the labeled molecules in the unlabeled matrix. The combination of SANS and intermediate-angle neutron scattering (IANS) with the labelling method gives information about the global and local conformations of polymer chains, respectively. Hence this method is very useful for solving various important problems in the physics of polymer solids. A partial labelling method is also important for investigation of the partial structure such as crosslinking in polymer networks. By means of the CEM heterogeneous structures due to density fluctuations such as long periods in semicrystalline polymers can be studied more easily. For example, the contrast enhancement is made for usual polymer materials by introducing a deuterated swelling agent into the amorphous or less ordered regions with lower density through absorption. The CVM reveals the internal structures of composite particles such as core-shell latex and many kinds of biological globular proteins without destroying the particles. Thus, by immersing the core-shell latex in the mixture of deuterated and hydrogenated media (usually the mixture of light and heavy water) and adjusting the fraction of the deuterated medium so as to match the scattering contrast of the shell part, we can 'see' only the core part, and vice versa. This method is not further described in this chapter, however.

In this chapter many interesting problems in the solid state physics of polymers are presented. These are applications of SANS, mainly with the labelling method and partly with the contrast enhancement method. Both are particularly useful for solving these problems. In some cases the IANS and the wide-angle neutron scattering (WANS) are as well employed in order to obtain information about more detailed local structures of the polymer chains. Some of the interesting problems are: Is a single polymer molecule Gaussian in the amorphous bulk state? How is a polymer chain incorporated in the single folded chain crystal? What is the effect of annealing on the chain conformation? Are the crystallization and the melting precisely inverse processes which occur during a conformational change? What are the differences in chain trajectory between single crystals and melt-grown crystals? How are polymer chains deformed by the external forces? We shall also investigate the problems of transesterification of polyesters, the structure of polymer networks and the long period structure in fibres. Before discussing these interesting issues we must of course acknowledge some experimental problems in the SANS labeling method, which merit some comments first.

For a reader keen to be acquainted with the fundamentals of neutron scattering of polymers the following textbooks are recommended in particular: Higgins and Benoit (1992), Kostorz (1979), and Glatter and Kratky (1982) (see also references in Chapter 1). The last book is limited to "Small Angle X-ray Scattering", but as the many of the underlying principles hold for neutron scattering as well, it is a very helpful reading for neutron scattering researchers.

## 5.2 EXPERIMENTAL PROBLEMS IN THE SANS LABELING METHOD

When the SANS labeling method is applied to the study of crystalline polymers in the solid state, several experimental problems should be taken into account. These are: exchange of H and D atoms during sample preparations, heterogeneities in the matrix, such as voids and amorphous-crystalline density difference, isotope fractionation effect, polydispersity effect of molecular weight, and so forth. These issues are important and have been examined in detail especially for polyethylene (PE), though in principle they must exist for all polymers, even if negligible.

### 5.2.1  Exchange of H and D Atoms

PE samples for SANS experiments are prepared by dissolving in boiling xylene at temperatures around 140°C, precipitating the solution into excess methanol and then pressing into bubble-free films. In this case prolonged boiling in xylene causes exchange of H and D atoms even when they are bonded to carbon atoms (Schelten *et al.*, 1974a). It is therefore necessary to determine the degree of deuteration (or hydrogenation) of the labeled PE by NMR.

On the other hand a catalytic exchange of H and D by a suspended rhodium catalyst is utilized to prepare partially deuterated samples; a labeled polymer can be then synthesized from the conventional polymer in one simple step (Tanzer & Crist Jr., 1985). The most important advantage of this preparation method is that exactly matched sets of the labeled and unlabeled species (in molecular size and chemical microstructure) are available. This makes interpretation of the SANS data easier.

### 5.2.2 Heterogeneities in the Matrix

For the analysis of the SANS data from labeled molecules in the crystalline matrix, we need to take account of the excess scattering due to heterogeneities in the matrix. The heterogeneities arise from the density difference between amorphous and crystalline regions and also from the presence of a small fraction of voids. Ullmann et al. (1977) examined such excess scattering. The mean square density fluctuation $\langle \delta\rho^2 \rangle$ due to the heterogeneities may be given by

$$\langle \delta\rho^2 \rangle / \langle \rho^2 \rangle = \phi_c \phi_a (\rho_c - \rho_a)^2 / \langle \rho^2 \rangle + \phi_v (1 - \phi_v) \tag{1}$$

where $\rho_c$ and $\rho_a$ are the densities of crystalline and amorphous regions, respectively, $\phi_c$ and $\phi_a$ are their volume fractions, $\langle \rho^2 \rangle$ is the mean-square density of the system, and $\phi_v$ is the volume fraction of voids. In the case of PE the contribution from the amorphous-crystalline contrast, i.e. the first term on the right side of equation (1), is estimated to be less than $6.03 \times 10^{-3}$. In this estimate, $\rho_c = 0.998$ g/cm$^3$ and $\rho_a = 0.855$ g/cm$^3$ (Zugenmeier & Cantow, 1969), because the product $\phi_c \phi_a$ cannot exceed 0.25 and when $\phi_c = \phi_a = 0.5$, then $\langle \rho^2 \rangle = 0.848$(g/cm$^3$)$^2$.

Experimentally the mean-square density fluctuation $\langle \delta\rho^2 \rangle$ in the sample can be estimated as follows. According to Debye and Bueche (1949), the scattering intensity $I_\delta(Q)$ due to the mean-square density fluctuation is described by

$$I_\delta(Q) = K\alpha^3 \langle \delta\rho^2 \rangle / (1 + Q^2 \alpha^2)^2 \tag{2}$$

where $Q = (4\pi/\lambda)\sin(\theta/2)$ is the magnitude of scattering vector, $\lambda$ and $\theta$ are neutron wavelength and scattering angle, respectively. $K$ is a constant depending on the experimental conditions and $\alpha$ is the characteristic correlation length in the sample when a Markoffian type of correlation function $\gamma(r) = \exp(-r/\alpha)$ is assumed. As seen from equation (2), $\alpha$ is obtained from the ratio of slope to intercept at $Q = 0$ of the $I_\delta(Q)^{-1/2}$ vs. $Q^2$ plot. The term $\langle \delta\rho^2 \rangle$ is also obtained from the intercept if the value of $K$ is given. In the experiments (Ullmann et al., 1977) the $K$ value was determined using the incoherent scattering data of liquid cyclohexane obtained under the same experimental conditions because it has the same chemical structure as PE and no structural density fluctuations. The SANS experiments for various commercial linear PE samples prepared by quenching from the melt gave the values of $\alpha = 100 \sim 150$ Å and $\langle \delta\rho^2 \rangle / \langle \rho^2 \rangle = 1.4 \sim 4.1 \times 10^{-2}$. This observed mean-square density fluctuation seems to be due entirely to the scattering from voids, since the contribution from the amorphous-crystalline density difference estimated above is one order of magnitude smaller.

Crystalline PE samples usually contain voids, and strong void scattering covers the scattering from chain conformations to be measured. Fortunately, in neutron scattering the void scattering is extremely weak when the protonated polyethylene (PEH) matrix is used (Schelten, 1974a). This is due to the cancellation between the positive scattering length of $^{12}$C atom ($b_C = +0.66 \times 10^{-12}$ cm) and the negative scattering length of H atom ($b_H = -0.37 \times 10^{-12}$ cm), resulting in $b_{CH_2} = -0.08 \times 10^{-12}$ cm for the coherent scattering length for the $CH_2$ basic unit. In this case, therefore, the correction of the void scattering is easy. By subtracting the scattering of the pure PEH matrix (blank scattering) from that of the labeled sample, both the void scattering at low $Q$ and incoherent scattering as $Q$-independent background can be removed simultaneously. On the other hand the deuterated polyethylene (PED) matrix should be avoided except when the system is confirmed to be free of voids because the scattering length of $^2$D atom ($b_D = +0.62 \times 10^{-12}$ cm) has almost the same value as that of $^{12}$C and the average for the $CD_2$ basic unit is $b_{CD_2} = +1.90 \times 10^{-12}$ cm. For samples with a large amount of voids the PED matrix gives strong void scattering. Hence the measurements of molecular dimensions of PEH in the solid PED matrix are therefore virtually impossible. For molten samples the void scattering clearly need not be considered, though a small correction for the blank scattering should be made.

### 5.2.3  Isotope Fractionation Effect (Segregation Effect)

The isotope fractionation effect for PE was first pointed out by Stehling et al. (1971), and thereafter it was studied in more detail by Schelten et al. (1974a, b). Figure 1 shows the schematic sketch comparing the unclustered and clustered, labeled molecules (Schelten et al., 1974b). They showed that if the clustering due to isotope segregation occurs, the scattering intensity at $Q = 0$ from an observed sample exceeds the intensity from an unclustered sample where the labeled molecules are completely randomly distributed in the matrix. This follows from the Guinier approximation (Glatter & Kratky, 1982). In this approximation, for $Q^2 R_{gz}^2 \ll 1$, the scattering intensity $I(Q)$ is generally given by

$$I(Q) \propto M_w (1 - Q^2 R_{gz}^2 / 3) \tag{3}$$

where $M_w$ is the weight-average molecular weight and $R_{gz}$ is the $z$-average radius of gyration (see subsequent subsection 5.2.4 and Section 5.3), and the clustering makes apparent values of $M_w$ and $R_{gz}$ larger than the true values. The analysis of the experimental data proceeds as follows. The scattering from the labeled molecules is first obtained by subtracting the matrix scattering from the sample scattering. Then a Guinier plot, $I(Q)^{-1}$ vs $Q^2$, is drawn for the difference scattering. The intercept at $Q = 0$ and the slope of this curve provide approximate values of $M_w$ and $R_{gz}$, respectively. In order to obtain more accurate $M_w$ and $R_{gz}$ the Zimm plot should be made as described later.

An example of the scattering patterns from a PED in PEH matrix (PED/PEH) is shown in Figure 2. Here (PED/PEH) contains 0.5% PED of $M_w = 150,000$ and $M_w/M_n = 2.2$ (open circles) while the scattering from a PEH matrix is denoted by solid circles. The upturn in the matrix scattering at low $Q$ is due to void scattering from the matrix. The difference intensity gives the scattering from the labeled molecules. The results obtained from the Guinier plots show that in the crystalline state after slow cooling from the melt there exist clusters of PED consisting of 24 molecules, but in the molten state at 150°C no clustering occurs. That clustering increased the aparent $R_{gz}$ from 250 Å to 550 Å. Due to an appreciable difference in melting temperatures $T_m$ of the PED and PEH samples PE is prone to the isotopic fractionation effect when it is slowly crystallized from the melt.

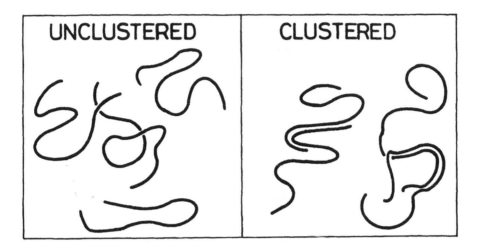

FIGURE 1. Schematic sketch for a clustering of labeled molecules in the matrix (Schelten *et al.*, 1974b. Reprinted with permission from *Colloid & Polymer Sci.*, Neutron small angle scattering by mixtures of H- and D-tagged molecules of polystyrene and polyethylene, J. Schelten, G.D. Wignall, D.G.H. Ballard and W. Schmatz, 1974, **252**, 749–752. Copyright © 1974 Steinkopff Publishers.)

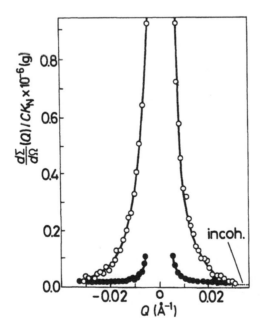

FIGURE 2. Scattering curves from PED in PEH matrix with 0.5% PED molecules (○) and from PEH matrix (●). (Schelten *et al.*, 1974a. Reprinted from *Polymer*, 15, J. Schelten, G.D. Wignall and D.G.H. Ballard, Chain information in molten polyethylene by low angle neutron scattering. Pages 682–685. Copyright © 1974 with permission from Elsevier Science.)

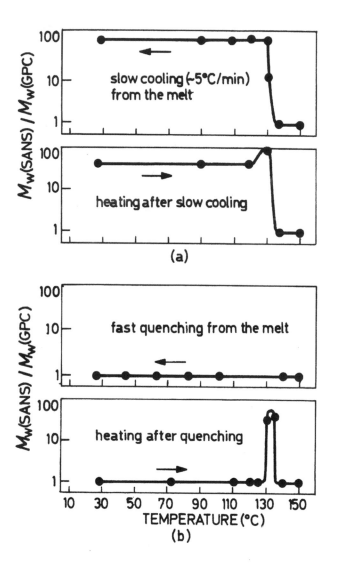

FIGURE 3. $M_w$ determined by SANS normalized by the true $M_w$ from GPC as a function of temperature for various cooling or beating conditions. (Wignall, 1993. Reprinted with permission from *Physical Properties of Polymer*, ed. by J.E. Mark, A. Eisenberg, W.W. Graessley, L. Mandelkern, E.T. Samulski, J.L. Koenig and G.D. Wignall, ACS, Washington, DC, pp. 313–378 (Chapter 7 by G.D. Wignall). Copyright © 1993 American Chemical Society.)

$T_m$(PED) $= 130.4°C$ is about $5°C$ lower than $T_m$(PEH) $= 135.2°C$ (Stehling *et al.*, 1971).

The conditions where the clustering occurs have been investigated in more detail (Schelten *et al.*, 1977). Figure 3 shows the results; the ratio of $M_w$ determined by SANS to the true $M_w$

determined by permeation chromatography (GPC) is plotted as a function of temperature (see Wignall, 1993). From this figure it is concluded that the aggregation occurs only when the sample is held for some time in the melting range 130 ~136°C. The slow cooling from the melt therefore causes clustering while fast quenching does not. This is the reason why only rapidly quenched PED-PEH blend samples are employed for the SANS labeling method, as shown later. It is also noticed that the aggregation is not affected by annealing below the melting range.

On the other hand, blends of deuterated and non-deuterated polystyrenes of sufficiently high molecular weights hardly exhibit any segregation effects owing to the small thermodynamic difference between them (Bates & Wignall, 1986) and may be treated as ideal mixtures.

### 5.2.4 Polydispersity

Polymers usually display *polydispersity* in their molecular weight. The effect of polydispersity on the scattering function has been discussed in detail by Higgins and Benoit (1994). Here only one limiting case (useful to estimate the radius of gyration $R_g$) is quoted, with some more details given in the next section. Thus, the values obtained from the the Guinier plot for SANS data are $M_w$ and $R_{gz}$. The conversion from $R_{gz}$ to the weight-average radius of gyration $R_{gw}$ is possible on the assumption of a Schultz distribution of molecular weight (Schultz, 1939) for a Gaussian coil:

$$R_{gw}^2 = R_{gz}^2(1 + U)/(1 + 2U) \tag{4}$$

where $U = M_w/M_n - 1$ (Zimm, 1948; Altgelt & Schulz, 1960; Oberthuer, 1978; Boothroyd, 1988). Of course, $R_{gw} = R_{gz}$ for monodisperse samples ($M_w/M_n = 1$). For other chain conformations see the literature (Sharp & Bloomfield, 1968; Glatter & Kratky, 1982).

### 5.2.5 Persistence Length

The straightness of polymer chains is an important parameter when we consider the structure and properties of polymer materials including liquid crystalline polymers (Donald & Windle, 1992). This parameter is usually represented by the persistence length, which was first defined by Porod (1949). For polymer chains in solutions it can be estimated using small angle x-ray scattering (SAXS) technique, but for those in the bulk state only a labeling SANS method is available. In what follows the definition of the persistence length and a precise evaluation method for amorphous bulks will be provided.

### Definition of Persistence Length

The freely rotating chain comprises $n$ bonds of fixed length jointed at fixed bond angles. For simplicity, all bond lengths $a$ and all bond angles $\psi$ are usually taken to be equal, respectively. The average projection of the $k$-th bond of a freely rotating chain on the direction of the first bond is given by $a(\cos \gamma)^{k-1}$ where $\gamma = \pi - \psi$ where the averaging is made by repeating the projection on the direction of the neighbouring bond by $(k - 1)$

times. The average sum of projections of these $n$ bonds on the direction of the first bond is given by

$$\langle X \rangle = a\Sigma(\cos\gamma)^k = a(1 - \cos^n\gamma)/(1 - \cos\gamma). \tag{5}$$

The persistence length $b$ is defined as the average sum of projections for the infinitely long chain ($n = \infty$) and hence

$$b = a/(1 - \cos\gamma). \tag{6}$$

Here it should be noted that the persistence length of the completely rigid chain ($\gamma = 0$) is infinite. Inserting equation (6) into equation (5), we obtain

$$\langle X \rangle = b[1 - (1 - a/b)^{L/a}] \tag{7}$$

where $L = na$ is the contour length of the polymer chain.

For the wormlike chain (Porod-Kratky chain), i.e. the chain with continuously varying direction, the persistence length becomes the limiting value of equation (6) for $a \to 0$ and $\gamma \to 0$:

$$b = \lim_{a\to 0, \gamma\to 0}[a/(1 - \cos\gamma)]. \tag{8}$$

In this case equation (7) reduces to

$$\langle X \rangle = b[1 - \exp(-L/b)]. \tag{9}$$

As seen from this equation, $\langle X \rangle = L$ for the completely rigid chain ($b \gg L$), while $\langle X \rangle = b$ for the flexible chain ($b \ll L$). The latter gives the basis for the persistence length being nearly equal to the average length of rigid segments in the flexible chain.

### Precise Determination of Persistence Lengths by SANS

As an example, let us show the change of the persistence length of poly(ethylene teraphthalate) (PET) during the induction period of crystallization when it was crystallized from the glass just above the glass transition temperature (Imai et al., 1995). In the glassy state polymer chains are Gaussian — or unperturbed — and three different analytical models have been proposed to describe the scattering function for the unperturbed wormlike chain. Yoshizaki and Yamakawa (1980) have developed the scattering function describing the low $Q$ range corresponding to the Guinier region, which is inapplicable to the determination of the persistence length. Des Cloizeaux (1973) has given the exact scattering function in a series expansion for the infinitely long chain. This expression is valid only in the high $Q$ range corresponding to the rod part of the real finite chain. In principle, the limiting form of this function for $Q \to \infty$ seems to be sufficient for the estimation of the persistence length, but in practice a fit with the limited observed data involves considerable uncertainty owing to the usual experimental error. In this study where a slight change of the persistence length is detected, this function could not be used. On the other hand, Sharp and Bloomfield (1968) have obtained the required scattering function in a simple analytical form covering both the Guinier and rod regions for the unperturbed wormlike chain. It is a special case of the general scattering function for perturbed wormlike chains with the excluded volume effect. It is confirmed that this function agrees well with the des Cloizeaux's function for

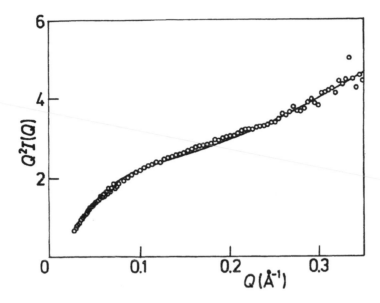

FIGURE 4. Fitting of the scattering function by Sharp and Bloomfield (1968) with the observed SANS data for PETD in PETH matrix, providing a reliable persistence length of 12.2 Å. (Imai *et al.*, 1995. Reprinted with permission from M. Imai, K. Kaji, T. Kanaya and Y. Sakai, *Physical Review B*, **52**, 1995, 12696–12704. Copyright © 1995 by the American Physical Society.)

$Q \to \infty$ (Kirste & Oberthuer, 1982). The scattering function developed by Sharp and Bloomfield (1968) includes the Debye function (see equation (22) below) to describe the lower $Q$ range:

$$I_n(Q) = 2(e^{-x} + x - 1)/x^2 + (2/15)[4 + 7/x - (11 + 7/x)e^{-x}](M_L/M)b \qquad (10)$$

with $x = LbQ^2/3$, where $M$ is the molecular mass of a polymer and $M_L = M_0/p_0$ is the mass per unit length. Here $M_0$ and $p_0$ are the mass of a monomer and the pitch of the monomer in the extended chain conformation, respectively. Figure 4 shows that the observed SANS data can be well fitted with the scattering function of Sharp and Bloomfield in the observed $Q$ range, providing a considerably reliable persistence length of 12.2 Å.

## 5.3 SANS DATA REDUCTION

A usual procedure is as follows (Kostorz, 1979; Cotton, 1991; Auvray, 1992; Higgins & Benoit, 1994). The differential scattering cross-section per unit solid angle of the sample per unit volume $d\Sigma(\theta)/d\Omega$ or $d\Sigma(Q)/d\Omega$ is calculated from the total experimental counting rate at an angle $\theta$ (or at a scattering vector $Q$). After the correction of instrumental background this reads

$$Z_{exp}(\theta) = I_0(\lambda)A_s D_s T_s(d\Sigma(Q)/d\Omega)\varepsilon_d A_d. \qquad (11)$$

Here $I_0(\lambda)$ is the neutron flux at a given wavelength $\lambda$ in neutrons/cm$^2$ s. $A_s$ and $D_s$ are the neutron beam-irradiated area and the thickness of the sample, respectively. $T_s = \exp[-\Sigma_t D_s]$ is the transmission factor of the sample, $\Sigma_t$ being the total cross-section per unit volume of the sample. $\varepsilon_d$ and $A_d$ are the efficiency and the area of a detector element, respectively. Then $\varepsilon_d$ is inversely proportional to the velocity of scattered neutrons (the reciprocal velocity law) and $A_d = R_{sd}^2 \Delta\Omega$, where $R_{sd}$ and $\Delta\Omega$ are the sample-to-detector distance and the solid angle of collection of neutrons, respectively. The main correction factor required to obtain the absolute values of differential scattering cross-sections is $I_0(\lambda)$ in equation (11). This factor can be determined by two methods: one is to measure the intensity of the direct beam using calibrated attenuators, and the other is to use standards. Usually the following standards are employed (Wignall & Bates, 1987; Russel et al., 1988; Cotton, 1991): vanadium and water as incoherent standards, dilute monodisperse suspensions and blends of well-characterized monodisperse ($M_w/M_n < 1.10$) labeled and unlabeled polymers, such as polystyrene and polyethylene, as coherent standards. The latter secondary coherent standards are calibrated against standard vanadium. In conventional experiments the latter standards are more frequently employed in order to save machine time.

The effect of multiple scattering should be kept in mind though it is assumed not to be large (less than 2%). The quantitative estimation of this correction is not easy and the best method is to reduce this scattering by making the sample as thin as possible. A criterion for it is length of the neutron mean free path $\Lambda$ in the sample; $\Lambda$ is between 0.5 and 10 mm for typical polymers (Cotton, 1991; des Cloizeaux & Jannink, 1987). Hence it is desirable for the sample thickness to be less than 0.5 mm for the severest case (also see refs Schelten & Schmatz, 1980, Goyal et al., 1983).

Even after these corrections, the noise scattering coming from the sample itself still remains; as described above, it contains the incoherent scattering $I_{inc}$ from all nuclear species, including impurities in the sample, and the scattering $I_\delta(Q)$ in equation (2) from heterogeneities in the matrix. Therefore, the observed differential scattering cross-section per unit volume $d\Sigma(Q)/d\Omega$ may be written as

$$d\Sigma(Q)/d\Omega = I(Q) + I_\delta(Q) + I_{inc} \tag{12}$$

where $I(Q)$ is the scattering intensity from the chain conformations to be measured. Further corrections for the incoherent scattering and the scattering from sample heterogeneities are made as below. Firstly, we measure the scattering intensities from both purely hydrogenated and purely deuterated samples prepared in the same way as in the case of the sample. Then, we can calculate the sum of $I_\delta(Q)$ and $I_{inc}$ from the D/H blend sample using the fraction of labeled molecules.

Let us assume a system composed of the mixture of labeled (deuterated) and unlabeled (hydrogenated) molecules with the same degree of polymerization $N$ where no interactions between the labeled and unlabeled molecules exist. As the scattering arises from the intra- and inter-molecular contributions, denoted respectively by $S_1(Q)$ and $S_2(Q)$, the scattering intensity $I(Q)$ is proportional to the sum of $[x_D S_1(Q) + x_D^2 S_2(Q)]$ where $x_D$ is the molar fraction of the labeled polymers. Here it is noted that $S_2(Q) = -S_1(Q)$ because no scattering should occur except at $Q = 0$ for the pure labeled molecular system ($x_D = 1$)

or the pure unlabeled one ($x_D = 0$), with the result that $I(Q) \propto x_D(1 - x_D)S_1(Q)$. Hence the scattering intensity per unit volume becomes

$$I(Q) = \rho_p N^2 K_N x_D (1 - x_D) S(Q) \qquad (13)$$

where $\rho_p$ is the number of moles of the total polymers per unit volume and $K_N$ is the contrast factor defined by

$$K_N = N_A (b_{mD} - b_{mH})^2, \qquad (14)$$

$b_{mD}$ and $b_{mH}$ being the scattering lengths of the labeled and unlabeled monomeric units, respectively, and $N_A$ the Avogadro number. Here,

$$b_{mD} = \sum_j^{n_D} b_j, \quad b_{mH} = \sum_k^{n_H} b_k \qquad (15)$$

where $b_j$, $b_k$ and $n_D$, $n_H$ are the coherent scattering lengths and the numbers of atoms in their monomers, respectively. Further, $S(Q)$ is the form factor of a single polymer molecule, which is the replacement of $S_1(Q)$ and given by

$$S(Q) = (1/N^2) \left\langle \left| \sum_{j=1}^N \exp(i\mathbf{Q} \cdot \mathbf{R}) \right| \right\rangle. \qquad (16)$$

The brackets denote an average over all orientations and conformations of the labeled polymers.

Equation (13) is valid only for monodisperse samples consisting of identical molecules, but the real samples are more or less polydisperse in molecular mass. As described before, the effects of polydispersity appear in the determinations of the molecular weight and the radius of gyration of labeled molecules (Cotton et al., 1974; Des Cloizeaux & Jannink, 1987; Higgins & Benoit, 1994). This effect on the molecular weight can be written as

$$M_w = M_0 \sum_\alpha (\rho_p x_D N^2)_\alpha / \sum_\alpha (\rho_p x_D N)_\alpha \qquad (17)$$

where $\sum (\rho_p x_D N)_\alpha \cong \rho_0 x_D$ and hence equation (13) becomes for $x_D \ll 1$

$$I(Q) \cong \rho_0 (M_w/M_0) K_N x_D S(Q) \qquad (18)$$

where $\rho_0$ is the sample density expressed in number of moles of the total monomers per unit volume, $M_w$ and $M_0$ are the weight-average molecular weight of labeled polymers and the labeled monomer mass, respectively. The effect on the radius of gyration should be contained in $S(Q)$ of equation (18). Generally $S(Q)$ can be expanded for the case $QR_g \ll 1$ as (Guinier & Fournet, 1955)

$$S(Q) = 1 - (1/3)Q^2 R_g^2 + \cdots \qquad (19)$$

where $R_g$ is the radius of gyration. For polydisperse samples $R_g^2$ should be replaced by an expression

$$R_{gz}^2 = \sum_\alpha (mN^2 R_g^2)_\alpha / \sum_\alpha (mN^2)_\alpha \qquad (20)$$

where $m$ is the number of molecules in each component. This $R_{gz}$ is conventionally called the *z-average radius of gyration* because $R_g^2$ is nearly proportional to $N$ for flexible chains. However, rigorously $R_g^2 \sim N^{1.0 \sim 1.2}$ as shown below (see equation (23)). The form factor in equation (19) is therefore modified as

$$S(Q) = 1 - (1/3)Q^2 R_{gz}^2 + \cdots \tag{21}$$

## 5.4 CHAIN CONFORMATION IN THE AMORPHOUS STATE

In polymer physics it is fundamentally important to know the conformational structure of polymer chains in the amorphous glasses and melts, which in turn is essential to understand the structure of semi-crystalline polymer materials and their physical properties.

The first theoretical attempt to tackle these issues is due to Flory (1949). He predicted that polymer molecules are shaped as Gaussian random coils in the homogeneous phase, i.e. have the same conformation as in the $\Theta$-solvent. A possible explanation is that polymer molecules dissolve in themselves and hence are shielded from either attractive or repulsive forces due to surrounding molecules. This model has been believed for a long time. The first doubts were voiced in 1970's: there was an evidence from the electron micrography of "nodule" structures in a melt-quenched amorphous film of poly(ethylene terephthalate) (PET) provided by Yeh and Geil (1972). In addition, Pechhold (1979) proposed a "meander" model which included regular structure on a very local scale. Many researchers investigated this problem from the viewpoints of long-range, intermediate-range and short-range order using various kinds of techniques. It was however the small angle neutron scattering technique which led to the conclusion that there exist neither the long range, nor the short range order. Therefore the Flory model of statistical random flight chains was widely accepted. However, current thinking and several experiments using wide angle x-ray and neutron scattering provided fresh evidence for the existence of the short range order in amorphous polymers. For more detailed discussion, see chapter by Young and Gabrys in this book.

In this section we show how the Flory model was confirmed using a SANS technique using the results obtained by Cotton *et al.* (1974) as an example. In order to avoid the problem of clustering due to the isotope segregation they used atactic polystyrene (aPS), a typical amorphous polymer. Sample was prepared by dispersing about 1% deuterated polystyrene (aPSD) with $M_w = 21,000$ and $M_w/M_n = 1.05$ into the hydrogenated polystyrene (aPSH) matrix with $M_w = 20,000$. This corresponds to a dilute system of aPSD, and hence the SANS measurement gives the scattering function for a single chain of aPSD. Figure 5 shows the scattering function $I(Q)$ of this sample after subtraction of the scattering intensity from the aPSH matrix. The solid circles are observed values, and the solid line is a Debye function fitted with the observed values. The Debye function represents the scattering function corresponding to the conformational structure of a statistically random flight chain. The fitted Debye function is given by

$$S(Q) = (2/X^2)(X - 1 + e^{-X}) \tag{22}$$

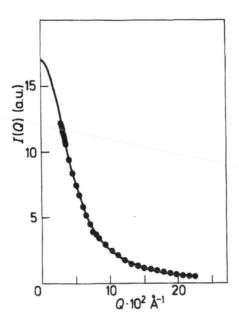

FIGURE 5. Scattering function of aPSD in aPSH matrix after subtraction of the scattering from the PSH matrix. The solid line is a Debye function. (Cotton *et al.*, 1974. Reprinted with permission from *Macromolecules*, **7**, 863–872. Copyright © 1974 American Chemical Society.)

where $X = Q^2 R_{gz}^2$. In this figure the solid line was calculated assuming $R_{gz} = 35$ Å, and a good agreement is obtained with the observed values. This strongly supports the Flory model.

Radius of Gyration

The Flory model was further tested on the basis of the relation between the molecular weight and the radius of gyration. Generally the following relationship holds (des Cloizeaux, 1970):

$$R_{gw} = K_w M_w \qquad (23)$$

where $K_w$ is a constant and $M_w$ is the weight-average molecular weight. $\nu$ is called the *characteristic critical exponent*, which depends on the chain conformation or the excluded volume. For example, $\nu = 0.5$ for the polymer chain in the $\Theta$-solvent, while $\nu = 0.6$ for that in the good solvent.

Experimentally $R_{gz}$ values are determined from the Zimm plots of observed SANS curves in order to remove the effect of intermolecular interactions. The Zimm relationship is given by (Zimm, 1948)

$$C_D K_c / I(Q) = (1/M_w)[1 + (1/3)Q^2 R_{gz}^2] + 2A_2 C_D. \qquad (24)$$

where $C_D = \rho_0 M_0 x_D$ is the concentration of polymer labeled by deuteration (PSD) expressed in weight per unit volume, $A_2$ is the second virial coefficient and $K_c = K_N/M_0$ is a constant. The first term of the right side of equation (24) comes from equation (21) with the condition $QR_{gz} \ll 1$ and the second term is due to the intermolecular interaction. If we plot $C_D/I(Q)$ against $Q^2$ for several values of $C_D$ and extrapolate to $C_D = 0$, we obtain $R_{gz}$ from the slope of the extrapolated curve, and $M_w$ from its intercept at $Q = 0$. The obtained values of $R_{gz}$ may be converted to $R_{gw}$ values through equation (4). Figure 6 shows the log $R_{gw}$ vs log $M_w$ plot for the values extracted this way. The open circles are the results for the glassy state; the slope of the line fitted with these data provides $\nu = 0.5$. This again confirms the Flory model. The crosses are the values for the $\Theta$-solvent when aPSD was dissolved in cyclohexane, which agree well with those obtained for the glassy state. The 'x' symbols denote data for the good solvent (sample dissolved in $CS_2$) yielding $\nu = 0.6$, which agrees well with the theoretical value. These results certify that the observed data are accurate enough to strongly support the Gaussian coil approximation.

Intermediate-angle neutron scattering (IANS)

However, even if the long-range order such as $R_g$ indicates a similar behaviour, the chain conformation is not always the same because of the possibility that different local conformations provide the same long-range order. In order to check such local conformation the scattering curves in the intermediate $Q$ range were also investigated. For $Q^2 R_{gz}^2 \gg 1$, equation (22) reduces to

$$S(Q) \cong 2/Q^2 R_{gz}^2. \tag{25}$$

This prediction can be confirmed from the Kratky plot, i.e. $Q^2 I(Q)$ vs $Q$ plot; as seen from Figure 7, $Q^2 I(Q)$ was almost constant in a range of $Q = 3 \sim 25 \times 10^{-2}$ Å$^{-1}$ for aPSD ($M_w = 3.3 \times 10^5$) in aPSH matrix. Note that a more realistic rotational isomeric state (RIS) model calculation for the random coil conformation often shows slight deviation from the Debye function (Yoon & Flory, 1976); for aPS the $Q^2 I(Q)$ somewhat increases with increasing $Q$ in the IANS range. Since the experimental data for aPS labeled only in the chain backbone agree well with the RIS model (Rawiso et al., 1987), the apparent agreement of the wholly deuterated aPS data with the Debye function may be due to the effect of side chain groups.

Other methods giving information about the short-range order such as depolarized light scattering, magnetic birefringence and Raman scattering as well supported the validity of the Flory theory (Fischer et al., 1976; Patterson, 1976). The confirmation of the Flory model means that polymer molecules penetrate deeply into one another and entangle themselves in the amorphous state. This concept is of a great importance when crystallization of flexible polymers from the bulk state is considered.

## 5.5  CHAIN CONFORMATION IN THE SEMICRYSTALLINE STATE

In the previous section it was clarified that the polymer chain in the amorphous state is Gaussian. When polymers are crystallized from the solution or from the amorphous bulk of

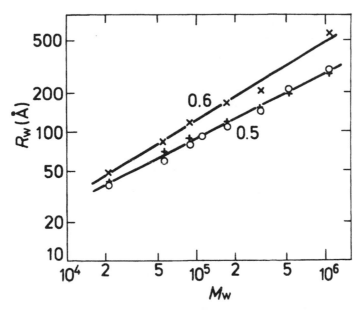

FIGURE 6. $R_{gw}$ vs $M_w$ plots in log scales for aPS in different environments. X: in a good solvent $CS_2$, +: in a $\Theta$ solvent, ○: in the bulk (Cotton *et al.*, 1974. Reprinted with permission from *Macromolecules*, **7**, 863–872. Copyright © 1974 American Chemical Society.)

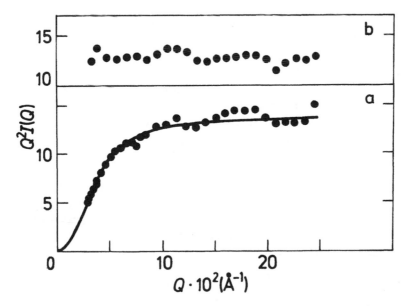

FIGURE 7. Kratky plots for aPSD in aPSH matrix. $M_w$'s of aPSD are $2.1 \times 10^4$ (a) and $3.3 \times 10^5$ (b). The solid line denotes the Debye function. (Cotton *et al.*, 1974. Reprinted with permission from *Macromolecules*, **7**, 863–872. Copyright © 1974 American Chemical Society.)

glass or melt, how does it change? This problem was also investigated by SANS. In what follows let us consider two distinct cases of solution-grown single crystals and bulk-grown crystals. For convenience we will roughly separate neutron scattering into three different $Q$ ranges. These are: (i) SANS range ($0.005$ Å$^{-1} < Q < 0.03$ Å$^{-1}$) giving molecular weight and radius of gyration, (ii) IANS range ($0.03$ Å$^{-1} < Q < 0.5$ Å$^{-1}$) giving information about the ribbon structure of the folded chain in the crystalline lamella, and (iii) WANS range ($0.5$ Å$^{-1} < Q < 5$ Å$^{-1}$) which yields crystal structure.

### 5.5.1  Chain Trajectory in Solution-Grown Single Crystals

Theoretical Scattering Functions for Possible Models

As will be described later, the molecular weight dependence of the radius of gyration of the labeled chains when they are crystallized under low super-cooling shows that the chain conformation in solution-grown single crystals is greatly extended compared to that in the amorphous state. Further, one may assume a sheet-like structure where a polymer chain molecule folds regularly in a fixed width though real systems may have more or less disordered chain conformations from such an ideal model. This assumption is based on the data from the intermediate $Q$ range (IANS) as well as the concept of chain folding in single crystals. Guenet *et al.* (1990) calculated the theoretical scattering functions for five simplified models. These models are reproduced in Figure 8, and their characteristic parameters and the calculated results are summarized below.

All the models consist of a chain-folded sheet or ribbon. In the first model (a) the chain folds regularly in the completely extended planar ribbon (sheet-like model) whose dimensions are defined by the length $L_c$ and the width $l_c$ of the ribbon. In the model (b) the ribbon is split into two parts at the center; these parts shift by the width of the ribbon $l_c$ to the chain axis (hinge-like model). The model (c) is the so-called "super-folding model" where the ribbon folds back on or near itself. If the number of super-folds is $n$, the actual length of the ribbon becomes $L'_c = L_c/(n + 1)$. The inter-ribbon distance between neighbouring super-folded parts is denoted by $\delta_c$. The model (d) is one of partial super-folding models and called the "Z-like model" because it has two consecutive partial super-folds. The actual length of the folded ribbon is given by $\mathcal{L}_c$ in this case and $\delta'_c = 2\delta_c$. The final model (e) is termed the "dilution model" where regular folding of polymeric chains is interrupted by non-adjacent folds; foreign molecules are partially incorporated in the labeled chain-folded ribbon. The lengths of the regular folding part and the interrupted part are given by $\lambda$ and $\alpha$, respectively. These two parameters are not fixed but have a distribution. If the number of $\lambda + \alpha$ sub-units is given by $p$, then the actual length $\Lambda_c$ of the ribbon becomes $p(\lambda + \alpha)$.

The scattering functions for these models may be summarized as follows. Here the effect of polydispersity is neglected and hence the discussion is valid when $R_{gz} = R_{gw} \equiv R_g$. In the Guinier range ($Q \ll R_g^{-1}$) the scattering function is approximately given by

$$I(Q) \propto M_w(1 - Q^2 R_g^2/3) \tag{26}$$

where $M_w$ is the weight-average molecular weight of the labeled species. The mean-square radius of gyration $R_g$ is theoretically given by

$$R_g^2 = (1/12)(l_c^2 + A^2), \tag{27}$$

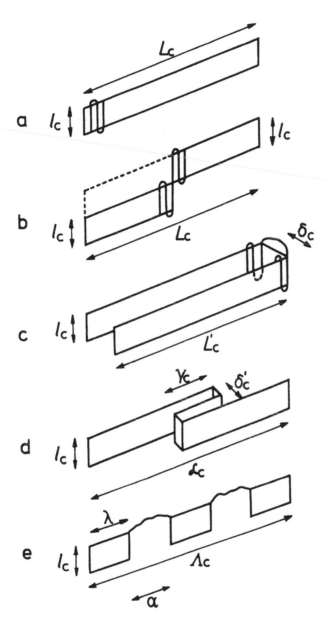

FIGURE 8. Possible models for chain trajectory in solution-grown crystals. (a) sheet-like model, (b) hinge-like model, (c) super-folding model, (d) Z-like model and (e) dilution model. (Guenet *et al.*, 1990. Reprinted from *Polymer*, **31**, J.M. Guenet, D.M. Sadler and S.J. Spells, Solution-grown crystals of isotactic polystyrene: neutron scattering investigation of the variation of chain trajectory with crystallization temperature. Pages 195–201. Copyright © 1990 with permission from Elsevier Science.)

for all the models except model (b), where $A$ is the actual length of the ribbon. Moreover, $A = L_c$, $L'_c$, $= \mathcal{L}_c$ and $\Lambda_c$ for the models (a), (c), (d) and (e), respectively. Here $\lambda_c = p(\lambda + \alpha)$, $p$ being the number of $(\lambda + \alpha)$ sub-units. For model (b)

$$R_g^2 = (1/12)[(5/2)l_c^2 + (5/8)L_c^2].\tag{28}$$

Equation (27) is reduced to $R_g^2 = (1/12)A^2$ for $l_c \ll A$, and $R_g^2 = (1/12)l_c^2$ for $l_c \gg A$. Similarly equation (28) is reduced to $R_g^2 = (5/96)L_c^2$ for $l_c \ll A$, and $R_g^2 = (5/24)l_c^2$ for $l_c \gg A$.

In the intermediate $Q$ range the scattering functions are approximated for the two $Q$ sub-ranges as follows. For $A^{-1} < Q < l_c^{-1}$ and $A > l_c \gg t$ where $t = n\delta_c$ is the super-folding thickness

$$I(Q) \propto M_w[(\pi f/QA)\exp(-Q^2l_c^2/24)] \quad \text{for models (a)(c)(d)(e)}$$
$$I(Q) \propto M_w[(3\pi/2QL_c)\exp(-Q^2l_c^2/28)] \quad \text{for model (b)}\tag{29}$$

where $f = 1$ for the models (a), (c) and (d), and $f = \lambda/(\lambda + \alpha)$ for the model (e). Except when $\ln QI(Q)$ vs $Q^2$ plot is made in order to estimate $l_c$, the exponent terms in equation (29) can be dropped because of the condition $Q < l_c^{-1}$, giving a $I(Q) \sim Q^{-1}$ behaviour. For $Q > l_c^{-1}$ the scattering function becomes

$$I(Q) \propto M_w(2\pi/Q^2l_cA)\exp(-Q^2t^2/\varphi).\tag{30}$$

Here the correction term for the super-folding thickness $t = n\delta_c$ appears. For the models (a), (b) and (e), $t = 0$ ($n = 0$); for model (c), $\varphi = 24$; and for model (d), $t = \delta_c(n = 1)$. For $t \cong l_c$ the effect of thickness $t$ becomes more serious, resulting in different scattering functions which are not shown here.

When the super-folding hardly occurs, i.e. $t \cong 0$, the exponent term of equation (30) may be dropped. In this case the behaviour of $I(Q)$ changes from $Q^{-1}$ to $Q^{-2}$ at around $Q = l_c^{-1}$. Thus the Kratky plot $Q^2I(Q)$ vs $Q$ gives two linear asymptotes $Q^2I(Q) \sim Q$ and $Q^2I(Q) = $ constant, which intercept at $Q^*$:

$$Q^* = 2/(l_c f) \quad \text{for models (a)(c)(d)(e)}$$
$$Q^* = 4/(3l_c) \quad \text{for model (b)}\tag{31}$$

From these $Q^*$ one can obtain $l_c$ for $f = 1$, but $l_c$ can also be estimated from the slope of $\ln QI(Q)$ vs $Q^2$ plot for equation (29) as described above. If these two estimated values of $l_c$ are close and also close to those determined by other techniques such as the lamellar thickness from electron microscopy (EM) and a long period from SAXS, the validity of this model may be ensured. If the apparent value of $l_c$ obtained from $Q^* = 2/l_c$ is smaller than those determined both from EM and from the $\ln QI(Q)$ vs $Q^2$ plot, this may be due to the dilution effect implied by $f$ in equation (31) because, strictly, the apparent value is $l_c f$ and $f \leq 1$.

Further, the slope of the Kratky plot is a function of $A$ and $f$,

$$\omega = \pi M_w f/A \quad \text{for models (a)(c)(d)(e),}$$
$$\omega = (3/2)\pi M_w/L_c \quad \text{for model (b),}\tag{32}$$

and the slope for model (b) is more than 1.5 times larger than those for models (a) and (e) of $f \leq 1$ and $A \geq L_c$.

## SANS Observations

As described before, polyethylene (PE) displays the isotope fractionation due to the difference in melting temperatures between PED and PEH. Hence when the chain trajectory is investigated, this polymer had better be avoided except when crystallized under high super-cooling conditions. For this reason, the results obtained for isotactic polystyrene (iPS), which virtually does not show the isotope fractionation due to negligibly small difference in melting temperatures between deuterated (iPSD) and hydrogenated (iPSH) one, will be explained first, and afterwards compared to those of PE.

*Isotactic Polystyrene (iPS):* The chain trajectory of iPS has been studied as functions of molecular weight and crystallization temperature by Guenet *et al.* (1980; 1990). Single crystals were prepared from 0.66% w/v solutions of the mixture (99:1 w/w) of iPSH with $M_w = 3.2 \times 10^5$ and $M_w/M_n = 2.8$ and iPSD with $M_w = 1.5 \sim 7 \times 10^5$ and $M_w/M_n = 1.1 \sim 1.3$ in dibutyl phthalate by quenching from 190°C to given temperatures.

First, iPS single crystals crystallized from 1.5% solutions in dibutyl phthalate at 130°C were investigated (Guenet, 1980). The $M_w$ dependence of $R_g$ of the labeled chains was obtained as $R_g \sim M_w^{0.91}$, which is in contrast with the result $R_g \sim M_w^{0.5}$ for the amorphous state (Cotton *et al.*, 1974). This suggests that the chains are greatly extended compared with the Gaussian coils. Then, the scattering function in the intermediate $Q$ range was examined, and it was found that $I(Q)$ changes from $Q^{-1}$ to $Q^{-2}$ at $Q^* \cong l_c^{-1}$. This means that the super-folding does not occur. Further, the values of $l_c$ obtained from $Q^*$ assuming $f = 1$ in equation (31) and from the Kratky plot were 75 and 85 Å, respectively, which agreed with that of 78 Å determined from electron microscopy. These observations lead to the conclusion that the sheet-like model with adjacent or near-adjacent reentry is valid. It is also reported that the direction of chain folding is seemingly along the {110} fold planes (Sadler *et al.*, 1984). In this case $L_c = 1140$ Å is obtained from equation (27) with $R_g = 330$ Å and $l_c = 80$ Å. These values give a reasonable $M_w$ near $5 \times 10^5$.

The crystallization temperature dependence of the chain trajectory was also studied (Guenet *et al.*, 1990). Figure 9 shows the Kratky plot, $Q^2 I(Q)$ vs $Q$ for $M_w = 5 \times 10^5$ as a function of crystallization temperature $T_x$. At $T_x = 115$°C (solid circles) the crossover behaviour of $I(Q)$ from $Q^{-1}$ to $Q^{-2}$ is still seen. The values $l_c \cong 50$ Å obtained by the three methods agreed with one another as well. At this $T_x$ the molecular chain is therefore considered to assume the sheet-like structure. However, at $T_x = 105$ and 85°C (triangles and squares, respectively) the $Q^{-1}$ to $Q^{-2}$ crossover is no longer seen. While the $Q^{-1}$ behaviour is still observed in a low $Q$ range, $I(Q)$ decreases more rapidly than $Q^{-2}$ with increasing $Q$ in a high $Q$ range owing to the exponent term in equation (30). This suggests that super-folding has taken place. Further, the slopes of the Kratky plot are higher than that for $T_x = 115$°C, and hence it is expected from equation (32) that the actual lengths $A$ of the ribbon are shorter than $L_c$, corresponding to the super-folding model or the Z-like model. Actually the values of $A$ calculated from $R_g = 290$ and 260 Å are ca. 1,000 and 900 Å, respectively, which are smaller than $L_c = 1,140$ Å for $T_x = 130$°C. On the other hand, there is also a possibility that the decrease of $A$, which is due to super-folding,

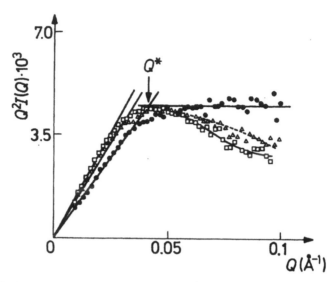

FIGURE 9. Kratky plots for iPSD with $M_w = 5 \times 10^5$, crystallized at 115°C (●), 105°C (△) and 85°C (□). (Guenet *et al.*, 1990. Reprinted from *Polymer*, **31**, J.M. Guenet, D.M. Sadler and S.J. Spells, Solution-grown crystals of isotatic polystyrene: neutron scattering investigation of the variation of chain trajectory with crystallization temperature. Pages 195–201. Copyright © 1990 with permission from Elsevier Science.)

is compensated by the dilution or the interruption of regular folding. It is therefore difficult to judge only from these data which model is more realistic, but nevertheless the super-folding model with dilution was considered most probable in the case of $T_x = 105°C$. In the case of $T_x = 85°C$, however, $l_c = 71$ Å determined from $\ln QI(Q)$ vs $Q^2$ plot is much larger than $l_c = 45$ Å obtained from electron microscopy. This suggests that crystallization occurs in more than one lamella, and the theoretical analysis supports the possible existence of a central core. Further, thermal observations indicating less organized crystals and low crystallinity suggest all types of deviations from the thin sheet at such low $T_x$.

In the case of lower $M_w$ of 1.5 or $3 \times 10^5$ the thin sheet structure was kept almost independent of $T_x$ until 85°C. The general conclusion is that the chain trajectory at relatively low under-cooling is regularly folded.

*Polyethylene (PE):* As was described previously, the chain trajectory in PE crystals can be investigated only in the case of crystallization at high super-cooling to minimize isotope fractionation. In the case of $T_x = 85°C$ it was shown (Sadler & Keller, 1976) that there is a possibility of partial segregation of the deuterated PE (PED). Therefore, crystallization was made at low temperatures such as 70°C, where it is anticipated from the results for iPS as mentioned above that the chain trajectory corresponds to the super-folding model with stem dilution or less ordered structures.

Figure 10 shows the $M_w$ dependence of $R_g$ determined from the Guinier range of SANS for both melt-crystallized sample and solution-grown single crystals (Sadler & Keller, 1979). In the former case (△, ×), the relation $R_g \sim M_w^{0.5}$ was obtained, which will be explained

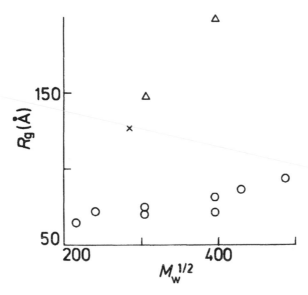

FIGURE 10. $M_w$ dependence of $R_g$ for melt-crystallized PE samples ($\triangle$, $\times$) and solution-grown PE crystals ($\bigcirc$). (Sadler & Keller, 1979. Reprinted with permission from "Neutron Scattering of Solution-Grown Polymer Crystals: Molecular dimensions are insensitive to molecular weight". D.M. Sadler and A. Keller, *Science*, **203**, 263. Copyright © 1979 American Association for the Advancement of Science.)

later, while in the latter case $R_g$ is only slightly dependent on $M_w$ in spite that $R_g$ should be nearly proportional to $M_w$ if the chains assume a sheet-like structure. Such results suggest the presence of super-folding.

IANS provides information about more detailed folding structure of the chains. Figure 11 shows a Kratky plot for a sample with $M_w = 2.9 \times 10^4$ crystallized from a dilute solution at 70°C (Spells & Sadler, 1984). This curve has a broad maximum at around $Q = 0.15$ Å$^{-1}$, suggesting the super-folding as previously described. This agrees with the results of $M_w$ dependence of $R_g$. However, Yoon and Flory (1979) pointed out that the compact adjacent stem arrangement in the super-folding model did not agree with experimental results. The general profile of scattering function was reproduced in terms of super-folding in two or three adjacent (110) growth layers, but the scattering intensity was higher by a factor of approximately two than the observed, and hence the super-folding model with stem dilution was proposed. More elaborate works by Spells and Sadler (1984) demonstrated that a strong statistical preference model is similar to a model having 75% adjacent re-entry and 50% dilution of a molecule along the 110 fold planes. Their computer simulation models are based on 50% stem dilution which was concluded from the difference, by the factor of two, in the scattering intensity between the compact adjacent stem model and the experimental data. The following scenarios were considered: (a) regular alternate re-entry model where the probability $p$ of the next stem coming from the same molecule is small, (b) block stem dilution model, where $p$ is very large such as $p = 0.75$ and there is a preference for

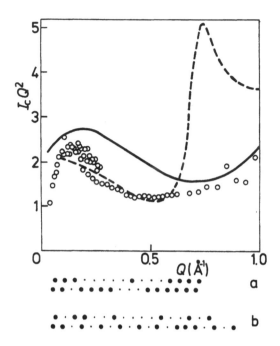

FIGURE 11. Kratky plot for a solution-grown PE sample with $M_w = 2.9 \times 10^4$, crystallized at 70°C. The solid and broken curves show calculations for 75% adjacent reentry (a) and 25% adjacent reentry (alternate reentry) (b), respectively. (Spells & Sadler, 1984b. Reprinted from *Polymer*, **25**, S.J. Spells and D.M. Sadler, Neutron scattering studies on solution-grown crystals of polyethylene: a statistical preference for adjacent re-entry. Pages 739–748. Copyright © 1984 with permission from Elsevier Science.)

the stems belonging to the same molecules to be grouped together by adjacent re-entry, and (c) random dilution model where $p = 0.5$; the stem deposition from either the same or a different molecule occurs equally. The random distribution model can be excluded because the scattering function in the Kratky representation increases monotonously with $Q$ and has no broad peak (Yoon & Flory, 1979). The calculated Kratky plot for the regular alternate re-entry model ($p = 0.25$) is indicated in Figure 11 by a broken line, producing a peak at $Q = 0.7 \text{ Å}^{-1}$ which corresponds to an additional periodicity of 8.8 Å. Such a peak is not observed, and hence this model is removed as well. The calculated results for the block stem dilution model with the number of super-folding sheets $n = 2$ is also shown by a solid line, which agrees more closely with the observed data. The $M_w$ dependence of $n$ was further investigated; Table 1 lists the values of the best fit for samples as a function of $M_w$. Although $n$ changes from 1 to 7, $M_w/n$ is roughly constant ($2.1 \times 10^4$), i.e. the average actual length of the ribbon does not change much as $M_w$ increases. These results, obtained from neutron scattering, were in agreement with those from FTIR study (Spells *et al.*, 1984a). The final conclusion is therefore that the most probable model for solution-grown crystals of PE under high super-cooling (crystallized at 70°C) is the super-folding model with block stem dilution (cf model (a) in Figure 11 for the case of $n = 2$).

TABLE 1. Molecular weight $M_w$ dependence of the average number of superfolding sheets $n$ in solution-grown crystals of PE, estimated from SANS. (Spells & Sadler, 1984b). Reprinted from *Polymer*, **25**, S.J. Spells and D.M. Sadler, Neutron scattering studies on solution-grown crystals of polyethylene: a statistical preference for adjacent re-entry. Pages 739–748. Copyright © 1984 with permission from Elsevier Science.)

| Matrix | $M_w \times 10^{-3}$ | average number of sheets $n$ | $M_w \times 10^{-3}/n$ |
|---|---|---|---|
| PEH | 216 | 7 | 31 |
| PEH | 118 | 5 | 24 |
| PED | 87 | 4 | 22 |
| PED | 70 | 4 | 17.5 |
| PED | 51 | 2 | 35.5 |
| PED | 29 | 2 | 14.5 |
| PED | 15 | 1.25 | 12 |
|  |  |  | average 21 |

### 5.5.1 Annealing of Solution-Grown Single Crystals

The change of the chain conformation in the solution-grown PE single crystals caused by annealing has also been investigated. In this case it should be noted that annealing causes lamellar thickening without partial melting at 123°C or lower, and with partial melting above 123°C (Sadler, 1985). In what follows the former case is discussed.

### Radius of Gyration

First the dimensions of radius of gyration along the normal ($z$-direction) and at right angles ($x$, $y$-directions) to the lamellar crystals were measured (Sadler, 1985) using an analytical method on oriented mats of crystals (Sadler, 1983). The results showed that the radius of gyration along the $z$-direction $R_z$ increases with annealing temperature $T_a$, but the value in the $x$, $y$-directions $R_{xy}$ hardly changes. For example, when the sample with $M_w = 103,000$ crystallized at 70°C was annealed at $T_a = 112 \sim 123$°C, $R_z$ increased from ca 40 Å to ca 80 Å and $R_{xy}$ remained almost constant, about 65 Å. The increase of $R_z$ was approximately in line with the lamellar thickness obtained from the SAXS long periods. However, how can we explain the values of $R_{xy}$? IANS should give an answer.

### IANS

Sadler and Spells (1989) investigated the change of chain trajectory by annealing. They examined the possible models, schematically shown in Figure 12. The starting model is the super-folded block-stem dilution one, mentioned in the previous subsection. It consists of four sheets, each having 20 stems; the stem dilution is 50% and a fraction of adjacent re-entry $p_0 = 0.75$. The conformation within each sheet is schematically drawn in Figure 12a; $p_0 = 0.75$ corresponds to four stems (17.6 Å) as the average group size of adjacent stems within one sheet. Annealing increases the average length of the stems and hence decreases their number. In this case two possible models can be considered. The simplest way of reducing the number of stems is the removal of stems randomly within each molecule (Figure 12b). This model does not contradict the observation that $R_{xy}$ hardly changes by

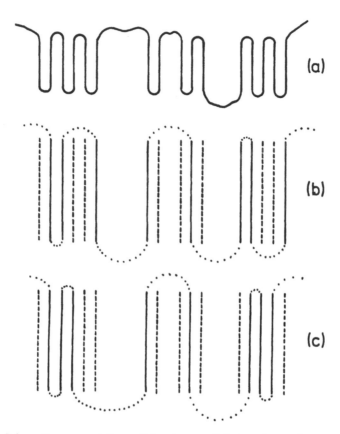

FIGURE 12. Schematic representations of the change of chain trajectory by annealing. (a) unannealed, (b) annealed: random removal of stems, (c) annealed: shrinkage of groups from their extremities. (Sadler & Spells, 1989. Reprinted with permission from *Macromolecules*, **22**, 3941–3948. Copyright © 1989 American Chemical Society.)

annealing. The second model proposes the removal of stems from the ends of adjacent stem groups, which leads to large gaps between adjacent stem groups. This in turn leads to the loss of overall sheet-like conformation (Figure 12c). Let us examine these models on the basis of the results of IANS data below.

Two IANS data are shown for $Q \leq 0.2$ Å$^{-1}$ and $Q \leq 0.8$ Å$^{-1}$ in Figure 13a and b, respectively. The former corresponds to the system of a concentration of 3% PED with $M_w = 93,000$ in PEH matrix and the latter to that of PEH with $M_w = 87,000$ in PED matrix. In the case of non-annealed solution-grown single crystals the Guinier plot shows a broad peak, indicating the super-folded ribbon structure as described previously. When they are annealed at $112 \sim 123°$C, the scattering intensity decreases for $Q < 0.2$ Å$^{-1}$, while it increases almost linearly with $Q$ for $Q > 0.4$ Å$^{-1}$. The decrease in intensity for low $Q$ suggests the decrease in the number of stems due to the increase in stem length. The linearity for $Q > 0.4$ Å$^{-1}$ reminds us of the scattering from the rod with intensity

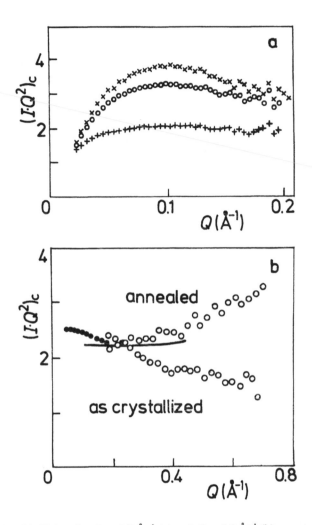

FIGURE 13. Two IANS data for $Q < 0.2$ Å$^{-1}$ (a) and $Q < 0.8$ Å$^{-1}$ (b) represented in Kratky plot. (a) PED in PEH with $M_w = 9.3 \times 10^4$ (PED). ×: unannealed, ○: annealed at 112°C, +: annealed at 123°C. (b) PEH in PED with $M_w = 8.7 \times 10^4$ (PEH). ●, ○ (lower): unannealed; ○ (upper): annealed at 120°C for 1 h. (Sadler & Spells, 1989. Reprinted with permission from *Macromolecules*, **22**, 3941–3948. Copyright © 1989 American Chemical Society.)

proportional to $Q^{-1}$; higher stem dilution by annealing may cause single rods. Such results support the random stem removal concept.

The solid lines in Figure 14 show calculated results of the random stem removal effect on the scattering intensity. The top line in the low $Q$ side is for the starting model with a total stem occupancy of 0.5 and a fraction of adjacent reentry $p_0 = 0.75$. The other three towards the down side are those for dilutions of the inter-stem correlations by factors of

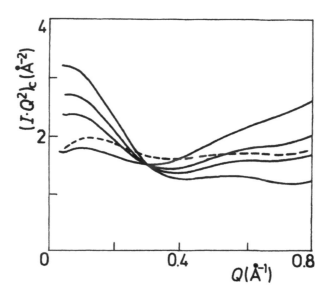

FIGURE 14. The Kratky plot calculated for the random stem removal model. In the low $Q$ side the top line is the curve for the starting model with total stem occupancy 0.5, fraction of adjacent reentry $p_0 = 0.75$ and number of superfolds 4, and towards the down side, stem dilutions by factors 1.2, 1.5 and 2. (Sadler & Spells, 1989. Reprinted with permission from *Macromolecules*, **22**, 3941–3948. Copyright © 1989 American Chemical Society.)

1.2, 1.5 and 2, respectively. The factor of 2 corresponds to the increase of the SAXS long period and $R_z$ values for $T_a = 123°C$ when the lateral dimensions are assumed to remain constant. These results roughly reproduce the observed SANS data at $T_a = 120 \sim 123°C$. The broken line is the calculation based on the end stem removal model; the total stem occupancy is 0.25 and the fraction of adjacent reentry is 0.57. As seen from the comparison between Figures 13 and 14, the SANS data favours the mechanism of the random removal of stems from groups of adjacent stems. The IR data at $T_a = 123°C$ is better fitted by the mechanism of the removal of stems from the ends of groups for $M_w \leq 100,000$, though they favour the random removal mechanism for $M_w > 100,000$, which is consistent with the SANS conclusion (Spells & Sadler, 1989).

### 5.5.3  *Melting of Solution-Grown Crystals*

The crystallization and melting of polymer crystals are not exactly inverse processes. Barham and Sadler (1991, 1993) raised an interesting problem: on melting, do the chains detach themselves individually from crystal surfaces and diffuse into the random melt or do they melt through, like in a cooperative bulk process? Thus, how do the molecules change their trajectories from folded chain ribbons to isotropic random coils? SANS experiments supports the co-operative melting process. In this subsection, the evidence for it is presented. To this end, the melting time dependence of the radius of gyration $R_g$ was investigated.

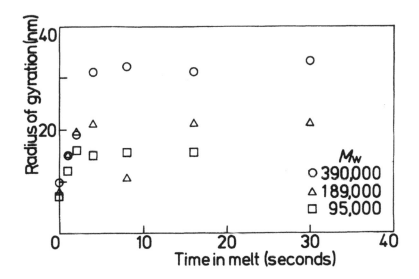

FIGURE 15. Time dependence of $R_g$ during melting of PE single crystal mats as a function of $M_w$. (Barham, 1993. Reprinted from *Crystallization of Polymers*, NATO ASI Series, Vol. 405, 1993, pages 153–158, ed. by M. Dosiere, in Chapter 3, "Melting of Polymer Crystals", Barham, Fig. 1. Copyright © 1993, with kind permission from Kluwer Academic Publishers.)

The molecular orientation in the original PE single crystals disappeared after melting for more than 1s at 140°C, and the long period increased from ca 120 Å to ca 250 Å which thereafter remained constant until complete melting. For the isotropic samples after melting $R_g$ was determined from the Zimm plots. For anisotropic samples before melting the molecular sizes $R_{xy}$ and $R_z$ (in the lamellar plane and along the lamellar normals of single crystals, respectively) were estimated according to the analytical method proposed by Sadler (1983). Subsequently $R_g$ was calculated from the equation

$$R_g^2 = R_z^2 + 2R_{xy}^2. \tag{33}$$

The results are shown in Figure 15. The initial radii of gyration $R_g^\circ = 72, 80$ and 98 Å for $M_w = 0.95, 1.89$ and $3.9 \times 10^5$, respectively, increase to the 'equilibrium' radii of gyration $R_g^\infty = 155, 210$ and 320 Å. For all the samples the radii of gyration reach their individual 'equilibrium' values within 4s, independently of their molecular weight. The fact that the randomization of the tightly packed compact molecules in the crystals occurs in very short time and almost at the same time may indicate that the melting involves a cooperative process. If reptation plays any role, then the randomization time would strongly depend on $M_w$ because the reptation time changes by two orders of magnitude, i.e. roughly from 0.1 to 10s in the present case (Klein & Briscoe, 1979). It might be natural that the melting time is much shorter than the reptation time because the entanglements in the single crystals are greatly reduced compared with those in the molten state. Figure 16 shows a computer simulation image of typical molecular trajectories

FIGURE 16. Computer simulation image of typical molecular trajectories of a polymer with $M_w = 3.9 \times 10^5$ before and after melting. Broken circle: final $R_g^\infty$ (diameter 64 nm). (Barham & Sadler, 1991. Reprinted from *Polymer*, **32**, P.J. Barham and D.M. Sadler, A neutron scattering study of the melting behaviour of polyethylene single crystals. Pages 393–395. Copyright © 1991 with permission from Elsevier Science.)

of a polymer with $M_w = 3.9 \times 10^5$ before and after melting; the difference in size is clear between the initial crystal lamella of $R_g^\circ = 98$ Å before melting and the sphere indicated by the broken circle of $R_g^\infty = 320$ Å after melting.

### 5.5.4 Chain Trajectory in Melt-Grown Crystals

The chain trajectory in melt-grown crystals has been studied for a long time, however the details remain still unclear. Roughly speaking, any hypothesis lies between the so-called switch-board model by Flory and the completely adjacent re-entry model by Keller. This problem was extensively studied by SANS for PE in particular, but in this case crystallization conditions were limited only to high super-cooling ones despite the chain trajectory dependence on the crystallization condition. This is of course due to the problem of tendency of clustering in mixtures of PED and PEH as mentioned before. Hence the results of PE at high super-cooling will first be presented, and subsequently those obtained for other polymers such as isotactic polystyrene (iPS) and isotactic polypropylene (iPP) will be commented upon, taking into account a degree of super-cooling.

Polyethylene (PE) — crystallization at high super-cooling

*Radius of Gyration.* In the case of polyethylene quenched from the melt the molecular weight $M_w$ dependence of the radius of gyration $R_g$ obtained from the measurements in low $Q$ ranges ($Q < 0.02$ Å$^{-1}$) surprisingly showed the relation $R_g = 0.46M_w^{0.5}$ (Lieser *et al.*, 1975; Schelten *et al.*, 1976). This result agrees with that for the $\Theta$-solvent where $R_g = 0.45M_w^{0.45}$. This finding is explained as follows. The key factor to determine the chain conformation in the crystalline state is the ratio of the relaxation time for a whole polymer chain to escape from the original tube in the reptation model, proposed by de Gennes (1971) and Doi-Edwards (1978), to the crystallization rate. The relaxation time $\tau$ does not much depend on temperature but it is strongly affected by molecular weight: $\tau \sim M_w^{3.4}$. On the other hand, the crystallization rate depends on the quenching depth $\Delta T = T_m - T_x$ where $T_m$ and $T_x$ are melting temperature and crystallization temperature, respectively; it increases exponentially with increasing $\Delta T$. Under the crystallization condition of quenching from the molten state to below room temperature the crystallization time is greatly shorter than $\tau$ for high $M_w$. Hence there is no time for polymer chains to escape from the tubes and deform before crystallization, with the result that the $R_g$ of polymer chains hardly changes from that in the molten state.

However, it was found that in a range of very small $M_w$ the $R_g$ levels off as $M_w$ decreases. Figure 17 shows the $M_w$ dependence of $R_g$ for PE and isotactic polypropylene (iPP) (Ballard *et al.*, 1979). For comparison, the result obtained for atactic polystyrene (aPS) in the amorphous glassy state is also shown. The polymer chain in the glassy state (aPS) indicates the size of Gaussian coils in the whole observed range of $M_w$, while the $M_w$ dependence of $R_g$ for the crystalline chains (PE and iPP) deviates from the Gaussian size in low $M_w$ ranges. This deviation may be caused by the reduction in number of inter-penetrations of a polymer chain into crystalline lamellae as well as by the decrease in escape time of the polymer chain from the original tube. The effect of the length of crystalline stem starts to appear below around a degree of polymerization 3,000 and 4,000. In this case the relationship between $R_g$ and $M_w$ is given by

$$R_g^2 = l^2/12 + \beta^2 M_w \qquad (34)$$

where $l$ is the stem length in the crystalline lamella and is constant. The values of $l$ for PE and iPP were estimated from Figure 17 using equation (34): they were about two times as large as values of lamellar thickness (SAXS long spacing $d$). This means that the polymer chain penetrates two crystalline lamellae in succession. Based on these findings, Fischer (1978) proposed a solidification model as shown in Figure 18. This model may, however, be modified regarding the detailed fold structure, as discussed below.

Finally let us refer to the entanglement which had existed in the melt. Since the whole conformation of polymer chains hardly changes on crystallization under high super-cooling as described above, the entanglement in the melt does not disappear on crystallization. As entanglement points cannot be incorporated into crystalline lattices, they must be excluded from the crystalline lamellae, resulting in the amorphous regions on the lamellar surfaces, i.e. the so-called long period structure with alternation of crystalline lamella and non-crystalline layer. This is the reason why polymers cannot crystallize completely under the usual conditions.

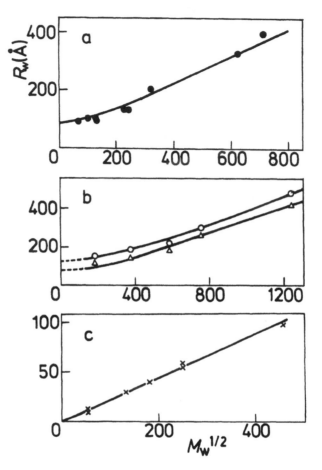

FIGURE 17. $M_w$ dependence of $R_g$ for polyethylene (a), isotactic polypropylene (b) and atactic polystyrene (c). For iPP(b), $\triangle$: quenched, $\bigcirc$: slow cooled. (Ballard *et al.*, 1979b. Reprinted from *Polymer*, **20**, D.G.H. Ballard, G.W. Longman, T.L. Crowly, A. Cunningham and J. Schelten, Neutron scattering studies of semicrystalline polymers in the solid state: The structure of isotropic polylefins. Pages 399–405. Copyright © 1979 with permission from Elsevier Science.)

*IANS — Chain folding structure.* The SANS in the intermediate $Q$ range (IANS) gives information about the chain-folding structure of lamellae. In such $Q$ ranges the isotopic concentration fluctuations (the segregation effect) is negligibly small, and hence the data analysis is more reliable.

As shown in Figure 10, $R_g$ of a chain in the melt-quenched crystals is considerably larger than that in the single crystals when $M_w$ is not so small. This suggests that the adjacent re-entry of the chain is greatly limited. In order to investigate this problem the comparison between the observed scattering intensities and computer-simulated functions in the IANS region has been made. Figure 19 shows the experimental IANS data (open circles) in the Kratky plot where a plateau and hence the $Q^{-2}$ dependence of the scattering intensity $I(Q)$

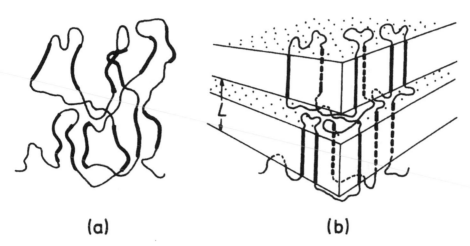

**(a)** **(b)**

FIGURE 18. Solidification model showing how a simple polymer chain in the melt (a) is incorporated into the growing lamellae (b) during the crystallization process. No significant reorganization of the chain conformation occurs. (Fischer, 1978. Reprinted with permission from *Pure & Applied Chemistry*, **50**, 1319–1341. Fig. 15. Copyright © 1978 IUPAC, Pergamon.)

is seen in the region of 0.03 to 0.3 Å$^{-1}$ (Schelten *et al.*, 1976; Sadler, 1982). Further the data for $Q > 0.3$ Å$^{-1}$ which lies on a straight line passing through the origin give the relation $I(Q) \sim Q^{-1}$, meaning that the scattering comes from isolated rod-like stems. Monte Carlo simulations for the 100% adjacent re-entry produce a plateau, but its intensity is about 1.7 times higher than the observed data (Figure 19a). Models for two-dimensional random walks of stems did not give the plateau in this region. Only nearby re-entry models can reproduce the observed data in the observed whole $Q$ range including the $Q > 0.3$ Å$^{-1}$ region. In Figure 19b the simulated result for one of the best fit nearby re-entry models is also shown. This model is called the modified row model where several sites occupied by successive stems are on a row with occasional stagger and the probabilities of their stems occupying each site decrease with increasing interstem distance $f$ or with a probability of $1/f$. In this model 46% of the folds are adjacent. Quench-crystallized iPS gives a similar result to that obtained for PE (Guenet, 1981).

### 5.5.5 Isotactic Polypropylene (iPP)

The crystallization of iPP was studied by Ballard *et al.* (1978, 1979a, 1997b, 1980). For this polymer no clustering occurs because the melting temperatures $T_m$ of deuterated and protonated polypropylenes (PPD and PPH) do not show measurable difference. However, different degrees of isotacticity and other characteristic quantities give different melting temperatures $T_m = 160 \sim 170°C$ for standard samples. Hence the effect of the degree of crystallinity $x$ on the chain conformation can be studied without considering the problem of clustering. The radius of gyration $R_g$ of iPP hardly changed from that of the melt by crystallization whether it was quenched, annealed or isothermally crystallized at 139°C.

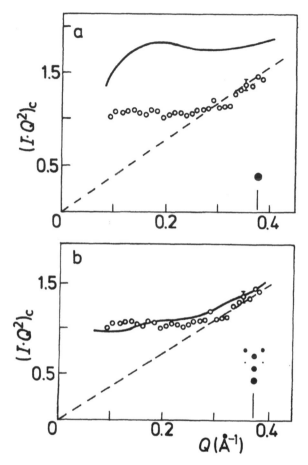

FIGURE 19. IANS data in the Kratky plot for melt-quenched polyethylene. The solid curves in (a) and (b) correspond to the theoretical calculations for the adjacent reentry model and the modified row model (46% adjacent re-entry), respectively. The insets indicate the sites of the crystalline stems; the area of a dot is proportional to the probability of successive stems occupying each site from the end of vertical line. (Sadler & Harris, 1982. A Neutron Scattering Analysis: how does the conformation in polyethylene crystals reflect that in the melt? D.M. Sadler and R. Harris, *J. Polymer Sci. Polym. Phys.* Ed., **20**, 561–578. Copyright © 1982 John Wiley & Sons, Inc. Reprinted by permission of John Wiley & Sons, Inc.)

The $R_g$ was almost independent of the degree of crystallinity within a range of $x = 0.5 \sim$ 0.7, and of the lamellar thickness within a range of $114 \sim 245$ Å. The SANS data gave a relation $R_g = 0.35 M_w^{0.5}$ for all cases which means a Gaussian conformation except for low $M_w$'s as described in the part of PE (crystallization at high super-cooling). Such a result conforms to that of quenched PE. Therefore it seems that the chain conformation of iPP having $x = 0.5 \sim 0.7$ is like the solidification model of Figure 18. However, the model proposed by Ballard *et al.* (1979) as shown in Figure 20 is different in one point,

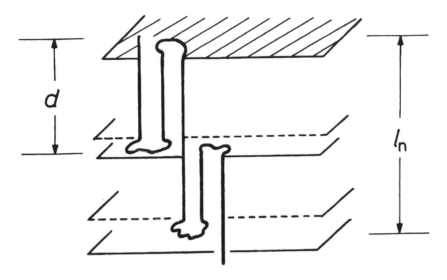

FIGURE 20. Schematic conformation model for melt-crystallized iPP with degrees of crystallinity 0.5–0.7. This is different from the solidification model only in one point that one of 3–5 stems has a length of twice the lamellar thickness and connects two lamellae. (Ballard *et al.*, 1979a. D.G.H. Ballard, A.N. Burgess, T.L. Crowley, G.W. Longman and J. Schelten, *Faraday Disc. Chem. Soc.*, **68**, 279–287. Copyright © 1979. Reproduced by permission of The Royal Society of Chemistry.)

namely one of $3 \sim 5$ stems has a length of twice the lamellar thickness and connects two lamellae. This is based on the experimental result that the SANS stem length $l_n$ is approximately twice the SAXS long spacing $d$. In the figure it should be noted that various combinations of short and long stems are possible, and the chain can assume adjacent or nearby re-entry into the lamella. The Kratky plot for these samples shows a plateau in the IANS region ($Q = 0.025 \sim 0.10$ Å$^{-1}$). Suprisingly the intensity level of this plateau did not change by crystallization conditions regardless a large change of the long period from 120 to 250 Å, corresponding to $x = 0.5 \sim 0.7$. The effect of crystallization appears only in the high $Q$ region extending to the WANS range such as 0.1 to 0.8 Å$^{-1}$; the scattering curves in this region are sensitive to the amount of chain folding within a lamella (Stamm *et al.*, 1981).

As iPP does not show the problem of clustering, we can investigate the conformation even in the case of extremely high crystallinity. Such a sample was prepared as follows. The rapidly quenched sample from the melt was annealed for 24h at 139°C and then cooled to the room temperature. After that the obtained sample was again melted just above $T_m$ for a short period and then cooled, indicating $x = 0.83$ and $d = 305$ Å. Such a high crystalline sample gave the $R_g$ some 60% larger than that in the melt. This result is similar to the iPS case — to be later described — but the important difference is that the former has another $R_g$ value, which was obtained from the intermediate $Q$ region. A model for iPP with an extremely high crystallinity proposed on the basis of such observations is shown in Figure 21.

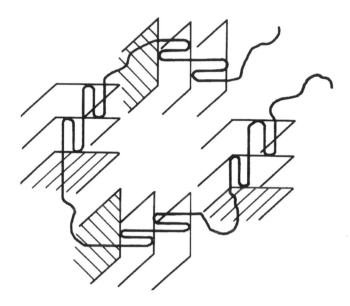

FIGURE 21. Schematic conformation model for melt-crystallized iPP with extremely high crystallinity (0.83). The chain consists of folded parts but assumes a disc as a whole. (Ballard *et al.*, 1980. Reprinted with permission from *Macromolecules*, **13**, 677–681. Copyright © 1980 American Chemical Society.)

Thus the polymer chain assumes a disc-like conformation as a whole, but it consists of folded chain parts being approximately 150 Å apart from each other and unfolded chain parts connecting the folded chain parts. The scattering intensity from the former parts is greatly higher than that from the latter ones. Further the chain molecule does not occupy the region around the center of its mass.

Isotactic Polystyrene (iPS) — Crystallization at low supercooling

As iPS neither causes clustering during crystallization, the fact confirmed by Guenet *et al.* (1979a), the evolution of the chain conformation can be studied as a function of the degree of crystallinity. Guenet *et al.* (1979a,b) have studied this question in detail in terms of $R_g$ using IANS. They found a very interesting result. In the case of solution-grown single crystals the $R_g$ of a single molecule in the solution greatly decreases after crystallization and it drastically increases after melting without solvent; such change in size is almost reversible. This observation suggests that even in the crystallization from the melt $R_g$ would decrease after crystallization. Surprisingly, actual experimental observations of the melt-crystallization show this is not the case. Let us consider this noteworthy problem while including the chain trajectory in the melt-grown crystals at slow crystallization.

*Radius of gyration.* For this purpose Guenet *et al.* (1979a) measured the radii of gyration $R_g$ for labeled iPSD with $M_w = 5 \times 10^5$ dispersed in an iPSH matrix ($M_w = 4 \times 10^5$) when

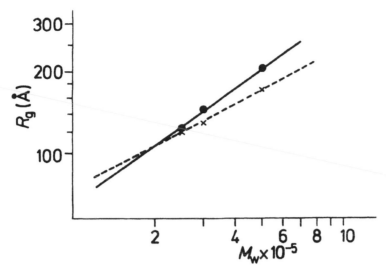

FIGURE 22. $M_w$ dependence of $R_g$ for iPS as a function of crystallinity $x$. $\times : x = 0$, $R_g \sim M_w^{0.5}$; $\bullet : x = 0.3$, $R_g \sim M_w^{0.78}$. (Guenet *et al.*, 1979b. Reprinted from *Polymer*, **20**, J.M. Guenet and C. Picot, Conformation of isotactic polystyrene (IPS) in the bulk crystallized state as revealed by small-angle neutron scattering. Pages 1483–1491. Copyright © 1979 with permission from Elsevier Science.)

it is crystallized at 185°C. The $R_g$ values were obtained from the initial slope of $I^{-1}(Q)$ plotted against $Q^2$ as a function of degree of crystallinity $x$. Contrary to the expectation $R_g$ increased from 165 Å to 228 Å with the value of $x$ increasing from 0 to 0.35. The molecular weight dependence was also investigated for $x = 0.3$, resulting in $R_g \sim M_w^{0.78}$ as shown in Figure 22.

These results show that slow crystallization changes the chain conformation from the Gaussian ($R_g \sim M_w^{0.5}$) to extended ones. Below let us examine what such extended chain conformations are like. Guenet *et al.* (1979a) first derived theoretical equations for possible models as a function of crystallinity and then compared with the observed data. In their theoretical approaches chains are assumed consisting of long amorphous (A) and crystalline (C) sequences having the same crystallinity as the bulk, and from the combination of their sequences with different incorporations of crystalline sequences in a monocrystal three models are considered, which are shown in Figure 23. In this figure, AC model comprises one crystalline and one amorphous sequences, ACA model of one crystalline and two amorphous sequences, and CAC model of two crystalline and one amorphous sequences. For the last model two types of combination can further be considered, depending on whether the amorphous sequence connects two crystalline sequences in the same monocrystal or in two different monocrystals. The restriction to these three models has been justified from the comparison with the Flory model of Gaussian chain statistics; beyond three sequences the Flory model discussed in the previous subsection of "crystallization at high supercooling" is rapidly recovered. The average total $R_g$ of a tagged chain can be written in terms of individual $R_g$'s of the sequences (Leng & Benois, 1962), and hence $R_g$ of crystalline and amorphous

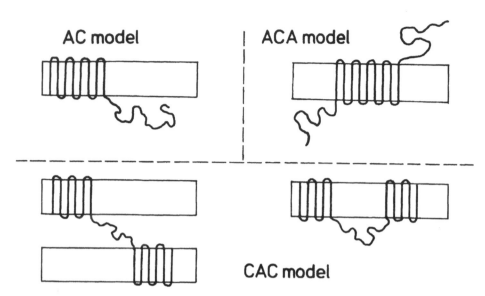

FIGURE 23. Three models for chain trajectory in melt-grown crystals at low super cooling, based on the combination of long amorphous (A) and crystalline (C) sequences. (Guenet & Picot, 1979a. Reprinted from *Polymer*, **20**, J.M. Guenet and C. Picot, Small-angle neutron scattering by semicrystallized chains: evaluation of the mean dimensions. Pages 1473–1482. Copyright © 1979 with permission from Elsevier Science.)

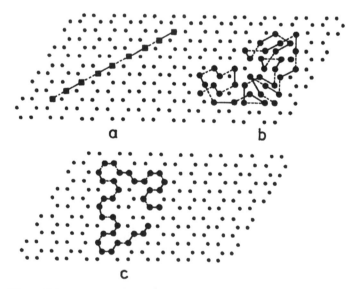

FIGURE 24. Three different types of chain incorporation in the monocrystal: (a) unidirectional crystallization (UC), (b) switch-board incorporation (SB) and (c) statistical adjacent reentry (SAR), ie a two-dimensional self-avoiding walk. (Guenet & Picot, 1979a. Reprinted from *Polymer*, **20**, J.M. Guenet and C. Picot, Small-angle neutron scattering by semicrystallized chains: evaluation of the mean dimensions. Pages 1473–1482. Copyright © 1979 with permission from Elsevier Science.)

sequences are first calculated, separately and then the total values. For these three models three different types of chain incorporation in one monocrystal are further considered. These are (i) unidirectional crystallization (UC) resulting in a sheet-like structure, (ii) switch-board incorporation (SB) corresponding to a two-dimensional random walk, and (iii) statistical adjacent reentry (SAR), a two-dimensional self-avoiding walk which is a special case of the SB incorporation. These chain incorporations are shown in Figure 24. In order to judge which model is the most probable one the crystallinity $x$ dependence on the $R_g$ was first examined. Model calculations showed the results that the UC incorporation in the AC, ACA and CAC models leads to an appreciable increase of $R_g^2$ with $x$ while the SB incorporation in these models gives a slight decrease of $R_g^2$ with increasing $x$. As the observed data indicated a considerable increase of $R_g^2$ with $x$, the SB model can be discarded. Of the UC incorporations the CAC model can also be removed, because it remains almost constant for $x < 0.5$ where the data were gathered while $R_g^2$ rises up for $x > 0.5$. Hence possible candidates are the AC and ACA models with UC. In order to determine which model of these two fits in with the observations, the weight fractions $w_c$ and $w_a$ of the amorphous material linked to the crystalline sequences and the residual amorphous material unlinked to crystalline sequences, respectively, were calculated. For $x = 0.35$, for example, $w_a$ is very small ($w_a = 0.07$) for the ACA model with tight loops ($z = 0.89$), while it is considerably large ($w_a = 0.3$) for the ACA model with loose loops ($z = 0.75$). Here $z$ is the fraction of stems in the crystalline sequence consisting of stems and loops. The AC models both with tight loops and with loose loops also indicated large calculated $w_a$ values. Optical microscopy observations, however, did not show any amorphous residual area and hence the ACA model with tight loops is considered the best model. At the same time the calculations of $R_g^2$ as a function of crystallinity $x$ supported this model as shown in Figure 25; the best fit is seen for the ACA model with tight loops ($z = 0.89$). In this best-fit model it is noted that $w_c$ is close to the linear crystallinity determined by X-ray diffraction and it remains roughly constant, independent of $x$, while $w_a$ decreases with increasing $x$. In conclusion the molecular weight $M_w$ dependence of $R_g$, i.e. $R_g \sim M_w^\nu$ was calculated for these models, providing $\nu = 0.79 \sim 0.80$ for the ACA models and $\nu = 0.65 \sim 0.67$ for the AC models. Since the experimental exponent was $\nu = 0.78$, these results again support the ACA model.

*IANS.* IANS studies have been carried out on both amorphous ($x = 0$) and crystalline ($x = 0.35$) samples having $M_w = 5 \times 10^5$ and a concentration of IPSD, $C_D = 0.51\%$. Figure 26 shows their Kratky representation [$Q^2 I(Q)$ vs $Q^2$]. Both curves have plateaus, but the values of the ordinate are different; the intensity for the amorphous sample is 1.14 times higher than the crystalline sample. All the crystallized samples exhibit the same behaviour and no peak has been observed. Referring to equation (30) for the solution-grown crystals, the plateau in the IANS region suggests that the labeled chain contains at least one folded chain sheet, and the scattering form factor $P_c(Q)$ for $Q > L_c^{-1}, l_c^{-1}$ is given by

$$P_c(Q) = 2\pi/Q^2 l_c L_c. \tag{35}$$

On the other hand the amorphous Gaussian chain gives the form factor

$$P_a(Q) = 2/Q^2 R_{ga}^2. \tag{36}$$

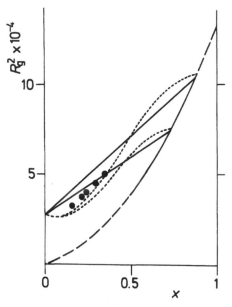

FIGURE 25. Crystallinity $x$ dependence of $R_g^2$ for iPS with $M_w = 5 \times 10^5$. Solid line: blend of pure amorphous and complete crystalline chains, broken line: ACA model; dashed line: limits of $x$ and $R_g$ caused by amorphous loops. Upper and lower curves correspond to tight ($z = 0.89$) and loose ($z = 0.75$) loops, respectively. (Guenet & Picot, 1979a. Reprinted from *Polymer*, **20**, J.M. Guenet and C. Picot, Conformation of isotactic polystyrene (IPS) in the bulk crystallized state as revealed by small-angle neutron scattering. Pages 1483–1491. Copyright © 1979 with permission from Elsevier Science.)

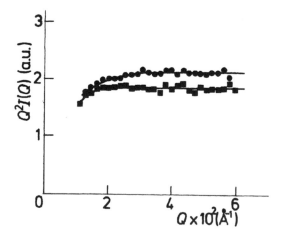

FIGURE 26. Kratky plots for amorphous and crystalline ($x = 0.35$) samples of iPS with $M_w = 5 \times 10^5$. The concentration of iPSD $C_D = 0.51\%$. (Guenet *et al.*, 1979a. Reprinted from *Polymer*, **20**, J.M. Guenet and C. Picot, Conformation of isotactic polystyrene (IPS) in the bulk crystallized state as revealed by small-angle neutron scattering. Pages 1483–1491. Copyright © 1979 with permission from Elsevier Science.)

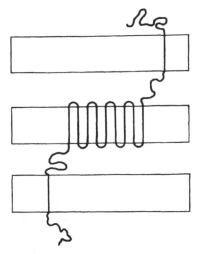

FIGURE 27. The most probable ACA model with small amorphous portions incorporated into different lamellae. This cannot be distinguished from the pure ACA model in Figure 21 by SANS. (Guenet *et al.*, 1979b. Reprinted from *Polymer*, **20**, J.M. Guenet and C. Picot, Conformation of isotactic polystyrene (IPS) in the bulk crystallized state as revealed by small-angle neutron scattering. Pages 1483–1491. Copyright © 1979 with permission from Elsevier Science.)

for $Q > R_{ga}^{-1}$ where $R_{ga}$ is the radius of gyration for the Gaussian chain. Hence the intensity ratio of the plateaus reduces to $P_a(Q)/P_c(Q) = l_c L_c/\pi R_{ga}^2$. If we assume $l_c = 100$ Å, such intensity ratios for the ACA models with tight and loose loops are obtained to be 1.1 and 1.05, respectively, supporting the former model. However, this model must be regarded as only a mean conformation; small fraction of the other model structures such as the CAC and a blend of pure A and C could exist.

Taking into account that iPS usually assumes a low crystallinity, the ACA model as shown in Figure 27, which cannot be distinguished from a pure ACA model by SANS, is considered the most probable where small portions of the amorphous sequences are incorporated into different lamellae. The incorporated amorphous parts prevent further crystallization. In conclusion it should be emphasized that with increasing $M_w$ the chain conformation approaches the Gaussian conformation because of decreasing diffusion rate of the chain molecule. If it is crystallized at 220°C near the equilibrium melting point $T_m = 242°C$, then it exhibits a sheet-like structure with adjacent folding along the (330) plane which is similar to that envisaged for solution-grown single crystal samples (Guenet & Picot, 1983).

## 5.6 DEFORMATION OF POLYMER CHAINS BY EXTERNAL FORCE

This section describes how polymer chains are deformed by external forces as observed in SANS experiments. Since it may be assumed that the deformation is different for amorphous and crystalline polymers, we consider these two cases separately.

## 5.6.1 Deformation of Amorphous Polymers

First the deformation of aPS by extrusion and then that of the same polymer during steady elongational flow in the molten state are explained according to Hadzijoannou *et al.* (1982) and Muller *et al.* (1990, 1996), respectively. These two SANS studies examine whether the deformation is affine-like.

### Extrusion

aPS samples were mixtures of 95% protonated (PSH) and 5% completely deuterated (PSD) polymers, both having $M_w = 6 \times 10^5$ and $M_w/M_n = 1.1$. Extrusion was carried out at 127°C under a pressure of 280 kg/cm$^2$; a deformation rate was 0.1 ∼ 0.2 cm/min. Figure 28a and b show the SANS contour diagrams for an unoriented sample ($\lambda_e = 0$) and an oriented one ($\lambda_e = 4.2$), respectively, where $\lambda_e = L/L_o$ is the macroscopic draw ratio or the effective draw ratio. Here $L_o$ and $L$ denote the length of the sample before and after extrusion, respectively. The elongated contour map for the oriented sample indicates that the molecular chain is greatly elongated in the drawing direction being perpendicular to the long axis of the elliptic contour map. In principle the radii of gyration $R_{g//}$ and $R_{g\perp}$ in the drawing and transversal directions, respectively, can be determined by the Zimm plot analysis for the corresponding directions. However, the intensity in the orientation direction rapidly falls at extremely small $Q$ values where the scattering is obscured by the beam stop, and so the measurement of $R_{g//}$ is actually impossible. For this reason $R_{g//}$ was calculated indirectly from the ratio of $Q_\perp/Q_{//}$ on the equal intensity contour by assuming an affine deformation with constant volume. Such assumption gives relations $R_{g//} = R_{go}\lambda$ and $R_{g\perp} = R_{go}/\lambda^{1/2}$ where $R_{go}$ is the radius of gyration for the unoriented sample as before, and $\lambda$ is the molecular draw ratio. Then the scattering laws become

$$S(Q_{//}) = 1 - Q_{//}^2 R_{g//}^2 = 1 - Q_{//}^2 R_{go}^2 \lambda^2$$
$$S(Q_\perp) = 1 - Q_\perp^2 R_{g\perp}^2 = 1 - Q_\perp^2 R_{go}^2/\lambda \tag{37}$$

where the total radius of gyration $R_g$ for the uniaxially oriented sample is given by $R_g^2 = R_{g//}^2 + 2R_{g\perp}^2$. Further, the condition of equal-intensity contour provides the relationship $\lambda = (Q_\perp/Q_{//})^{2/3}$, from which $R_{g//} = R_{go}\lambda$ can be calculated.

The experimental results are plotted as a function of $\lambda_e$ in Figure 29. The solid lines represent the theoretical prediction when an affine transformation is assumed. The agreement between the theoretical and observed values is fairly good.

### Elongational flow

As described above, the large deformation of aPS by extrusion up to draw ratios 10 was confirmed to follow the affine transformation model. Here it is examined whether the affine model works also in the case of the small deformation of aPS in the molten state under the steady elongational flow condition according to Muller *et al.* (1990, 1996). This is done in

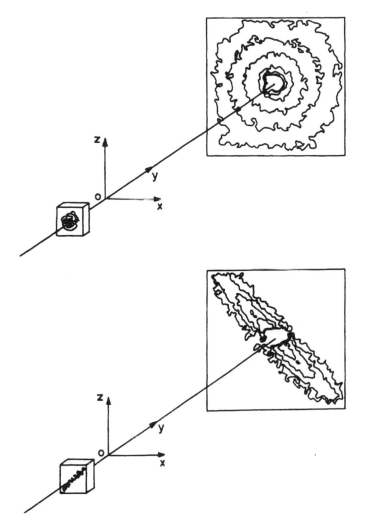

FIGURE 28. SANS experimental geometries of unoriented (a) and oriented (b) amorphous aPS with scattering intensity contour maps. (Hadziionannou *et al.*, 1982. Reprinted with permission from *Macromolecules*, **15**, 880–882. Copyright © 1982 American Chemical Society.)

comparison with the temporary network model which is well-known in the field of rheology. For the temporary network model one obtains

$$R_{g//} = R_{go}[1 + (\lambda^2 - 1)(1 - 1/n_e - 7/2n_e^2 + 3/n_e^3)]^{1/2}$$
$$R_{g\perp} = R_{go}[1 + (\lambda^{-1} - 1)(1 - 1/n_e - 7/2n_e^2 + 3/n_e^3)]^{1/2} \quad (38)$$

where $n_e = M/M_e$, $M$ and $M_e$ being the molecular weight of the polymer chain and the average molecular weight between entanglements determined from the plateau modulus.

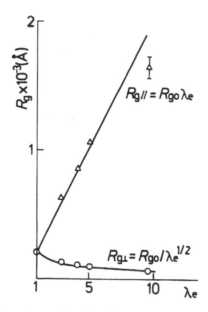

FIGURE 29. Radius of gyration as a function of the macroscopic draw ratio $\lambda_e$. $\triangle$: longitudinal radius of gyration $R_{g//}$, $\bigcirc$: transversal radius of gyration $R_{g\perp}$. Solid lines represent an affine deformation with constant volume: $R_{g//} = R_{go}\lambda_e$ and $R_{g\perp} = R_{go}/\lambda_e^{1/2}$, $R_{go}$ being the radius of gyration for the unoriented sample. (Hadziionannou *et al.*, 1982. Reprinted with permission from *Macromolecules*, **15**, 880–882. Copyright © 1982 American Chemical Society.)

When $n_e$ is infinite, these equations reduce to $R_{g//} = R_{go}\lambda$ and $R_{g\perp} = R_{go}\lambda^{-1/2}$, which correspond to the affine model; it is therefore possible to judge from $n_e$ values obtained by fitting these equations to experimental data whether or not the affine model is valid.

In order to examine the validity of the affine model the form factors of a deformed polymer chain in the directions parallel and perpendicular to the elongation axis were calculated for various $n_e$ values using the $\lambda$ values obtained from rheological experiment, and then compared with the SANS data. Figure 30 shows the SANS results for aPS with PSH/PSD=90/10 in the form of the Kratky plot as a function of the recoverable strain $\lambda_r$ corresponding to $\lambda$ in equation (38); $\lambda_r$ is defined as the ratio of stretched and quenched specimen length to the length of the specimen after recovery. Figure 30 (a) and (b) are the results for $\lambda_r = 1.2 \sim 2.2$ with $M_w = 95,000$ and $M_w/M_n = 1.13$ and for $\lambda_r = 2.5 \sim 2.6$ with $M_w = 1,170,000$ and $M_w/M_n = 1.2$, respectively. As is seen from the slopes of the curves in the low $Q$ range, $R_{g//}$ increases with increasing $\lambda_r$ while $R_{g\perp}$ decreases; for $M_w = 95,000$, $R_{g//}$ changes from 82 Å to 165 Å and $R_{g\perp}$ Å from 82 Å to 56 Å as $\lambda_r$ increases from 1.0 to 2.2. Qualitatively, these results are reasonable.

In what follows let us compare the data with some models. The Rouse model where the monomeric friction enhancement by entanglements is not taken into account could not explain the data (the calculated curves based on this model are not shown here). On the other hand the temporary network model satisfactorily describes the SANS data as is clear

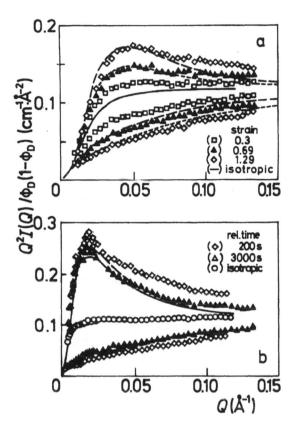

FIGURE 30. Kratky plots for aPS. (a) Samples with $M_w = 9.5 \times 10^4$ and $M_w/M_n = 1.13$, deformed at a constant shear rate $2.3 \times 10^{-2}$ s$^{-1}$ at 123°C. Recoverable strains $\lambda_r$: 1.2 (○), 1.6(△) and 2.2(×). Isotropic specimen (——). The temporary network model: $\lambda_r = 1.6$ (broken line) and 2.2 (dashed line). (b) Samples with $M_w = 1.17 \times 10^5$ and $M_w/M_n = 1.2$, deformed rapidly to $\lambda_r = 3$ at 140°C and then relaxed for given time before quenching. $\lambda_r = 2.45(△)$ and 2.63(×). Isotropic specimen (○). The temporary network model: $\lambda_r = 2.45$ (solid line) and 2.63 (dashed line). (Muller & Picot, 1996. Reprinted from R. Muller and C. Picot, Chain conformation in elongational and shear flow as seen by SANS, in *Rheology for Polymer Melt Processing*, Rheology Series 5, ed. by J.-M. Pia and J.F. Agassant, Copyright © 1996, Pages 65–94. with permission from Elsevier Science.)

upon inspection of Figure 30; the model calculations are indicated with solid and dashed lines. When $\lambda_r$ is large, there exists some disagreement in high $Q$ ranges, but this is due to the change of local chain conformation. The $n_e$ values obtained from the best fit with the SANS data are 5, 10 and 60 for $\lambda_r = 1.2 \sim 1.4$, $1.6 \sim 2.2$ and $2.5 \sim 2.6$, respectively. These results indicate that the affine model is clearly invalid when $\lambda_r$ is less than 2.2, but its validity recovers steadily as $\lambda_r$ exceeds 2.5. This is due to a rapid rise of $n_e$ when $\lambda_r$ is larger than 2.5, and this fact supports the results of extrusion with high draw ratio as described above.

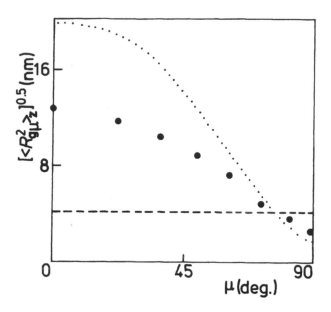

FIGURE 31. Azimuthal angle $\mu$ dependence of radius of gyration $R_{g\mu}$ of the cold-drawn PET determined by SANS. ●: observed values for drawn samples, $\cdots$: calculation assuming affine deformation, −−−: values for the isotropic sample. (Gilmer *et al.*, 1986. Reprinted from *Polymer*, **27**, J.W. Gilmer, D. Wiswe, H.-G. Zachmann, J. Kugler and E.W. Fischer, Changes in molecular conformations in poly(ethylene terephthalate) produced by crystallization and drawing as determined by neutron scattering. Pages 1391–1395. Copyright © 1986 with permission from Elsevier Science.)

### 5.6.2 *Deformation of Crystalline Polymers*

The problem of the affine deformation has been examined also for crystalline polymers such as polyethylene (PE), poly(ethylene terephthalate) (PET) and isotactic polypropylene (iPP). In addition to this the relation between the macroscopic draw ratio and the molecular draw ratio is also addressed.

### Problem of the Affine Deformation

Sadler and Barham (1983) raised an interesting question regarding the deformation of a molecule in crystalline polymers. For drawing process, the deformation over the dimension of spherulites is affine (Hay & Keller, 1965), but crystalline lamellae are not deformed affinely. Then, what is the deformation over the scale of a molecule? SANS experiments provided the answer (Sadler & Barham, 1983).

The melt-quenched samples of 3% PED with $M_w = 6.2$, 10.6 and $21.6 \times 10^4$ in PEH matrix were drawn by draw ratios of 10, 5.4 and 6.5, respectively, at room temperature with the speed of 1cm/min. The radii of gyration of the PED molecules normal to the draw direction were determined from the SANS patterns using the Zimm relation

$$I(0)/I(Q_\perp) = 1 + Q_\perp^2 R_\perp^2 \tag{39}$$

where $Q_\perp$ and $R_\perp$ are the length of scattering vector and the radius of gyration in the vertical direction, respectively. The scattering results have revealed that the deformation of the radius of gyration in the transversal directions $R_{g\perp}$ approximately follows the affine relation $R_{g\perp} = R_{go}\lambda^{-1/2}$ and the distribution of the molecule keeps Gaussian in the transversal plane. The deformation in the draw direction $R_{g//}$ never agrees with the affine relation $R_{g//} = R_{go}\lambda$; the molecular extension after drawing is much less than that predicted from the affine relation by a factor of two or less.

Such a deformation mechanism of crystalline polymers was further investigated in detail for poly(ethylene terephthalate) (PET) by Gilmer et al. (1986). PET samples were cold-drawn by a draw ratio of 4.9 at 45°C at 2mm/min and the two-dimensional SANS data were sliced into small sectors with an angle $\mu$ relative to the draw direction. Figure 31 shows the azimuthal angle $\mu$ dependence of radius of gyration $R_g\mu$ of the cold-drawn PET. As seen from Figure 31, the observed values depend on $\cos^2_\mu$ and are between those for the isotropic sample and the affine deformation. Here it is again confirmed that $R_{g\perp}(\mu = 90°)$ follows the affine relation while $R_{g//}(\mu = 0°)$ greatly deviates from it. The non-affine deformation suggests chain slippage during draw.

Relation between Macroscopic and Molecular draw ratios

The practical properties of fibres such as tensile modulus depend on the molecular draw ratio and do not directly on the macroscopic draw ratio. This has been clearly shown for polyethylene fibres using the SANS method by Sadler and Barham (1990). Various samples were prepared from blends of hydrogenated and deuterated polyethylenes (98% PEH and 2% PED) with molecular weights in a range of $5 \sim 30 \times 10^4$ and of $3 \sim 20 \times 10^4$, respectively. Fibres were drawn through a neck at various temperatures ranging from 70 to 118°C. Figure 32a shows the measured molecular draw ratio as a function of the macroscopic draw ratio after necking. As clearly seen from Figure 32a, the molecular draw ratio after necking is very close to the macroscopic neck draw ratio for the drawing temperatures below ca. 80°C, whereas above 80°C the former is much lower than the latter, suggesting local melting during necking. When we plot the final modulus of all the fibres as a function of the molecular draw ratio all the data are in a single curve as shown in Figure 32b. This indicates that the fibre modulus is uniquely determined by the molecular draw ratio, and not by the macroscopic draw ratio.

The taken-up velocity $v_a$ dependence of isotactic polypropylene chain conformation in fast melt-spun fibres has also been investigated by Hahn et al. (1988). In spite that the stress at the solidification point is proportional to $v_a^2$, the molecular elongation $\lambda = R_{g//}/R_{go} = 4.4 \sim 5.0$ is practically independent of $v_a$ for $v_a = 2,500$ and $4,000$ m/min, being identical with the value $\lambda = 4.8$ for the cold-drawn sample of a draw ratio 4.

## 5.7 TRANSESTERIFICATION IN POLYESTERS

The phenomenon of transesterification in polyesters is well known to industrial chemists, and it is often used to produce copolyesters by heating a mixture of two different compatible polyesters. Usually the characterization of composition and sequence length of such copolymers is made by analytical methods, but the SANS technique is more powerful and can

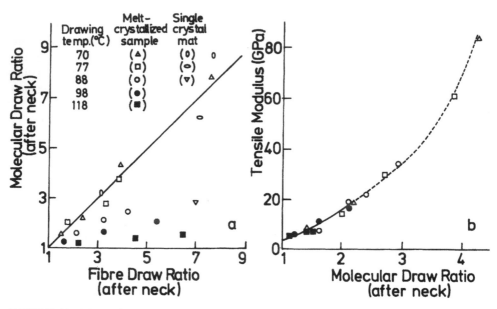

FIGURE 32. Molecular draw ratio $\lambda$ in PE fibres determined by SANS as a function of macroscopic draw ratio after necking (a) and the tensile modules as a function of $\lambda$ (b). $T_d$ is drawing temperature. (Sadler & Bahram, 1990. Reprinted from *Polymer*, **31**, D.M. Sadler and P.J. Barham, Structure of drawn fibres: 3. Neutron scattering studies of polyethylene fibres drawn beyond the neck. Pages 46–50. Copyright © 1990 with permission from Elsevier Science.)

determine these parameters directly. Kugler *et al.* (1987) and Benoit *et al.* (1989) have studied the kinetics of transesterification in poly(ethylene terephthalate) (PET), i.e. the influence of temperature and time on the amount of transesterification using the deuterium labeling SANS technique. Amorphous blends of deuterated and non-deuterated PET were annealed for different times at 280 and 250°C. During the annealing, block copolymers consisting of deuterated and non-deuterated sequences are formed by transesterification. To obtain the block molecular weight, absolute intensity measurements for SANS were carried out. The scattering function $S(Q)$, available from the absolute intensity $I(Q)$ through the relation $I(Q) = K\rho_N S(Q)$ where $K$ is the contrast factor between protonated and deuterated monomer units, $\rho_N$ is its number density per unit volume and $S(Q)$ is approximately represented by

$$S^{-1}(Q) = (1 + Q^2 R_g^2/3)/[x_D(1 - x_D)P_{SANS}], \qquad (40)$$

according to the random-phase approximation theory where $x_D$ is the fraction of deuterated material. $P_{SANS}$ is the weight-average degree of polymerization as determined by SANS and $R_g$ is its radius of gyration. Figure 33 shows an example of the Zimm plot which serves to determine the block molecular weight and the block radius of gyration from the extrapolated value at $Q = 0$ and the ratio of the slope to the intercept, respectively. The results obtained for non-crystalline PET blends with $M_w = 23,000$ melt-pressed at 280°C are summarized in Table 2. As is seen from this table, transesterification takes place very fast during melt

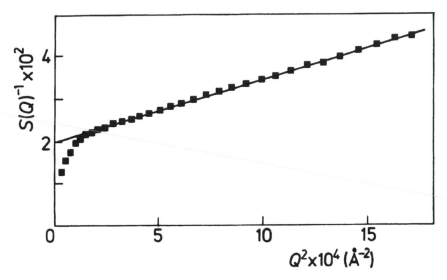

FIGURE 33. Zimm plot of noncrystalline PET blend (50% PETD) melt pressed at 280°C for 10s and quenched. (Kugler *et al.*, 1987. Reprinted with permission from *Macromolecules*, **20**, 1116–1119. Copyright © 1987 American Chemical Society.)

pressing at 280°C; in 3 min just after the beginning of pressing the molecular weight of PET greatly decreases from the original value $M_w = 23,000$ to only about 900 as an average block molecular weight. In this case no degradation should be expected judging from other measurements such as GPC and light scattering. The average number $u$ of intermolecular transesterification reaction per molecule can be calculated from the equation

$$u = (P_{LS}/P_{SANS}^{-1})/[x_D^2 + (1 - x_D)^2] \tag{41}$$

$P_{LS}$ being the original degree of polymerization measured by light scattering, and it is also listed in Table 2. This value $u$ increases linearly with pressing time, and hence the rate constant can be defined. The temperature dependence of the rate constant per monomer unit provides an activation energy of $E_a = 152$ kJ/ mol. Further, the authors suggested a possibility that transesterification decreases the number of entanglements.

TABLE 2. Block molecular weight $M_w$, $Z$-average radius of gyration $R_g$ and average number of intermolecular reaction $u$ as a function of melt-pressing time $t$ at 280°C. (Kugler *et al.*, 1987. Reprinted with permission from *Macromolecules*, **20**, 1116–1119. Copyright © 1987 American Chemical Society.)

| $t$/s | $M_w$ | $R_g$/nm | $u$ |
|---|---|---|---|
| 0 | 23,000 | – | 0 |
| 10 | 8,200 | 3.9 | 3 |
| 30 | 5,000 | 3.2 | 7 |
| 180 | 900 | 1.5 | 38 |

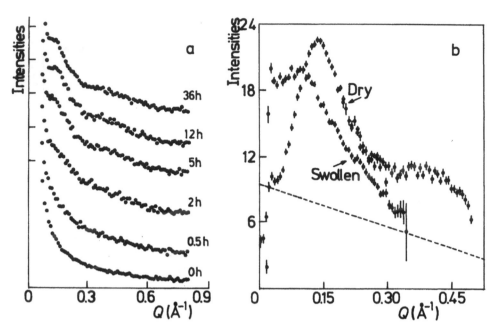

FIGURE 34. SANS curves of epoxy as a function of cure time (a), and those of dry and acetone swollen epoxy (b). The broken line in (b) indicates the estimated background for the dry sample. (Wu & Bauer, 1986a. Reprinted from *Polymer*, **27**, W. Wu and B.J. Bauer, Network structure of epoxies — a neutron scattering study: 2. Pages 169–180. Copyright © 1986 with permission from Elsevier Science.)

## 5.8  STRUCTURE OF POLYMER NETWORKS

### 5.8.1  Partial Labelling

Thermoset polymers are widely used as matrix for structural composites. Most of them are amorphous and highly cross-linked. It is therefore very difficult to characterize such materials. Equilibrium swelling ratio, elastic moduli and glass transition temperature have been employed to characterize them, but they give information on only the average density of the crosslinks. Such indirect measurements never give information about the distribution of the crosslinks or the homogeneity of the networks. Wu and Bauer (1985; 1986a, 1986b) have investigated the network structure of epoxy resins using the SANS technique.

Figure 34a shows the SANS curves of the partially deuterated diglycidyl ether of bisphenol A (DGEBA) cured with Jeffamine D-2000 [difunctional primary amines linked by poly(propylene oxide), $M_w \cong 2,000$] as a function of cure time. The deuterated parts are only those of bisphenol A. As cure proceeds, a pronounced peak at $Q = 0.13$ Å$^{-1}$, a broad peak at around $Q = 0.4$ Å$^{-1}$ and a weak shoulder at $Q = 0.24$ Å$^{-1}$ begin to appear. These scattering maxima were assigned to the intra-network correlations when the network is ideal. The peak at 0.13 Å$^{-1}$ and the shoulder at 0.24 Å$^{-1}$ are the first and the second order reflections for the average distance (43 Å) along the curing agent molecules between two adjacent deuterated bisphenol A groups. Finally, the peak at 0.4 Å corresponds to the

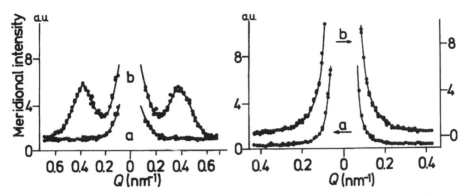

FIGURE 35. Meridional SANS curves of Fortisan (left) and Ramie (right). (a) Untreated, (b) exposed to heavy water vapor for 5 h. (Fischer *et al.*, 1978. Reprinted with permission from *Macromolecules*, **11**, 213–217. Copyright © 1978 American Chemical Society.)

average distance (16.5 Å$^{-1}$) along the epoxy chain between two such groups, suggesting that a rather ideal network is formed. However, the observed scattering intensities at zero angle are much higher than the theoretically calculated intensities for the ideal network, and the swelling effect increases the observed zero-angle intensities, contrary to the theoretical predictions (see Figure 34b). These results lead to the two-phase model consisting of the regular network portion and the disordered portion caused by certain (curing) side reactions other than the amine-epoxide reaction. The latter portion provides a background which is indicated with a broken line in Figure 34b for the dry sample. This two-phase model gives ca. 60% as an index of order in this epoxy resin.

## 5.9 CONTRAST ENHANCEMAENT METHOD (CEM)

### 5.9.1 Detection of Long Periods

Most semicrystalline polymers indicate the long-period structure where the crystalline and amorphous layers are arranged alternatively. The detection of long periods is carried out usually by small-angle X-ray scattering (SAXS). However, when the density difference between the crystalline and amorphous parts is too small, they cannot be detected by SAXS. On the other hand the SANS provides a possibility to detect them by introducing a deuterated solvent into the amorphous or disordered regions selectively. In what follows the results for native and regenerated cellulose investigated by Fischer *et al.* (1978) are explained as an example. In these cases the deuteration was carried out by exposing the samples into vapor of heavy water, resulting in hydrogen-deuterium exchange preferentially in the less ordered region. Figure 35 shows the meridional SANS curves before and after deuteration for regenerated (Fortisan) and native (Ramie) cellulose fibres. Before deuteration no peaks are observed for both celluloses, but through deuteration a clear strong meridional long spacing reflection with the spacing of 165 Å appears for Fortisan while such a reflection is not observed for Ramie. This reveals the difference in textural structure between native

TABLE 3. Estimated Chain Conformations in Various States of Amorphous and Crystalline Polymers.

| State | Chain Conformation |
| --- | --- |
| **AMORPHOUS** | |
| melt | Gaussian (random coil) |
| glass | Gaussian (random coil) |
| **CRYSTALLINE** | |
| *Single Crystals* | |
| As grown | |
| slowly at low supercooling | Sheet (ribbon) with adjacent or near-adjacent re-entry |
| at intermediate supercooling | Superfolded or Z-like sheet with block-stem dilution |
| rapidly at high supercooling | Hinge-like sheet (a molecular chain is incorporated in more than one lamella) |
| Annealed without partial melting | $R_g$ increases along the chain axis but hardly changes in the transverse directions. Stem length increases with random removal from adjacent stem groups in the sheet. |
| Melting process | $R_g$ increases to the equilibrium value rapidly in short time, independent of $M_w$. Tightly packed compact molecules randomize through a cooperative process. |
| *Melt-Grown Crystals* | |
| Generally | Between switch-board model (Flory) and sharp adjacent re-entry model (Keller) |
| Crystallized | |
| slowly at low supercooling | $R_g$ increases with increasing crystallinity. Tight loop sheet with long amorphous sequences at both ends incorporated into different lamellae, which prevents further crystallization.(ACA model with UC incorporation) |
| rapidly at high supercooling | Totally Gaussian. A single molecule is incorporated into two successive lamellae (Solidification model by Fischer). Near-by re-entry of several stems on a row with occasional stagger (Modified row model by Guenet *et al.*). |
| **MECHANICAL DEFORMATION** | |
| *Amorphous Polymers* | |
| under low strain | Non-affine deformation |
| under large strain | Affine deformation |
| *Crystalline Polymers* | |
| Radius of gyration in the transversal directions $R_{g\perp}$ | Keeping Gaussian, approximately deforms affinely. |
| Radius of gyration in the draw direction $R_{g//}$ | Non-affine deformation, the molecular extension is less by factors two or less than that predicted from the affine relation because of chain slippage. |

and regenerated celluloses; the former has a periodic structure of alternative crystalline and disordered regions along the fibre axis while the latter structure is almost homogeneous in this direction. These results make it clear that the CEM is useful for detecting weak density fluctuations, considering that SAXS could not detect a long period reflection for Fortisan.

## 5.10 CONCLUSION

In this chapter it has been shown that the SANS methods combined with labeling are very useful for investigation of both the total and local chain conformations in the bulk states such as melt, glassy and semicrystalline; and how the fundamentally important, interesting problems in polymer physics have been solved. However, these are but a few examples and the readers would find many other applications of neutron scattering in literature. Especially the author regrets that he could not touch the problems of dynamics in bulk polymers e.g. melting and glass transition because of time and space restrictions.

For the readers' reference the chain conformations in the amorphous and crystalline polymers as functions of the conditions for sample preparation are summarized in Table 3.

## Acknowledgements

The author wishes to thank Dr. Barbara Gabrys for giving him a chance to review the interesting problems written here, and reading through whole the chapter critically. He is indebted to the authors of literature cited in the text, especially for allowing him to reproduce their figures and tables.

## References

Akcasu, A.Z., Summerfield, G.C., Jahshan, S.N., Han, C.C., Kim, C.Y. and Yu, H. (1980) *J. Polymer Sci. Polym. Phys. Ed.*, **18**, 863.

Algelt, K. and Schultz, G.V. (1960) *Makromol. Chem.*, **36**, 209.

Anderson, J.E. and Bai, S.J. (1978) *J. Appl. Phys.* **49**, 4973.

Auvray, L. (1992) Contrast Methods in Small-Angle Neutron Scattering. Applications to Polymer Systems, in *Industrial and Technological Applications of Neutrons, Proceedings of the International School of Physics Enrico Fermié, Course CXIV*, Fontana, M., Rustichelli, F., Coppola, R., Terenzo Di Lerici, S. (Eds), North Holland, Amsterdam.

Barham, P.J. and Sadler, D.M. (1991) A neutron scattering study of the melting behaviour of polyethylene single crystals *Polymer* **32**, 393.

Barham, P.J. (1993) *Crystallization of Polymers*, M. Dosi?re (Ed), Kluwer, p.153.

Ballard, D.G.H., Cheshire, P., Longman, G.W. and Schelten, J. (1978) Neutron scattering studies of semicrystalline polymers in the solid state: the structure of isotopic polyoleofins *J. Polymer*, **19**, 379.

Ballard, D.G.H., Burgess, A.N., Crowley, T.L., Longman, G.W. and Schelten, J. (1979a) Structure of polyolefins in the solid state as revealed by small-angle neutron scattering *Faraday Disc. Chem. Soc.*, **68**, 279.

Ballard, D.G.H., Longman, G.W., Crowley, T.L., Cunningham, A. and Schelten, J. (1979b) Small angle studies of isotropic polypropylene *Polymer*, **20**, 399.

Ballard, D.G.H., Burgess, A.N., Nevin, A., Cheshire, P., Longman, G.W. and Schelten, J. (1980) Small-angle neutron scattering studies on polypropylene *Macromolecules*, **13**, 677.

Bates, F.S. and Wignall, G.D. (1986) Non-ideal mixing in binary blends of perdeuterated and protonated polystyrene *Macromolecules*, **19**, 932.

Benoit, H.C., Fischer, E.W. and Zachmann, H.G. (1989) Interpretation of neutron scattering results during transesterification reactions *Polymer*, **30**, 379.

Boothroyd, A.T. (1988) Polydispersity corrections to small-angle scattering data from polymers in the melt and in dilute solution *Polymer*, **29**, 1555.

Cotton, J.P., Decker, D., Benoit, H.C., Farnoux, B., Higgins, J.S., Jannink, G., Ober, R., Picot, C. and des Cloizeaux, J. (1974) Conformation of polymer chain in the bulk *Macromolecules* **7**, 863.

Cotton, J.P. (1991) Initial Data Treatment, in *Neutron, X-ray and Light Scattering: introduction to an investigative tool for colloidal and polymeric systems*, ed. by Lindner, P., Zemb, Th., North Holland, Amsterdam.

Des Cloizeaux, J. (1970) The statistics of long chains with non-markovian repulsive interactions and the minimal Gaussian approximation *J. Phys.* **31**, 715.

Des Cloizeaux, J. and Jannink, G. (1987) *Polymers in Solution — Their modelling and structure*, Oxford University Press, Oxford.

De Gennes, P.G. (1971) Reptation of a polymer chain in the presence of fixed obstacles *J. Chem. Phys.*, **55**, 572.

Doi, M. and Edwards, S.F. (1978) *J. Chem. Soc., Faraday II*, **74**, 1789, 1802, 1818.

Donald, A.M. and Windle, A.H. (1992) *Liquid Crystalline Polymers*, Cambridge Univ. Press, Cambridge.

Fischer, E.W. (1978a) Studies of structure and dynamics of solid polymers by elastic and inelastic neutron scattering *Pure Appl. Chem.*, **50**, 1319.

Fischer, E.W., Herchenršder, P., Manley, R.St.J. and Stamm, M. (1978b) Small-angle neutron scattering of selectively deuterated cellulose *Macromolecules*, **11**, 213.

Flory, P.G. (1949) The configuration of real polymer chains *J. Chem. Phys.*, **17**, 303.

Geil, P.H. (1972) *Ind. Eng. Chem. Prod. Res. Dev.*, **14**, 59.

Gilmer, J.W., Wiswe, D., Zachmann, H.-G., Kugler, J. and Fischer, E.W. (1986) Changes in molecular conformations in poly (ethylene terephthalate) produced by crystallization and drawing as determined by neutron scattering *Polymer*, **27**, 1391.

Glatter, O. and Kratky, O. (1982) Small Angle X-ray Scattering, Academic Press, London.

Goyal, P.S., King, J.S. and Summerfield, G.C. (1983) Multiple scattering in small-angle neutron scattering measurements on polymers *Polymer*, **24**, 131.

Guenet, J.M. and Picot, C. (1979a) Small-angle neutron scattering by semicrystalline chains: evaluation of the mean dimensions *Polymer*, **20**, 1473; ibid: Conformation of isotactic polystyrene (iPS) in the bulk crystallized state as revealed by SANS, 1483.

Guenet, J.M., Picot, C. and Benoit, H. (1979b) Chain conformation of semicrystalline isotactic polystyrene by small-angle neutron scattering *Faraday Disc. Chem. Soc.*, **68**, 251.

Guenet, J.M. (1980) A neutron scattering study of the chain trajectory in isotactic polystyrene single crystals *Macromolecules*, **13**, 387.

Guenet, J.M. (1981) Neutron scattering investigations on the effect of crystallization temperature and thermal treatment on the chain trajectory in bulk-crystallized isotactic polystyrene *Polymer*, **22**, 313.

Guenet, J.M. and Picot, C. (1983) Bulk crystallization of isotactic polystyrene near its melting point: a neutron scattering study of the chain trajectory *Macromolecules* **16**, 205.

Guenet, J.M., Sadler, D.M. and Spells, S.J. (1990) Solution-grown crystals of isotactic polystyrene: neutron scattering investigation of the variation of the chain trajectory with crystallization temperature *Polymer*, **31**, 195.

Guinier, A. and Fournet, G. (1995) *Small Angle X-Ray Scattering*, Wiley, New York.

Hadziioannou, G., Wang, L., Stein, R.S. and Porter, R.S. (1982) Small-angle neutron scattering studies on amorphous polystyrene oriented by solid-state coextrusion *Macromolecules*, **15**, 880.

Hahn, K., Kerth, J., Zolk, R., Schwahn, D., Springer, T. and Kugler, J. (1988) Determination of the chain conformation in fast-spun polypropylene fibers by small angle neutron scattering *Macromolecules*, **21**, 1543.

Higgins, J.S. and Benoit, H.C. (1994) *Polymers and Neutron Scattering*, Oxford University Press, Oxford.

Imai, M., Kaji, K., Kanaya, T. and Sakai, Y. (1995) Ordering process in the induction period of crystallization of poly(ethylene terephthalate) *Phys. Rev. B*, **52**, 12696.

Kirste, R.G., Kruse, W.A. and Ibel, K. (1975) Determination of the conformation of polymers in the amorphous solid state and in concentrated solution by neutron diffraction *Polymer*, **16**, 120.

Kirste, R.G. and Oberthuer, R.C. (1982) Synthetic Polymers in Solution, in *Small Angle X-ray Scattering*, ed. by O. Glatter and O. Kratky, Academic Press, London.

Klein, J. and Briscoe, B.J. (1979) *Proc. Roy. Soc. Lond.* **A365**, 53.

Kostorz, G. (1979) *Treatise on Materials Science and Technology*, **15** (Neutron Scattering), Academic Press, N.Y.

Kugler, J., Gilmer, J.W., Wiswe, D., Zachmann, H.-G., Hahn, K. and Fischer, E.W. (1987) Study of transesterification in poly(ethylene teraphthalate) by small-angle neutron scattering *Macromolecules*, **20**, 1116.

Leng, M. and Benoit, H. (1962) *J. Polym. Sci.*, **57**, 263.

Lieser, G., Fischer, E.W. and Ibel, K. (1975) *J. Polymer Sci.* B **13**, 39.

Muller, R., Picot, C., Zang, Y.H. and Froelich, D. (1990) Polymer chain conformation in the melt during steady elongational flow as measured by SANS. Temporary network model *Macromolecules*, **23**, 2577.

Muller, R. and Picot, C. (1996) *Rheology for Polymer Melt Processing* (Rheology Series 5), ed. by J-M. Piau and J.F. Agassant, Elsevier, p.65.

Oberthuer, R.C. (1978) *Makromol. Chem.*, **179**, 2693.

Pechhold, W.R. and Grossmann, H.P. (1979) Meander model of condensed polymers *Faraday Disc. Chem. Soc.*, **68**, 58.

Porod, G. (1949) *Monatsh. Chem.*, **2**, 251.

Rawiso, M., Duplessix, R. and Picot, C. (1987) Scattering function of polystyrene *Macromolecules* **20**, 630.

Russel, T.D., Lin, J.S., Spooner, S. and Wignall, C.D. (1988) Intercalibration of small angle x-ray and neutron scattering data *J. Appl. Cryst.* **21**, 629.

Sadler, D.M. and Keller, A. (1976) Trajectory of polyethylene chains in single crystals by low angle neutron scattering *Polymer*, **17**, 37.

Sadler, D.M. and Keller, A. (1979) *Science*, **203**, 263.

Sadler, D.M. and Harris, R. (1982) A neutron scattering analysis: how does the conformation in polyethylene crystals reflect that in the melt? *J. Polym. Sci. Polym. Phys. Ed.*, **20**, 561.

Sadler, D.M. (1983a) Analysis of anisotropy of small-angle neutron scattering of polyethylene single crystals *J. Appl. Cryst.*, **16**, 519.

Sadler, D.M. and Barham, P.J. (1983b) Neutron scattering measurements of chain dimensions in polyethylene fibers *J. Polym. Sci. Polym. Phys. Ed.*, **21**, 309.

Sadler, D.M., Spells, S.J., Keller, A. and Guenet, J.M. (1984) Wide-angle neutron scattering from isotactic polystyrene: the fold arrangement in solution grown crystals *Polymer Comm.*, **25**, 290.

Sadler, D.M. (1985) Neutron scattering of annealed solution grown crystals: molecular dimensions *Polymer Comm.*, **26**, 204.

Sadler, D.M. and Spells, S.J. (1989) Neutron scattering of annealed solution-grown crystals of polyethylene *Macromolecules*, **22**, 3941.

Sadler, D.M. and Barham, P.J. (1990) Structure of drawn fibres: 1. Neutron scattering studies of necking in melt-crystallised polyethylene *Polymer*, **31**, 36; ibid Structure of drawn fibres: 2. Neutron scattering and necking in single-crystal mats of polyethylene 43; ibid Structure of drawn fibres: 3. Neutron scattering studies of polyethylene fibres drawn beyond the neck 46; *"Integration of Fundamental Polymer Science and Technology — 4"*, ed. by P.J. Lemstra and L.A. Kleintjens, Elsevier Appl. Sci., London, 1990, p.304.

Schelten, J., Wignall, G.D. and Ballard, D.G.H. (1974a) Chain conformation in molten polyethylene by low angle neutron scattering *Polymer*, **15**, 682.

Schelten, J., Wignall, G.D., Ballard, D.G.H. and Schmatz, W. (1974b) Neutron small angle scattering by mixtures of H and D-tagged molecules of polystyrene and polyethylene *Colloid & Polymer Sci.* **252**, 749.

Schelten, J., Ballard, D.G.H., Wignall, G., Longman, G.W., Schmatz, W. (1976) Small-angle neutron scattering studies of molten and crystalline polyethylene *Polymer* **17**, 751.

Schelten, J., Wignall, G., Ballard, D.G.H. and Longman, G.W. (1977) Small-angle neutron scattering studies of molecular clustering in mixtures of polyethylene and deuterated polyethylene *Polymer*, **18**, 1111.

Schelten, J. and Schmatz, W. (1980) Multiple-scattering treatment for small-angle scattering problems *J. Appl. Cryst.*, **13**, 385.

Schultz, G.V. (1939) *Z. Physik. Chem.* **B43**, 25.

Sharp, P. and Bloomfield, V.A. (1968) *Biopolymers*, **6**, 1201.

Spells, S.J., Keller, A. and Sadler, D.M. (1984a) I.r. study of solution grown crystals of polyethylene: correlation with the model from neutron scattering *Polymer*, **25**, 749.

Spells, S.J. and Sadler, D.M. (1984b) Neutron scattering studies on solution grown crystals of polyethylene: a statistical preference for adjacent re-entry *Polymer*, **25**, 739.

Spells, S.J. and Sadler, D.M. (1989) Mixed-crystal infrared spectroscopy of annealed solution grown crystals of polyethylene *Macromolecules*, **22**, 3948.

Stamm, M., Fischer, E.W., Dettenmaier, M. and Convert, P. (1979) Chain conformation in the crystalline state by means of neutron scattering methods *Faraday Disc. Roy. Soc. Chem.*, **68**, 263.

Stamm, M., Schelten, J. and Ballard, D.G.H. (1981) Determination of the chain conformation of polypropylene in the crystalline state by neutron scattering *Colloid &se Polymer Sci.*, **259**, 286.

Stehling, F.S., Ergos, E. and Mandelkern, L. (1971) Phase separation in n-Hexatriacontane-n-Hexatriacontane-$d_{74}$ and polyethylene — poly(ethylene-$d_4$) systems *Macromolecules*, **4**, 672.

Ullmann, R., Summerfield, G.C. and King, J.S. (1977) Investigation of heterogeneities in solid polyethylene by small angle neutron scattering *J. Polymer Sci. Polym. Phys. Ed.*, **15**, 1641.

Wignall, G.D. and Bates, F.S. (1987) Absolute calibration of small angle neutron scattering data *J. Appl. Cryst.*, **20**, 28.

Wignall, G.D. (1993) Scattering Techniques with Particular Reference to Small-Angle Neutron Scattering, in *Physical Properties of Polymers*, ed by J.E. Mark, A. Eisenberg, W.W. Graessley, L. Mandelkern, E.T. Samulski, J.L. Koenig and G.D. Wignall, ACS, Washington, DC, p. 313 (Chapter 7).

Wu, W. and Bauer, B.J. (1985) Network structures of epoxies: 1. A neutron scattering study *Polym. Comm.*, **26**, 39.

Wu, W. and Bauer, B.J. (1986a) Network structures of epoxies — neutron scattering study: 2. *Polymer*, **27**, 169.

Wu, W. and Bauer, B.J. (1986b) Epoxy network structure. 3. Neutron scattering study of epoxies containing monomers of different molecular weight *Macromolecules*, **19**, 1613.

Wu, W. and Bauer, B.J. (1988a) Network structure in epoxies. 5. Deformation mechanisms in epoxies *Macromolecules*, **21**, 457.

Wu, W., Hunston, D.L., Yang, H. and Stein, R.S. (1988b) Epoxy network structure. 4. A neutron scattering study of epoxies swollen in a deuteriated solvent *Macromolecules*, **21**, 756.

Wu, W., Bauer, B.J. and Su, W. (1989) Network structure in epoxies: 6. The growth process investigated by neutron scattering *Polymer*, **30**, 1384.

Yeh, G.S.Y. (1972) *Crint. Rev. Macromol. Chem.*, **1**, 173.

Yoon, D.Y. and Flory, P.J. (1976) Small-angle x-ray and neutron scattering by polymethylene, polyoxyethylene, and polystyrene chains *Macromolecules*, **9**, 294.

Yoon, D.Y. and Flory, P.J. (1979) Molecular morphology in semicrystalline polymers *Faraday Disc. Chem. Soc.*, **68**, 288.

Yoshizaki, T. and Yamakawa, H. (1980) Scattering functions of wormlike and helical wormlike chains *Macromolecules*, **13**, 1518.

Zimm, B.H. (1948) The scattering of light and the radial distribution function of high polymer solutions *J. Chem. Phys.*, **16**, 1093 Ibid. Apparatus and methods for measurement and interpretation of the angular radiation of light scattering; preliminary results on polystyrene solutions, 1099.

# 6. Organisation and Dynamics of Amphiphilic Polymers at Fluid Interfaces

STELLA K. PEACE* and RANDAL W. RICHARDS

*Interdisciplinary Research Centre in Polymer Science and Technology, University of Durham, Durham, DH1 3LE, UK*

## 6.1 INTRODUCTION

Molecules at fluid interfaces are difficult to investigate by means which are non-perturbative and which provide molecular level information. ATR FTIR can be used to identify the nature (and perhaps the orientation of surface molecules), but making conclusions relevant to the global molecular properties is often difficult. The need for non-perturbative methods of investigation usually means that radiation of some form is used as the probe and the reflected or scattered radiation is analysed in some way to give the desired insight. We discuss here the application of neutron reflectometry to polymers at fluid interfaces and demonstrate that by judicious use of deuterium labelling a complete description of the organisation of the polymer is accessible. Along with this, we also describe how the long wavelength dynamics of polymers at fluid interfaces are obtainable using surface quasi-elastic light scattering. Although this technique has less complicated requirements in terms of equipment than neutron reflectometry, the analysis and interpretation of the data is considerably more demanding. Nonetheless, the information obtained complements the structural description and demonstrates the unusual visco-elastic behaviour of polymers at fluid interfaces.

Although some work on polymers at liquid-liquid interfaces has been discussed, this aspect is as yet a very small part of the overall activity. The majority of the work discussed below is concerned with polymers at the air-water interface. We begin with a precis of results concerning the most basic property accessible, the surface pressure.

[1]Current Address: Unilever Research, Colworth Laboratory, Sharnbrook, Bedford, MK44 1LQ, UK.

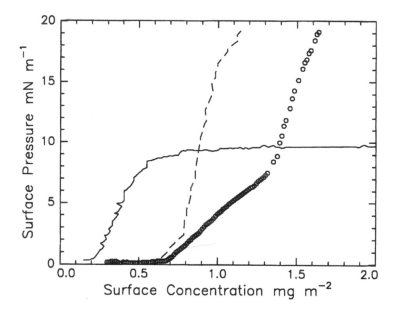

FIGURE 1. Surface pressure isotherms for spread films on water of polyethylene oxide (solid line), polymethyl methacrylate (dashed line) and an equimolar linear diblock copolymer of the two (○).

## 6.2 SURFACE PRESSURE ISOTHERMS

The surface pressure (i.e. difference in surface tension of the pure liquid and the pure liquid surface supporting a polymer film) of a spread polymer film is perhaps the simplest and easiest property to determine experimentally. All that is required is some form of Langmuir film balance and there are numerous reports in the literature of isotherms for various polymers. However the interpretation of isotherm data in molecular or thermodynamic terms is not clear cut. Figure 1 reproduces isotherms for a variety of polymers and copolymers, clearly the range of behaviours observed suggests that any completely general theory may have to be rather complex. Note that these isotherms are plotted as surface pressure ($\pi$) as a function of surface concentration rather than molecular (or monomer unit) area. This is the more commonly observed plot because much of the earlier work on spread polymer films used polydisperse (i.e. broad molecular weight distribution) polymers. From the shapes of the isotherms they may be classified into various types but generally for polymers, surface pressure isotherms are either liquid expanded (e.g. PEO) or liquid condensed films (e.g. PMMA). The wide variety of isotherm behaviour observed is a testimony to the interplay between enthalpic influences and entropic restrictions when the polymer is spread on the liquid surface. Interactions between subphase and polymer determine the enthalpic contribution to the excess surface free energy, the entropy is determined by the configurations available to the polymer molecule now it is confined to a (pseudo) two dimensional state. Thermodynamically the surface pressure is defined as

$$\pi = (\partial \Delta G_s / \partial A)_{N,T} \tag{1}$$

i.e. the change in excess surface Gibbs free energy per unit surface area at constant temperature and number of molecules. Over the years attempts have been made to calculate $\Delta G_s$ for spread polymers and thus obtain an equation for $\pi$ as a function of surface coverage (i.e. an equation of state). All of these equations have the form

$$\pi = \frac{RT}{A_o}\theta(1 + A_{2s}\theta + A_{3s}\theta^2 + \ldots) \tag{2}$$

where $\theta$ is the fraction of the surface covered by polymer, $A_o$ the molar area of the polymer and $A_{is}$ are the $i$th surface virial coefficients. Since $\theta = \Gamma_s A_o / \overline{M}_n$ ($\Gamma_s =$ surface concentration in mass per unit area, $\overline{M}_n =$ number average molecular weight) then interpretation of surface pressure isotherms via such equations of state should allow a value of $\overline{M}_n$ to be obtained. However these equations of state are only valid at very low values of $\Gamma_s$ where the surface pressure is small and data of high accuracy are required. They have been little used. In recent years the most commonly used means of interpreting surface pressure isotherms is the application of scaling law relations (de Gennes, 1979; Vilanove and Rondelez, 1980) to the data in the 'semi-dilute' region. These scaling laws are an extension of those developed for bulk solutions and for a fixed temperature the surface pressure is related to the surface concentration by

$$\pi \sim \Gamma_s^y \tag{3}$$

where $y = 2\nu/(2\nu - 1)$ and $\nu$ is the excluded volume exponent in the relation between radius of gyration and degree of polymerisation of the polymer. For favourable thermodynamic interactions between polymer and subphase, $\nu = 0.75$, if the interaction is such that ideal thermodynamic behaviour prevails (i.e. surface $\theta$-state) then $\nu = 0.571$. Consequently in a double logarithmic plot of $\pi$ as a function of $\Gamma_s$ a slope ($y$) of 3 is expected for good solvent conditions and a much higher value should be observed as $\theta$-conditions are approached. Very high values of $y$ should be observed if the spread film collapses because $\nu$ is predicted to have a value of 0.5 under these circumstances. Poupinet and Rondelez (Poupinet, Vilanove *et al.*, 1989) have used $\pi$ data at low concentrations and in the 'semi-dilute' region to establish the applicability of scaling laws to spread films of polymethyl methacrylate. Values of $\nu$ obtained for polymethyl methacrylate (PMMA) spread at the air-water interface are very dependent on the tacticity of the polymer. Syndiotactic PMMA has a $\nu$ value of 0.53 whereas isotactic PMMA exhibits a $\nu$ value of 0.7.

Scaling relations can only be used to deduce the thermodynamic state of spread homopolymer films. The surface pressure isotherms of copolymers are much more difficult to interpret because of the different interactions of the constituent parts of the copolymer with the subphase. One aspect of the spread copolymer films is to use them to generate polymer brush layers which can then be characterised by other techniques. We do not have the space to set out the detailed aspects of polymer brush theory as first developed by Alexander (Alexander, 1977) and de Gennes (de Gennes, 1980) and subsequently elaborated on by

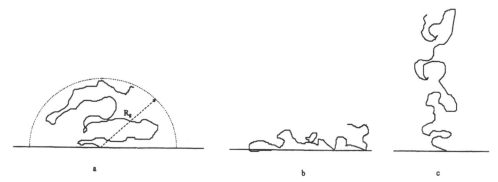

FIGURE 2.   Schematic arrangements of polymer molecules at a surface arranged as a mushroom (a), a pancake (b) and brush (c).

others (Milner, Witten *et al.*, 1988). Figure 2 summarises the situations dealt with. Polymer molecules are attached to an interface by one end. If the grafting density, $\sigma$ (number of molecules per unit area), is small the molecules exist either as mushrooms (Figure 2a) or pancakes (Figure 2b) depending on the magnitude of the interaction parameter (or 'sticking' energy) between the interface and the segments of the polymer molecule. As the grafting density increases, excluded volume effects between the grafted molecules causes a stretching of the molecules normal to the grafted surface to minimise the free energy increase due to the polymer-polymer interactions. Opposing this stretching is the entropic penalty associated with the less probable configurations of the stretched molecule. The equilibrium configuration is that which balances these enthalpic and entropic contributions. Typically the surface pressure isotherms of block copolymer spread films exhibit a pseudo plateau region before a further rapid increase in surface pressure is observed at higher surface concentrations. Bijsterbosch et al have used Scheutjens-Fleer (Fleer, Cohen-Stuart *et al.*, 1993) self consistent field theory to calculate the surface pressure isotherm by numerical methods. Favourable interactions between the segments of the solvated block and the surface result in a pancake configuration and a rise in surface pressure. At the inflection point before the pseudo plateau, the surface is completely filled with polymer segments. The surface pressure does not increase markedly as more polymer is added until the excluded volume interactions are such that the solvated blocks begin to stretch and the surface pressure now increases again due to contributions from the brush like layer (Figure 2c). Bijsterbosch *et al.* (Bijsterbosch, de Haan *et al.*, 1995) cite data for asymmetric block copolymers of polystyrene and ethylene oxide as support for this view, and such copolymers have been frequently used as models for brush like layers. However it must be pointed out that such copolymers form micelles at very low concentrations and the shape of the surface pressure isotherms observed is exactly that predicted by Israelachvili (Israelachvili, 1994) for systems where surface micelles appear. Consequently, clear cut conclusions from surface pressure isotherms still remain elusive and what is required is some means of determining the organisation and composition of the surface region occupied by spread polymers. It is also noteworthy that the form of the surface pressure isotherm attributed to brush formation is also observed for some homopolymers, e.g. PMMA (Henderson, Richards *et al.*, 1991).

## 6.3 NEUTRON REFLECTOMETRY

We provide the simplest outline of the basis of neutron reflectometry here, sufficient to cope with the examples to be discussed subsequently. More detailed treatments of the subject are available elsewhere (Thomas, 1995) as well as discussions specific to reflectometry from polymer systems (Russell, 1990). The aim of a neutron reflectometry experiment is to measure the intensity of the specularly reflected neutron beam from a surface or interface as a function of the scattering vector ($Q$) normal to the interface on which the beam is incident. The reflectivity, $R(Q)$, is the ratio of the reflected intensity to the incident intensity, $I(Q)/I_o$. The scattering vector is determined by the grazing angle of incidence, $\theta$, and the wavelength of the neutron beam, $\lambda$.

$$|Q| = Q = (4\pi/\lambda)\sin\theta$$

The reflectivity is expressed by Fresnel's law with the replacement of the optical refractive index by the neutron refractive index of the reflecting substance which, ignoring absorption effects, is

$$n = 1 - \frac{\lambda^2}{2\pi}\rho \qquad (4)$$

and $\rho$ is the scattering length density of the substance. The incident neutron beam is both reflected and refracted and there will be a critical angle of incidence where the beam is totally reflected. This angle can be determined using Snell's law and the critical value of $Q$ obtained is given by

$$Q_c = 4\pi^{1/2}\rho^{1/2} \qquad (5)$$

Hence the reflectivity and the critical scattering vector are dependent on the scattering length density, $\rho$. This parameter is determined by the composition of the reflecting material and by the variation in composition normal to the surface. In principle therefore, the composition distribution normal to a surface or interface can be obtained from the variation of the reflectivity as a function of $Q$, i.e. the reflectometry profile. In practice because phase information is lost in obtaining the reflectometry data and because the accessible $Q$ range may be limited because of incoherent background scattering, the data cannot be Fourier transformed to give the composition distribution directly. Neutron reflectometry data are frequently interpreted using models, i.e. a model composition distribution normal to the interface is generated and its reflectivity calculated and compared with the experimental data. The parameters of the composition distribution are iteratively refined until acceptable agreement between data and simulation is obtained. This process is aided by the optical matrix description of the reflectivity. In the model the specimen is described by a series of lamellar sheets parallel to the surface on which the beam is incident which reproduce the actual composition distribution. Within each layer the composition is constant and the scattering length density is calculated from this composition. Thus if there are two components in the system (e.g. polymer, $p$, and subphase, $s$) then in layer $n$;

$$\rho^n = \phi_s^n \rho_s + \phi_p^n \rho_p$$
$$= n_s^n b_s + n_p^n b_p \qquad (6)$$

where $\phi_i^n$ is the volume fraction of species $i$ in layer $n$ which have a scattering length density $\rho_i$, a coherent scattering length $b_i$ and a number density, $n_i^n$ in the layer.

Each layer is describable by a characteristic 'optical' matrix;

$$[M^n] = \begin{bmatrix} \exp(i\beta_{n-1}) & r_n \exp(i\beta_{n-1}) \\ r_n \exp(-i\beta_{n-1}) & \exp(-i\beta_{n-1}) \end{bmatrix} \tag{7}$$

where $\beta_{n-1}$ is the optical path length of the neutron beam in layer $(n-1)$

$$\beta_{n-1} = \frac{2\pi}{\lambda} n_{n-1} d_{n-1} \sin\theta_{n-1} \tag{8}$$

with $n_{n-1}$, $d_{n-1}$ and $\theta_{n-1}$ being the neutron refractive index, the thickness and the angle of refraction in layer $(n-1)$. The Fresnel reflectivity coefficient of layer $n$, is given by an expression involving the wave vector of the neutron beam in layer $n$. For a lamellar assembly the overall optical matrix is the product of all the individual matrices

$$M = [M^1][M^2]\dots\dots[M^n]$$

$$= \begin{bmatrix} M_{11} & M_{21} \\ M_{12} & M_{22} \end{bmatrix} \tag{9}$$

and the reflectivity is given by

$$R(Q) = \frac{M_{21}M_{21}^*}{M_{11}M_{11}^*} \tag{10}$$

Such optical matrix calculations are *exact* and can be done rapidly on a PC. Care has to be exercised in using optical matrix methods to analyse reflectivity data. Including more layers will *always* give a better fit to the data but not necessarily be a true description of the specimen. Likewise the layer thickness and scattering length density (and thus layer composition) are directly coupled, a reduction in one can be compensated by an increase in the other and values obtained from least squares fitting need to be examined critically.

Optical matrix type calculations of reflectivity are exact in all circumstances i.e. from total reflection to reflectivities of the lowest magnitude currently measurable ($\sim 10^{-7}$). For polymers at liquid-air interfaces the reflectivity over most of the accessible $Q$ range is at least two orders of magnitude smaller than total reflectivity ($R(Q) = 1$). In these circumstances the kinematic approximation describes the reflectivity with acceptable accuracy (Thomas, 1995). The assumptions of the kinematic approximation are that the interaction of the neutron with the reflecting substance is weak and there is no multiple scattering of neutrons before the reflected neutron is detected. Under these circumstances

$$R(Q) = \frac{16\pi^2}{Q^4} \rho(Q)^2 \tag{11}$$

where $\rho'(Q)$ is the one dimensional Fourier transform of the *gradient* in scattering length density distribution normal to the surface i.e.

$$\rho'(Q) = \int_{-\infty}^{\infty} \exp(iQz)\left(\frac{\partial\rho}{\partial z}\right) dz \tag{12}$$

At any distance $z$ from the interface

$$\rho(z) = \sum_i n_i(z)b_i \qquad (13)$$

where $n_i(z)$ and $b_i$ are the local number density of species $i$ with a coherent scattering length $b_i$. Hence

$$\rho(Q) = \sum b_i \int_{-\infty}^{\infty} \exp(iQz)n_i(z)$$
$$= \sum b_i n_i(Q) \qquad (14)$$

Now $(n_i(Q))^2$ is the partial structure factor of species $i$ represented by $h_{ii}(Q)$ and

$$h_{ii}(Q) = Q^2 h'_{ii}(Q) \qquad (15)$$

with

$$|\rho'(Q)|^2 = \sum b_i^2 h'_{ii}(Q)$$

and $h'_{ii}(Q)$ is the differential of the partial structure factor. Generalising for all possible species which may be present then

$$R(Q) = \frac{16\pi^2}{Q^4} \sum b_i b_j h'_{ij}(Q) \qquad (16)$$

To illustrate the potential of this formulation of the reflectivity we consider a polymer with constituent parts $A$ and $B$ on a subphase $S$, the reflectivity is given by

$$R(Q) = \frac{16\pi^2}{Q^4} \left[ \begin{array}{l} b_A^2 h'_{AB}(Q) + b_B^2 h'_{BB}(Q) + b_S^2 h'_{SS}(Q) + 2b_A b_B h'_{AB}(Q) \\ + 2b_A b_S h'_{AS}(Q) + 2b_B b_S h'_{BS}(Q) \end{array} \right] \qquad (17)$$

The self partial structure factors describe the composition and thickness of the regions containing the species whereas the cross partial structure factor terms $((h'_{ij}(Q))$ contain information about the separation between the two species. The application of this description of reflectivity will be illustrated later.

### 6.3.1 Scattering Lengths and Scattering Length Density

From the above brief description of the theoretical basis of neutron reflectometry, the importance of scattering length and scattering length density is clear. Just as in small angle neutron scattering, detection of a polymer molecule requires a *contrast* to be generated between the molecule of interest and its surroundings. The scattering lengths of hydrogen and deuterium are significantly different and thus deuterium labelling is sufficient to generate contrast or scattering length density difference. Table 2 in Chapter 1 reports $b$ values for atoms frequently found in polymers, Table 1 gives scattering lengths and scattering length densities of common polymers and solvents or subphases. For polymers the scattering length densities are calculated for the monomer unit of which the molecule is formed and

$$\rho = \frac{\sum b_i \, dens \, N_A}{m} \qquad (18)$$

TABLE 1.  Scattering Lengths and Scattering Length Densities at 298 K.

| Species | $b/10^{-4}$ Å | $\rho/10^{-6}$ Å$^{-2}$ |
| --- | --- | --- |
| methyl methacrylate-h | 1.49 | 1.07 |
| methyl methacrylate-d | 9.82 | 7.04 |
| ethylene oxide-h | 0.41 | 0.68 |
| ethylene oxide-d | 4.58 | 7.1 |
| styrene-h | 2.33 | 1.4 |
| styrene-d | 10.66 | 6.5 |
| $H_2O$ | −1.68 | −0.56 |
| $D_2O$ | 1.92 | 6.35 |

where *dens* is the physical density of the polymer (or solvent), *m* the monomer molecular weight and $N_A$ is Avogadro's number. For reflectivity investigations of polymers at the air-water interface, the particularly important scattering length densities are those for $H_2O$ and $D_2O$. The opposite signs mean that they can be mixed together to form an aqueous subphase with a zero scattering length density. There will be no specular reflection from this null reflecting water (NRW) mixture but there will be a background signal which arises from the incoherent scattering of $H_2O$.

## 6.4 APPLICATIONS OF NEUTRON REFLECTOMETRY TO POLYMERS AT AIR-WATER INTERFACES

The organisation of amphiphilic polymers at the air-water interface is of interest for several reasons. Academically one may be interested in creating a (hopefully) brush like layer of solvated polymer where the grafting density can be controlled. There may be intrinsic interest in determining the balance between enthalpic interactions (polymer-subphase) and entropic restrictions for polymers in a pseudo two dimensional state. Does the presence of the polymer 'structure' the water surface in some way and thus influence any extraction processes? Are surface micelles evident in the polymer layer which may affect subsequent use of the solution or dispersion? Does the polymer influence the dynamics of the liquid surface and if so, is this beneficial? From an industrial viewpoint, the situation of most interest is probably polymers at the liquid-liquid interface. Unfortunately, the use of neutron reflectometry to probe such interfaces is fraught with difficulty (mainly due to beam attenuation effects) and thus far most of the work reported is for polymers at the air-fluid interface. The majority of studies have used water as the fluid and the surface tension of pure water is so high that there is usually a strong interaction between polymer and surface so that even if part of the polymer is solvated, there is little tendency to form an extended layer.

An exhaustive discussion is precluded by space limitations, we cite examples which illustrate the technique, provide some new insight, and pose some questions. To this end we discuss polyethylene oxide and polymethacrylate polymers and their copolymers almost exclusively. These provide a sufficient range of hydrophobic-hydrophilic behaviour and the prospect of full use of deuterium labelling to illustrate all of the potential of neutron reflectometry. Finally we give a brief résumé on neutron reflectometry applied to other systems where the intention was to investigate brush formation and behaviour.

### 6.4.1  Polymethyl Methacrylate and Polylauryl Methacrylate

Syndiotactic polymethyl methacrylate (PMMA) spread at the air-water interface has been investigated by neutron reflectometry (Henderson, Richards *et al.*, 1991; Henderson, Richards *et al.*, 1993) over the surface concentration ($\Gamma_s$) range of 0.1 to 2.0 mg m$^{-2}$. A positive surface pressure is evident only when $\Gamma_s$ exceeds 1.0 mg m$^{-2}$ and from a scaling law analysis of the surface pressure data, the value of $\nu$ obtained was 0.52 indicating that the molecules were approaching the collapsed state. Neutron reflectometry data from the deutero polymer spread on null reflecting water and for the hydrogenous polymer on D$_2$O were analysed by the optical matrix method. The polymer film thickness was roughly constant at $20 \pm 2$ Å over the whole concentration range and there was a constant volume fraction of water ($\sim 0.1$) in the polymer film. At low values of $\Gamma_s$ ($< 1.0$ mg m$^{-2}$) the air content of the film was high (volume fraction of air stated at $\sim 0.8$) but fell to an approximately constant value of 0.2 for $\Gamma_s \geq 1.0$ mg m$^{-2}$. This finding supports earlier evidence that at low surface concentrations the PMMA exists as 'islands' on the surface. These data were also analysed using a simplified kinematic approximation (Henderson, Richards *et al.*, 1993; Henderson, Richards *et al.*, 1993) method and using a uniform layer model the layer thickness for the PMMA was identical with that obtained by the optical matrix method. Both optical matrix and kinematic approximation analyses gave a surface concentration which agreed with the amount dispensed except at the highest surface concentration investigated where consistently low values were obtained from the NR data. It is noteworthy that low values of $\Gamma_s$ were also calculated from the neutron reflectivity data for an atactic sample of PMMA. This indicates that some of the polymer is distant from where the majority is located and is diluted to such an extent that it does not contribute to the reflectivity. This 'absent' polymer is probably the loop and tail excursions into the air phase. Certainly one would anticipate a greater proportion of loops in the spread polymer at higher concentrations. A significant fact in this respect is that for the atactic polymer over a dispensed $\Gamma_s$ range of 2.0 to 4.0 mg m$^{-2}$, the value of $\Gamma_s$ calculated from the neutron reflectivity was constant at 1.8 mg m$^{-2}$.

Although the expectation for PMMA is that the polar carbonyl groups are in the main located immediately at the water surface, no direct evidence is obtainable from the reflectometry data. An attempt to provide evidence for this has been made using polylauryl methacylate (PLMA) (Reynolds, Richards *et al.*, 1995) which also demonstrates the power of the kinematic approximation in providing detailed information. Four deuterium labelled isomers of the polymer were synthesised as shown below in the schematic formulae of the repeat unit.

Each of these polymers was spread on both null reflecting water and D$_2$O and the reflectivity data obtained. We discuss here only the data obtained at the highest surface concentration used, 1.0 mg m$^{-2}$ corresponding to a surface pressure of 9 mN m$^{-1}$. The influence of the deuterium labelling on the reflectivity is shown in Figure 3, note that due to the intrinsically high reflectivity of D$_2$O, the influence of the different deuterium labelling is rather subtle when the polymers are spread on D$_2$O. However close inspection does reveal

$$\begin{array}{c} \text{CH}_3 \\ | \\ -\text{CH}_2-\text{C}- \\ | \\ \text{C}=\text{O} \\ | \\ \text{O-(CH}_2)_{11}\text{CH}_3 \end{array} \qquad \text{HMHL}$$

$$\begin{array}{c} \text{CD}_3 \\ | \\ -\text{CD}_2-\text{C}- \\ | \\ \text{C}=\text{O} \\ | \\ \text{O-(CD}_2)_{11}\text{CD}_3 \end{array} \qquad \text{DMDL}$$

$$\begin{array}{c} \text{CD}_3 \\ | \\ -\text{CD}_2-\text{C}- \\ | \\ \text{C}=\text{O} \\ | \\ \text{O-(CH}_2)_{11}\text{CH}_3 \end{array} \qquad \text{DMHL}$$

$$\begin{array}{c} \text{CH}_3 \\ | \\ -\text{CH}_2-\text{C}- \\ | \\ \text{C}=\text{O} \\ | \\ \text{O-(CD}_2)_{11}\text{CD}_3 \end{array} \qquad \text{HMDL}$$

that the trend in reflectivity is as expected, i.e. the reflectivity is reduced as the hydrogen content of the spread polymer increases. A propos of these reflectivity profiles on $D_2O$, it should be noted that the use of such data in the partial structure factor description of the reflectivity demands that they be corrected as suggested by Crowley (Crowley, Lee *et al.*, 1990). Figure 4a shows the self partial structure factors for the methacrylate backbone ($M$), the lauryl substituent ($L$) (Figure 4b) and the aqueous subphase (Figure 4c). The $Q$ range is much reduced compared to that over which data were collected (e.g. Figure 3) because for $Q$ values greater than circa 0.3 Å$^{-1}$ the reflectometry signal is essentially due to background.

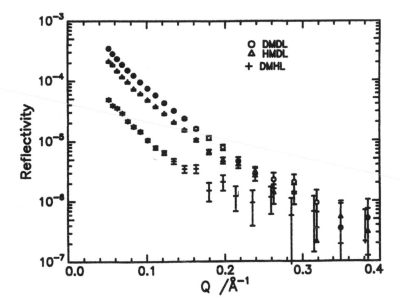

FIGURE 3. Reflectivity from partially deuterium labelled polylauryl methacrylate spread on null reflecting water. D=deutero; H=hydrogenous; M=methacrylate; L=lauryl

The solid lines through the data points are non-linear least squares fits of model partial structure factors to the data. In both cases the model used is that of a Gaussian distribution of segments normal to the water surface. The expression for the number density distribution is

$$n_i(z) = n_{1i} \exp(-4z^2/\sigma^2) \tag{19}$$

where $n_{1i}$ is the number density at the maximum of the Gaussian and $\sigma$ the full width of the distribution where the amplitude has fallen to $n_{1i}/e$. The one dimensional partial structure factor of this Gaussian distribution is

$$h_{ii}(Q) = n_{1i}^2 \frac{\sigma^2 \pi}{4} \exp\left(-\frac{Q^2 \sigma^2}{8}\right) \tag{20}$$

and both $n_{1i}$ and $\sigma$ are adjusted to obtain the fits shown.

The distribution of water molecules in this near surface layer has been modelled by a hyperbolic tangent profile

$$n_{wi}(z) = n_{wo}[0.5 + 0.5 \tanh(z/\xi)] \tag{21}$$

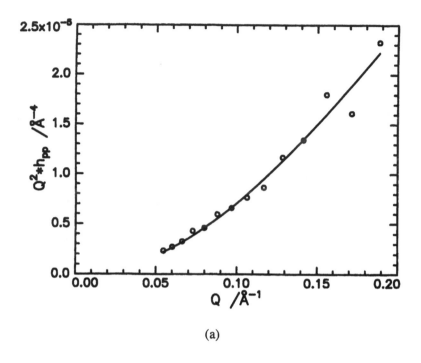

(a)

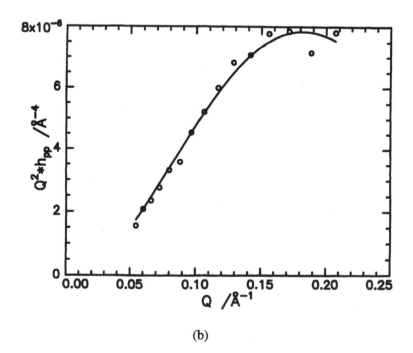

(b)

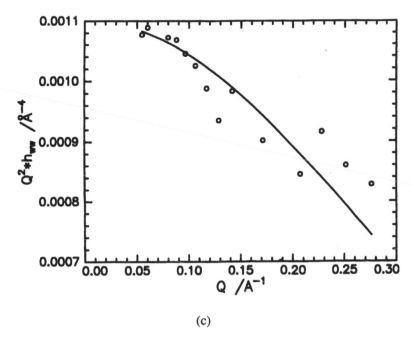

(c)

FIGURE 4. Self partial structure factors for (a) the methacrylate backbone, (b) lauryl substituent and (c) near surface water layer.

where $n_{wo}$ is the bulk number density of $D_2O$ and $\xi$ is the width of the diffuse interfacial region. From the best fits to these self partial structure factors, the composition and thickness of the $M$ and $L$ rich regions can be defined. However, it is only from the cross partial structure factors that we can say anything about the separations between these layers. Figure 5a–c shows the three experimentally obtained cross partial structure factors and the solid lines are least squares fits of the functions which theoretically describe them. Thus for the cross partial structure factor between backbone and lauryl substituent one has

$$h_{ML}(Q) = \pm[h_{MM}(Q)h_{LL}(Q)]^{1/2}\cos(Q\delta_{ML}) \qquad (22)$$

where $\delta_{ML}$ is the centre-to-centre separation of the two distributions and the $\pm$ sign reinforces the fact that we have lost all phase information, i.e. which layer is uppermost. For the cross partial structure factor between aqueous subphase and either backbone or substituent then

$$h_{M(L)S} = \pm[h_{MM(LL)}(Q)h_{SS}(Q)]^{1/2}\sin(Q\delta_{M(L)S}) \qquad (23)$$

The parameters obtained produced an arrangement of polymer with respect to the water surface which is described in the number density distribution profiles of Figure 6. One of the notable features is the almost complete immersion of the backbone in the aqueous

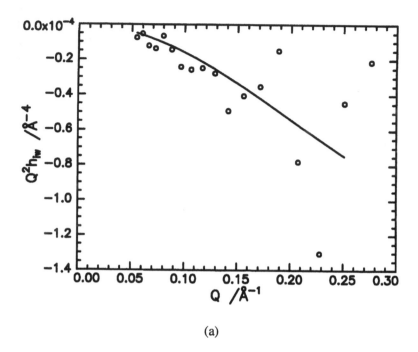

(a)

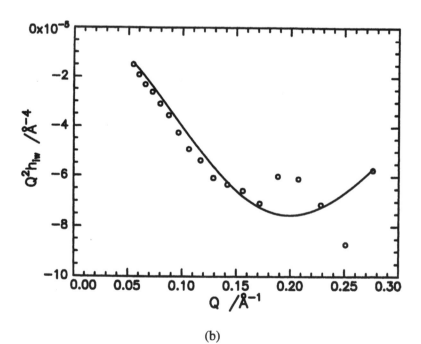

(b)

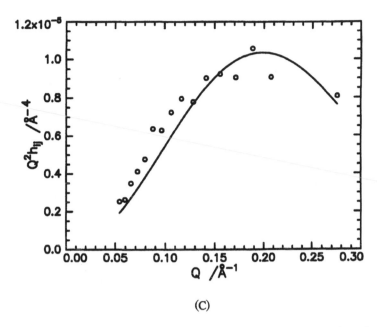

(C)

FIGURE 5. Cross partial structure factors and best fits (solid lines) for (a) methacrylate – water; (b) lauryl – water; and (c) lauryl – methacrylate backbone.

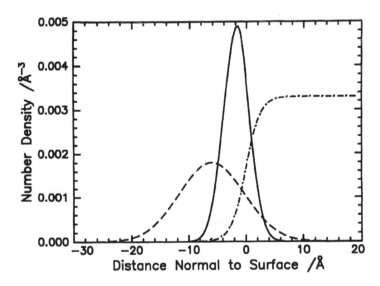

FIGURE 6. Number density distributions for the components of polylauryl methacrylate spread at the air – water interface. Solid line = methacrylate backbone; dashed line = lauryl substituent; dashed-dot line = water.

phase, this is undoubtedly due to the very hydrophobic nature of the lauryl substituents and their desire to reduce contact with water which forces the backbone into the aqueous phase. Secondly, although the majority of the lauryl unit is out of the aqueous phase, what is immersed actually extends to a greater depth than the backbone. This was rationalised by restrictions due to stereochemistry and the number of rotations about bonds in the lauryl side chain needed to ensure protrusion of the majority of the side chain into the air phase.

### 6.4.2  Polyethylene Oxide

Polyethylene oxide (PEO) is a water soluble polymer but one which also is able to be spread at the air-water interface to form a stable monolayer (Granick, 1985). Its solubility is attributed to the strong hydrogen bonding between the ether oxygen in the repeat unit of the polymer and a priori one would anticipate an extended configuration in solution to maximise such interactions. However, the polymer melt has a surface tension of circa 30 mN m$^{-1}$, far less than that of water, consequently in aqueous solution a surface excess layer of PEO of modest dimensions is formed. Lu *et al.* (Lu, Su *et al.*, 1996) used neutron reflectometry on solutions of deutero PEO in null reflecting water with polymer concentrations between $10^{-4}\%$ $w/w$ to 0.1% $w/w$. These reflectivity data were analysed by the optical matrix method and for two very different molecular weights the concentration profile of PEO normal to the surface was essentially identical. Rather more surprising was the result that adding magnesium sulphate to the solution produced no change in the concentration profile. Aqueous magnesium sulphate is a $\theta$-solvent for PEO and generally polymers form a denser, more compact layer when adsorbed to an interface from a solution under $\theta$-conditions. Including data for hydrogenous PEO in D$_2$O, the reflectivities for 0.1% $w/w$ solutions were analysed using the kinematic approximation. A Gaussian distribution fitted the self partial structure factor data well, the thickness of the layer being circa 17 Å which was also close to that of the near surface water layer thickness. Since the separation between these two layers was $1.5 \pm 1$ Å, it was concluded that near surface water layer and PEO layer were associated with each other. Both optical matrix analysis and the kinematic approximation gave the concentration of PEO in the surface excess region as 0.50 mg m$^{-2}$.

Preceding this work on the surface excess layer in PEO solutions, Henderson et al (Henderson, Richards *et al.*, 1993) had reported on spread films of PEO on water. From scaling law analysis of the surface pressure isotherm, the value of $\nu$ obtained was 0.75, i.e. a thermodynamically favourable situation. Again both optical matrix and kinematic approximation analyses were used, although the latter was the simplified approximation rather than making use of *all* the contrasts necessary. The surface concentration calculated from fits to neutron reflectometry data were in excellent agreement with the amounts spread up to a spread surface concentration of 0.4 mg m$^{-2}$. For greater surface concentrations the calculated values were consistently low and roughly constant at circa 0.45 mg m$^{-2}$. This change takes place in the region of surface concentration where it had been proposed that the PEO began to explore the aqueous subphase significantly, probably by looping of parts of the molecules. A two layer model could also fit the reflectivity data well with the second layer being thick ($\sim 45$ Å) but very dilute in PEO (volume fraction circa 0.02). This second layer is essentially a diffuse polymer layer and one can speculate that this is where the majority of the loops are contained. Notwithstanding the presence of a second diffuse layer, the calculated surface concentration was still considerably less than that actually spread, which leads one to conclude that either there is a third yet more diffuse and dilute layer (containing tails?) or that some of the polymer has dissolved into the bulk. Given that the surface pressure was constant over long times the dissolution of PEO into the bulk does not seem sustainable. In contrast to the influence of salt on the surface excess layer dimensions, Henderson et al noted a distinct effect on the characteristics of the concentration profile of the PEO spread at the air-water interface. As the concentration of salt in the aqueous phase increased, the surface layer of polymer became thinner but increased in its density.

### 6.4.3 Copolymers

Linear block copolymers which have amphiphilic characteristics can exhibit a range of behaviour depending on their molecular weight and relative proportions of hydrophobic and hydrophilic component in the copolymer. At one extreme (low molecular weight, high content of hydrophilic component), they can behave like a surfactant when dissolved in water and at high concentrations micelles are formed. At the other extreme (high molecular weight, low hydrophilic content) they may be totally insoluble and unable to be dispersed in water. Consequently what is set out here should not be interpreted as generic behaviour for all copolymers.

In principle copolymers with two chemically distinct components should be ideal vehicles to explore the full use of deuterium labelling and the use of the kinematic approximation to obtain highly detailed information regarding the arrangement of amphiphilic copolymers at air-water interfaces. In practice it is this need to have variously labelled specimens which may be a stumbling block because each of the copolymers so prepared should have *identical* compositions and molecular weights for the results to be truly meaningful in view of the dependence of properties on molecular weight and composition. If the molecular weight of the copolymer is sufficiently high then small variations in molecular weight from sample to sample may not be too important. A linear block copolymer of methyl methacrylate and ethylene oxide spread at the air water interface has been the most thoroughly examined copolymer thus far. Unfortunately the composition and molecular weight variation between the various deutero isomers was wider than is desirable because of the poor control that is able to be exercised over the polymerisation of ethylene oxide. Nonetheless some interesting insights were obtained. Perhaps the most significant was the change in organisation at two different surface concentrations. The copolymers were linear diblocks with circa 50 mol % ethylene oxide and a molar mass of $55000 \pm 10000$. Partial structure factor plots (Figure 7) clearly show marked differences between the two surface concentrations of 0.6 mg m$^{-2}$ and 1.2 mg m$^{-2}$; the lower concentration being where the surface pressure becomes just measurable, the higher concentration is just into the plateau of the surface pressure. Models used to interpret the partial structure factors showed that the ethylene oxide block was coincident with the near surface water and that at the lower concentration the methyl methacrylate block was intimately mixed with this layer. At the higher concentration there was some stretching of the ethylene oxide layer (the centre of which remained coincident with that of the surface water layer). This stretching was *not* significant, a change in $\sigma$ from 9 Å to 12 Å being noted. The major effect of increasing the block copolymer surface concentration was to increase the thickness of the methyl methacrylate layer

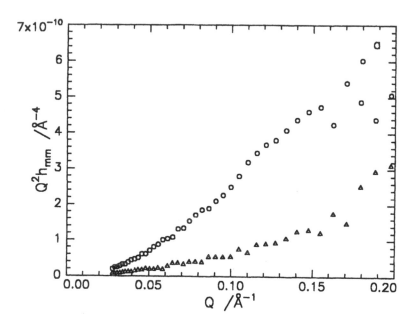

FIGURE 7. Partial structure factor for the methacrylate block of a polyethylene oxide – polymethyl methacrylate linear diblock copolymer spread at the air water interface at concentrations of 1.2 mg m$^{-2}$ ($\bigcirc$) and 0.6 mg m$^{-2}$ ($\triangle$).

(almost doubling it) but more significantly the separation between the two block layers was now considerably increased so that some of the methacrylate block was in the air phase. Although the segregated amount in the air phase was not large, there is clear evidence for a surface segregation but the dimensions of the two layers are such that one cannot claim that a brush like layer of the solvated ethylene oxide block is becoming evident at these concentrations. In this respect it is also noteworthy that an earlier optical matrix analysis of the higher surface concentration data had also resulted in an ethylene oxide layer whose dimensions were 15 Å, far short of anything expected for a brush like layer and indeed much smaller than the radius of gyration of the block length. A pancake configuration of the ethylene oxide block at the interface is thus indicated.

An interesting comparison with the behaviour of the linear block copolymer can be made with a graft copolymer of the same two chemical species. Peace et al examined graft copolymers with a polymethyl methacrylate backbone and polyethylene oxide grafts placed randomly along the backbone. The PEO grafts were all exactly 54 monomer units long. Because they were unable to obtain graft copolymers with deuterated PEO grafts, the full kinematic approximation could not be applied to the neutron reflectivity data. Optical matrix analyses was used together with a forceful application of Occam's razor to obtain a description of the surface layer. Figure 8a–c summarises the findings, as the surface concentration of graft copolymer increases more layers are needed to reproduce the observed neutron reflectivity data. At the lowest surface concentration, the grafts and backbone are mixed with the PEO grafts extending a little deeper in the subphase. The fact that the near surface water region requires a two layer model suggests that some PEO actually penetrates deeper, and this is confirmed by the observation that the surface concentration calculated from the reflectometry fits was less than that spread on the water surface. Increasing the surface concentration from 0.5 mg m$^{-2}$ to 1.5 mg m$^{-2}$ still requires a two layer model to fit the reflectivity data. The three components (water, PMMA and PEO) remain mixed at the surface with PMMA being the major constituent. The second layer is a thick ($\sim 30$ Å) layer mostly containing water but also some PEO and PMMA which is now the minor component. At 2.0 mg m$^{-2}$, three layers are needed to describe the surface structure. The top layer contains only PMMA and air, suggesting a looped arrangement of the PMMA backbone. Below this is a 20 Å thick layer containing water, PEO and PMMA as a minor constituent. Finally there is a third layer also some 20 Å thick which is dilute in PEO alone. The increased depth penetration of the PEO grafts as surface concentration increases is reminiscent of the behaviour predicted for grafted brush layers. Both original scaling law theory and the analytical theory of such layers predicts that the brush layer thickness (or height) has the following dependence on the grafting density, $\sigma$, (number of molecules attached per unit area of surface)

$$\text{thickness} \propto \sigma^{0.3} \tag{24a}$$

On the assumption that the total thickness of near surface water layer was associated with PEO, the water layer thicknesses obtained from kinematic approximation analysis were examined for agreement with equation 24. Figure 9 shows these data and the best linear fit to the data which gives

$$\text{thickness} \propto \sigma^{0.21} \tag{24b}$$

This is considerably different from the theoretical predictions but the size of the PEO grafts is much smaller than the high molecular weights which are assumed in the theories to observe limiting behaviour.

Bijsterbosch et al. (Bijsterbosch, de Haan et al., 1995) have used styrene-ethylene oxide block copolymers spread at the air water interface as systems to investigate brush layer formation. These hydrogenous copolymers were spread on the surface of D$_2$O and the neutron reflectivity data over a $Q$ range of 0.08 Å$^{-1}$ to 0.5 Å$^{-1}$ collected as a function of surface concentration. The parabolic profile expected for brush like layers was then fitted to the reflectivity using four adjustable parameters. In the one example shown the fit appeared to be reasonable and the variation of brush height with surface grafting density obtained for one of these copolymers is shown in Figure 10 together with the line predicted by theory. Given the molecular weight of the PEO block ($\sim 31000$) the thickness of the PEO layer is greatly extended over the dimensions anticipated for the end-to-end dimensions of the equivalent homopolyethylene oxide at the excluded volume limit. It should also be pointed out that styrene-ethylene oxide block copolymers of the compositions used by Bijsterbosch et al form micelles at extremely small concentrations $\sim 10^{-5}$ mg ml$^{-1}$. Admittedly the micellisation data refer to bulk dispersions, but the evidence from the MMA-EO blocks regarding water penetration suggests that it may be prudent to consider surface micelle formation in these systems. Evidently, what is needed is data for a block copolymer spread on null reflecting water where the EO block is deuterated.

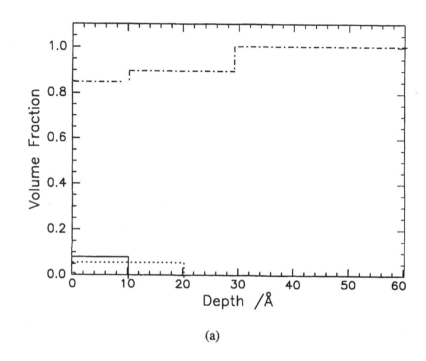

(a)

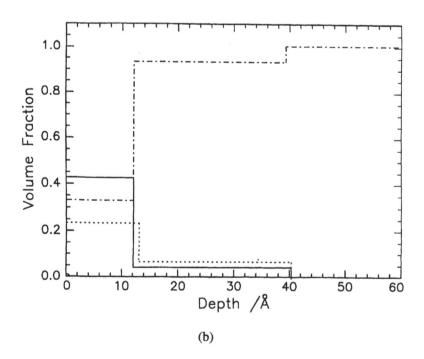

(b)

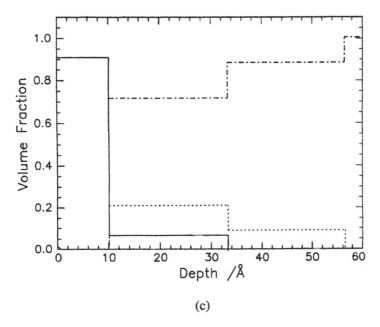

(c)

FIGURE 8. Volume fraction distribution for a graft copolymer of polymethyl methacrylate and polyethylene oxide spread at the air–water interface at different concentrations. Solid line = PMMA; dashed line = PEO, dashed-dot line = water.

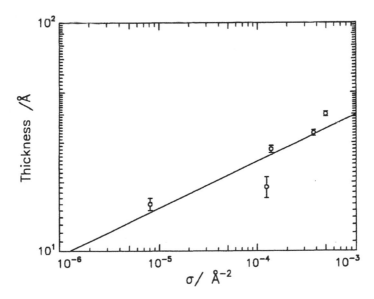

FIGURE 9. Thickness of near surface water layer as a function of 'grafting' density of the air–water interface by the polyethylene oxide grafts of a polymethyl methacrylate–polyethylene oxide graft copolymer spread at the interface.

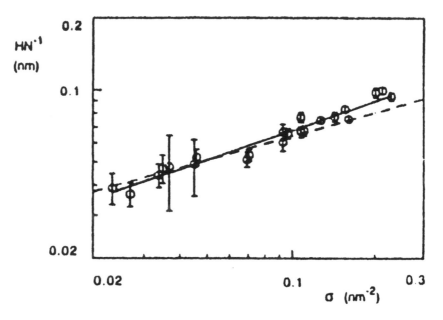

FIGURE 10. Brush height normalised by degree of polymerisation of polyethylene oxide block as a function of the surface grafting density.

Perhaps the only spread block copolymer system which unequivocally forms a brush like solvated layer and which has been examined by neutron reflectometry is a polydimethyl siloxane-polystyrene linear diblock copolymer spread on ethyl benzoate (Kent, Lee *et al.*, 1992). The polystyrene block becomes solvated in the subphase and both block molecular weight and grafting density were explored. A large number of functional forms describing the composition profile normal to the surface were explored (Kent, Lee *et al.*, 1995). The best fit profile was one which had a variable exponent in the dependence on the distance from the surface and included an exponential tail distant from the surface and depletion layer near the surface. Apart from the brush layer height there appear to be at least four other variables to fit in this model. The agreement between the best fitting profile and that predicted by a self consistent field theory was quite good but the dependence of brush height on grafting density was weaker than the theoretical prediction that it should scale as $M\sigma^{1/3}$.

## 6.5 SURFACE QUASI-ELASTIC LIGHT SCATTERING

### 6.5.1 Background

Quasi-elastic light scattering is a well established technique that has been applied extensively to the study of polymer dynamics, particularly to dilute solutions of polymers. In recent years (Langevin, 1992) an allied technique, surface quasi-elastic light scattering or SQELS, has been developed enabling the dynamic behaviour of polymers at interfaces to be probed in a novel way offering unique information. The basic principle of the technique is simple and relies on thermally generated capillary waves which continually roughen any fluid interface. Whilst the dimensions of these waves are microscopic, their root mean square amplitude being about 2.5 Å, their evolution across the surface depends explicitly on the physical properties of the fluid and, more importantly, its interface. The SQELS experiment exploits the fact that these waves scatter light, the spectrum containing information on the evolution of the waves and the properties of the interface.

This feature of capillary wave behaviour was predicted early this century by Smoluchowski and later verified by Mandelstamm. Despite this early work full development of the technique awaited the advent of the laser in the 1960s. The technique has been developed extensively since then, particularly in the last ten to fifteen years by Earnshaw (Earnshaw, McGivern *et al.*, 1988) and others (Hard and Neumann, 1981). Whilst the scattering of light by pure fluid surfaces has rightly received attention our particular interest lies in its application to monolayer covered surfaces.

The knowledge that spreading a film on a liquid surface damps surface waves predates even the early prediction of surface light scattering. The Romans, as recorded by Pliny, were exploiting these principles when oil was poured onto rough seas to calm the waves. Exactly the same behaviour is studied in SQELS where the presence of a monolayer at the fluid surface modifies capillary wave evolution, and thus the spectrum of the scattered light. It transpires that the spectrum of the scattered light contains information on the surface visco-elastic properties of the monolayer which are responsible for damping the surface waves.

Although the principles of SQELS are relatively straightforward in practise its application is both experimentally and analytically demanding and it is justifiable to consider its worth as opposed to other surface sensitive techniques. The abilities of neutron reflectivity have been set out above and SQELS cannot compete with the structural detail accessible from that technique. Other, perhaps more standard, surface techniques that spring to mind include ellipsometry, FTIR, and Brewster angle microscopy, BAM.

The application of ellipsometry to monolayers at liquid surfaces is not yet well established. The principle of the experiment is to measure the reflectivity of elliptically polarised light and relate this to film properties, in particular the composition and thickness. However, the underlying mathematics are those for $s$ and $p$ electromagnetic radiation, (see the section on neutron reflectometry), and consequently the film thickness and composition are coupled via the refractive index. Separation of these terms invariably requires some kind of assumption. In an attempt to avoid such assumptions most data have been interpreted in terms of a comparison to the bare interface and thus the conclusions are predominantly qualitative. Application of FTIR, in particular polarised FTIR, affords some information on the orientation of functional groups at the surface. Microscopic techniques, including fluorescence microscopy as well as BAM, differ from other approaches by being sensitive to in plane structure. However, both are sensitive over a micron length scale rather than molecular dimensions.

Although all the techniques discussed above are well established and offer valuable information, when compared to SQELS it is clear that none can rival the visco-elastic information offered by SQELS. By the very nature of the SQELS experiment we are studying the response of monolayers to perturbations, rather than static equilibrium properties. In fact in many applications of polymeric amphiphiles, i.e. as emulsion stabilisers, it is dynamic properties that will be vital to applications rather than static behaviour. A concept central to SQELS as opposed to similar techniques where waves are generated is that capillary waves occur naturally. This has two important consequences, firstly the accessible frequency range is much higher than for generated waves and, in addition, the entire experiment is non perturbative.

In summary it should be emphasised that while SQELS has been compared to structural characterisation techniques rather than rivalling such methods it is best viewed as complementary. Indeed one of the most promising approaches to establishing the molecular origins of surface visco-elasticity is extensive correlations with structural information. It is the great potential of SQELS data, allied to detailed structural understanding of polymeric amphiphiles that we wish to convey here.

### 6.5.2  Theory

Although up to five hydrodynamic modes can be supported by a monolayer covered surface only two, the transverse shear mode and compression mode, concern us here. As described above these surface wave modes are such that the fluid surface is agitated continually. Treatment of such a randomly rough fluid surface is simplified by Fourier decomposition of each perturbation into a set of modes. Each mode is characterised by an interfacial wave number, $q = 2\pi/\lambda$, see Figure 11, and gives rise to a spatial perturbation described by

$$\zeta = \zeta_o \exp(iqx + \omega t) \tag{25}$$

The temporal wave frequency, $\omega$, is a complex term that can be expressed as a spatial fluctuation with frequency $\omega_o$, decaying at a rate $\Gamma$ and $\omega = \omega_o + i\Gamma$.

Whilst equation 25 describes the spatial propagation of a capillary wave, provided $q$ and $\omega$ are known, a method of relating the two parameters is required. This is provided by the dispersion equation for surface waves, $D(\omega)$, which for a monolayer at the air-liquid interface is given by (Lucassen-Reynders and Lucassen, 1969);

$$D(\omega) = [\varepsilon q^2 + i\omega\eta(q+m)]\left[\gamma q^2 + i\omega\eta(q+m) - \frac{\omega^2 dens}{q}\right] - [i\omega\eta(q-m)]^2 = 0 \tag{26}$$

where $\eta$ and $dens$ are the liquid viscosity and density respectively and $m = \left(q^2 + \frac{i\omega dens}{\eta}\right)^{1/2}$.

The quantities $\gamma$ and $\varepsilon$ are the interfacial tension and elastic moduli respectively. These are expanded as linear visco-elastic response functions to include energy dissipation.

$$\gamma = \gamma_o + i\omega\gamma' \qquad \varepsilon = \varepsilon_o + i\omega\varepsilon' \tag{27}$$

$\Gamma_o$ and $\varepsilon_o$ are the interfacial tension and elastic modulus respectively. The primed quantities are the surface viscosities; $\gamma'$ is the transverse shear viscosity and governs response to shear transverse to the surface whilst $\varepsilon'$, the dilational viscosity, influences in plane dilation. The nature of these parameters is not clear; interpretations include viewing these as microscopic properties of the film or surface excess quantities. Perhaps the best way to consider these, especially in the context of recent results is in terms of their dissipative natures; $\varepsilon'$ representing dissipative forces transverse to the surface, $\gamma'$ dissipation in the surface plane.

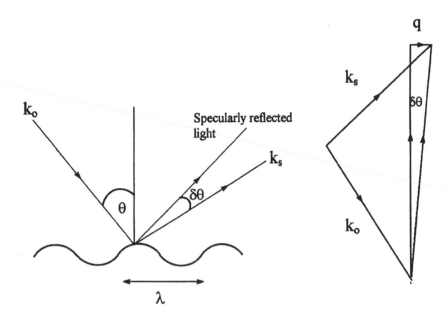

FIGURE 11.  Scattering of light by capillary waves and vector diagram identifying $q$.

Whilst by no means a simple expression, the dispersion equation enables us to predict the range of wave frequencies supported by a surface with particular properties. It has long been realised (Lucassen-Reynders and Lucassen, 1969) that for a surface supporting a monolayer, where $\varepsilon$ is non zero, there are two realistic roots for $D(\omega)$. One of these roots corresponds to the capillary waves, the other to the dilational waves. Capillary waves are predominantly transverse in nature. However, due to local variations in surface tension associated with an undulating surface, a tangential surface stress arises imparting some horizontal motion to the wave mode. Likewise the dilational waves, associated with in plane concentration fluctuations, have a vertical component. The implications of these features is that the surface waves are coupled, neither exhibiting pure transverse or longitudinal characteristics. It is because of this coupling that although experimentally we can assess only the capillary waves, they not only depend on $\gamma_o$ and $\gamma'$, but $\varepsilon_o$ and $\varepsilon'$ also have an influence on wave propagation albeit weak.

Whilst only the capillary waves scatter light the presence of the two wave modes can have a marked influence on behaviour of the frequency and damping of the capillary waves as a function of surface concentration. A particular example is the resonance which can occur between the two modes. Resonance occurs when the real frequencies of the two modes coincide: at modest $q$ values this is achieved when $\varepsilon_o/\gamma_o = 0.16$. Latterly this resonance condition has been associated with complex behaviour where the damping of the two modes converge, then modes exhibiting neither pure capillary or dilational characteristics are observed. This mode mixing has been extensively discussed by Earnshaw (Earnshaw and McLaughlin, 1991; Earnshaw and McLaughlin, 1993).

Having established the form of the dispersion equation we now consider the relationship between $D(\omega)$ and the scattered light. The power spectrum of the scattered light, $P(\omega)$, derived by Bouchiat and Meunier is given by;

$$P(\omega) = -\frac{k_B T}{\pi \omega} \text{Im} \left[ \frac{i\omega\eta(q+m) + \varepsilon q^2}{D(\omega)} \right]$$  (28)

$P(\omega)$ is comprised of a doublet, approximately Lorentzian in shape, disposed about the frequency of the incident light as in Brillouin scattering. An alternative to recording the power spectrum of the scattered light is provided by the correlation function, $g(\tau)$ which is the Fourier transform of $P(\omega)$

$$g(\tau) = FT[P(\omega/\gamma_o, \gamma', \varepsilon_0, \varepsilon')]$$  (29)

### 6.5.3  Data Analysis

As the previous discussion has demonstrated, once $P(\omega)$ or $g(\tau)$ have been measured interpretation of these data in terms of the surface visco-elastic parameters via $D(\omega)$ is not easy. In addition to quantifying the surface parameters it is also highly informative to determine the propagation characteristics of the surface waves. The latter problem can be addressed with relative ease and is considered first.

If the spectrum of the scattered light is recorded, $P(\omega)$, the shape of the pseudo – Lorentzian that results is defined by $\omega_o$ and $\Gamma$. The central frequency gives $\omega_o$, the line width $\Gamma$. These two quantities are more evident in the correlation function of the scattered light. Although both data collection methods are equivalent, emphasis will be placed on the latter method as the major developments in data analysis have been associated with correlation function detection methods.

It can be shown that a Lorentzian power spectrum corresponds to an exponentially damped cosine in the time domain, a feature clearly seen in Figure 12. The correlation function is described by a function of the form

$$g(\tau) \propto \cos(\omega_o \tau + \phi) \exp(-\Gamma \tau)$$  (30)

where $\phi$ is a phase term arising from the skewed nature of the power spectrum.

In practise additional terms are incorporated into the exact analytical form to account for instrumental effects and building vibrations. However, for the present purposes we need not concern ourselves unduly with the details of equation (30) but simply appreciate that by fitting it to experimental data the capillary wave frequency and damping can be obtained. Computer routines perform this fit relatively fast and the method is well established as returning non biased values of $\omega_o$ and $\Gamma$. Whilst knowledge of $\omega_o$ and $\Gamma$ are clearly essential to understanding behaviour there is no way of interpreting them in terms of the four surface parameters that we wish to determine. Formerly, the only way to extract any of the surface parameters involved some radical, unjustified assumptions. The approach generally adopted was to assume that $\gamma_o$ takes the same value as the static Wilhelmy surface tension and that $\gamma' = 0$: by this means values for $\varepsilon_o$ and $\varepsilon'$ were extracted. The only justification offered was that the values obtained were consistent. Such methods are far from ideal and with the development of a new approach to the analysis they are now redundant.

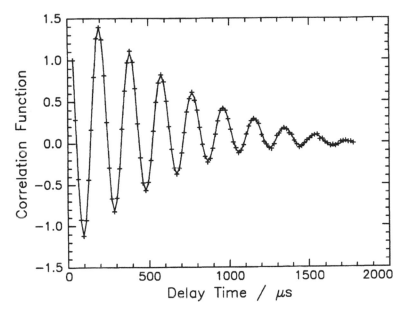

FIGURE 12. Heterodyne correlation function of light scattered from a liquid surface (+). The line is a fit of the cosine function as described in the text.

In, 1990 Earnshaw (Earnshaw, McGivern *et al.*, 1990), one of the main developers of the SQELS technique, published a method of data analysis that has been shown to yield the four surface visco-elastic parameters requiring no *a priori* assumptions. This spectral fitting method relies upon the realisation that the shape of the correlation function depends upon each of the surface parameters independently. In an extensive discussion of the approach he demonstrated successfully the independence of $\gamma_o$, $\gamma'$, $\varepsilon_o$ and $\varepsilon'$ and their influence on $g(\tau)$. The analysis requires no arbitrary assumptions because the entirety of the light scattering data are used, not just the parameters $\omega_o$ and $\Gamma$ extracted from the observed data.

Just as in obtaining values of $\omega_o$ and $\Gamma$ the method involves fitting of an equation to the data. The equation used is similar to equation 30; but the damped cosine function is replaced by the Fourier transform of the power spectrum.

$$g(\tau) \propto FT[P(\omega, q/\gamma_o, \gamma', \varepsilon_o, \varepsilon')] \qquad (31)$$

Analysing the data in this way requires estimates of $\gamma_o$, $\gamma'$, $\varepsilon_o$ and $\varepsilon'$ from which a power spectrum is generated, this is then Fourier transformed to give the correlation function. The generated correlation function is compared to the experimental function and a least squares fitting routine evoked to minimise the difference. The values of $\gamma_o$, $\gamma'$, $\varepsilon_o$ and $\varepsilon'$ required for the best fit are returned by the routine. This analysis is unique in yielding values for $\gamma_o$, $\gamma'$, $\varepsilon_o$ and $\varepsilon'$ and is now generally accepted. It should be noted that the principles of the analysis are not specific to data recorded in the time domain. Application of this approach to data recorded in the frequency domain, $P(\omega)$, should be straightforward, although as yet there are no instances of this in the literature.

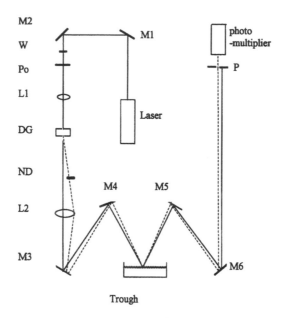

FIGURE 13. Schematic sketch of surface quasi-elastic light scattering apparatus. Mn = mirrors, Ln = lenses, W = quarter wave plate, Po = polariser, DG = grating, ND = neutral density filter, P = pinhole.

### 6.5.4 Experimental Details

The bare essentials of a SQELS experiment are a light source, a detector and a fluid surface on which the light is incident. Whilst conventional light sources suffice to demonstrate the phenomenon of scattering by capillary waves measurement of the spectral properties of the scattered light requires the coherence properties of the laser. Experimentally the incident light is guided to the fluid surface along an optical train using mirrors. The scattered and reflected light are collected in a similar manner and directed to the detector, a photomultiplier tube PMT, (see Figure 13). If data are collected in the time domain a correlator is used to collect the data, for experiments in the frequency domain a spectrum analyser is used. The final form of the raw data, either as $P(\omega)$ of $g(\tau)$, is stored on a PC. Most of the light incident upon the fluid surface is specularly reflected and the actual scattering is weak. In addition the frequency shift caused by the scattering is small, only circa 10–100 kHz requiring a more sensitive detection method than standard light scattering experiments. The technique employed to detect these frequency shifts relative to the frequency of the incident light is heterodyne detection. In heterodyne detection scattered light is mixed with a reference beam of light at the detector. The reference beam frequency is the same as the incident light so that the scattered component beats against it. (A reference beam is generated by placing a transmission grafting before the liquid surface.) In this manner a series of diffracted beams are generated which, by a combination of lenses, are focused into a single spot at the surface. Subsequently, the beams diverge so that a series of separate beams are available at the detector. Each beam corresponds to a different scattering angle, $\theta$, and is coincident with light scattered at that same angle and thus a range of scattering vector, $q$, can be explored. A pinhole is placed in front of the PMT and by adjustment of mirrors M3 and M4 the diffracted beams can be focused onto the detector in turn. This provides a means of probing behaviour over a range of $q$, and therefore wave frequencies. The $q$ values corresponding to the diffracted beams may be calculated using geometric principles and measuring $\theta$. Alternatively the apparatus may be calibrated by measuring the capillary wave frequency and damping for a well characterised liquid. The dispersion equation is then solved for $q$ using the appropriate physical properties. The limitations of a SQELS experiment are almost universally defined by the available laser power and vibration isolation. Laser power restricts the frequency range available, high laser powers are required to access high frequencies. This is a manifestation of the inverse squared dependence of the scattered intensity on $q$. Recent studies by Earnshaw (Earnshaw and McLaughlin, 1991; Earnshaw and McLaughlin, 1993) demonstrate that frequency dependent studies, extending to high frequencies, are necessary to identify complex behaviour of the surface modes. In such instances studying a limited frequency range would lead to important features of behaviour being overlooked. The problems of vibration - isolation require that experiments are mounted on isolation units and positioned carefully within the building.

Having achieved the optimum SQELS set up there are two types of experiment that may be performed. In monolayer experiments our variables are $q$ and $\Gamma_s$. By fixing $q$ and varying $\Gamma_s$, the composition dependence of the surface visco-elastic parameters may be studied by tracking along the isotherm. Alternatively $\Gamma_s$ may be fixed and $q$ varied, providing a frequency dependent study. The most successful way to understand a system is to combine both of these two approaches.

## 6.6 APPLICATIONS OF SQELS TO POLYMERS AT AIR-WATER INTERFACES

Attention is now turned to specific examples of polymeric amphiphiles that have been studied. The relatively small field of SQELS narrows further when attention is focused on polymeric systems. The workers most active in this area are the groups of Richards and Yu. Unfortunately these two groups have used different data analyses, a problem alluded to previously. Yu has interpreted data in terms of $\varepsilon_o$ and $\varepsilon'$ only assuming $\gamma' = 0$ and $\gamma_o$ takes the static value. In contrast Richards has adopted Earnshaw's spectral fitting approach. These basic differences make comparison of the results of these groups difficult, only $\omega_o$ and $\Gamma$ being truly comparable. The influence of Yu's adopted values of $\gamma_o$ and $\gamma'$ on the dilational surface parameters cannot be inferred due to the complexity of $D(\omega)$.

The following polymer systems are considered here, with comparisons where possible: polymethyl methacrylate (PMMA), polyethylene oxide (PEO) spread and adsorbed films and copolymers of these two.

### 6.6.1  Poly Methymethacrylate

Two SQELS studies of PMMA have been published, initially Yu (Kawaguchi, Sauer *et al.*, 1989) reported data collected at a fixed $q$ value at points along the surface pressure isotherm. More recently Richards *et al.* (Richards and Taylor, 1996) reconsidered the system, applying a different analysis to extract the visco-elastic parameters and extending to a frequency dependent study.

The tendency of PMMA to exist as discrete islands on the water surface was acknowledged before Yu's study, and was subsequently confirmed by both neutron reflectometry and ellipsometry. Yu commented on this in the discussion of wave frequencies and damping reported for surface concentrations below 0.5 mg m$^{-2}$, which were essentially the same as for water. In contrast Richards $et\,al.$ (Richards and Taylor, 1996) reported that at such surface concentrations the wave frequency and damping exhibit oscillations, this was demonstrated most clearly by monitoring $\omega_o$ and $\Gamma$ as a function of time. This behaviour was attributed to islands of PMMA moving in and out of the laser footprint on the water surface. At surface concentrations above 0.5 mg m$^{-2}$ both groups reported a decrease in the wave frequencies consistent with the surface pressure isotherm; to a first order approximation $\omega_o \propto \gamma^2$, the capillary damping being essentially constant over the concentration range studied.

Monolayers of PMMA have long been identified as adopting a condensed type structure. Irrespective of the ambiguities associated with the data analysis, Yu's conclusion that $\varepsilon_o$ is too high to determine both $\varepsilon_o$ and $\varepsilon'$ accurately is entirely consistent with the monolayer classification. Condensed type films are associated with large dilational moduli, reducing the horizontal motion associated with the capillary waves with the consequence that the sensitivity to the dilational parameters is poor. As a consequence of the inaccessibility of $\varepsilon_o$ and $\varepsilon'$ and the assumption that $\gamma' = 0$, Yu's study failed to quantify any of the surface visco-elastic parameters, leaving just the variation in $\omega_o$ and $\Gamma$ as a function of surface concentration.

Like Yu, Richards $et\,al.$ (Richards and Taylor, 1996) failed to extract $\varepsilon_o$ and $\varepsilon'$, however values of $\gamma_o$ and $\gamma'$ were reported both as a function of $\Gamma_s$ and $\omega_o$. The values of $\gamma_o$ at $q = 220$ cm$^{-1}$ followed the static isotherm closely. Like the capillary wave damping $\gamma'$ oscillated at low concentrations, reaching a plateau at a value of around $1 \times 10^{-5}$ mN sm$^{-1}$ for a uniform film. Values of $\gamma_o$ and $\gamma'$ as a function of frequency for a fixed film concentration of 2 mg m$^{-2}$ were also reported, and a negligible frequency dependence was demonstrated. Phenomenological visco-elastic models, comprised of springs and dashpots (Ferry, 1980), is one approach that has been adopted to interpret frequency dependent behaviour. The capillary waves are considered to impart an oscillatory stress and strain to the monolayers. These parameters are linked via a complex dynamic modulus, $G'(\omega) + i\,G''(\omega)$, where $G'(\omega)$ is the storage modulus and $G''(\omega)$ is the loss modulus. In the current application storage modulus corresponds to the surface tension, the loss modulus is equal to the product $\omega_o \gamma'$. Two simple models that describe the frequency dependence of $G'(\omega)$ and $G''(\omega)$ are the Voigt and Maxwell models. The former model does not predict a frequency dependence for either $\gamma_o$ or $\gamma'$, somewhat unrealistic since eventually the loss term, $\omega_o \gamma'$, would become infinitely large. Ignoring the limitations of the model it was observed that for PMMA the variation in $\gamma_o$ and $\gamma'$, both as a function of $q$ and $\Gamma_s$, was consistent with a Voigt response. Using the relations $G'(\omega) = G_i$ and $G''(\omega) = G_i \omega_o \tau$, where $G_i$ is the modulus contribution of element $i$ and $\tau$ is the relaxation time, and substituting for the appropriate surface visco-elastic parameter enabled $G_i$ and $\tau$ to be determined as a function of $\Gamma_s$. Relaxation times were reported and were found to be constant at circa $0.5 \times 10^{-6}$ s, up to $\Gamma_s$ of around 1 mg m$^{-2}$. At higher surface concentrations the values fell to $0.2 \times 10^{-6}$. No rationale was provided for these values.

In summary phenomenological models describe the frequency dependent behaviour exhibited by PMMA. Richards et al's study also provided further evidence that $\gamma'$ is non zero for polymers and its inclusion was absolutely necessary to fit the correlation functions. This result means that Yu's approach inevitably over simplifies the problems leading to discrepancies in the surface visco-elastic parameters reported.

### 6.6.2 Polyethylene Oxide

The ability of aqueous solutions of PEO to form a surface excess has been known for some time. In 1970 it was realised that PEO also spreads as a film and since then the two systems have received attention as workers have sought to investigate any differences between them. These studies have included both structural elucidation using NR and, more recently, investigation of their dynamic properties using SQELS.

Like PMMA spread films of PEO have been studied by both Yu (Sauer, Yu et al., 1987) and Richards (Richards and Taylor, 1996). Yu reported data at a single $q$ value whereas Richards et al., whilst repeating the earlier work, extended their study to encompass the frequency dependence. The broad features in the variation in $\omega_o$ and $\Gamma$ with surface concentrations, at low $q$ values, reported by the two groups are similar. The wave frequency decreases with increasing surface concentration, the wave damping increases from the value for water passes through a maximum, then falls back to the original values once more. On a more detailed comparison of the data certain differences are clear, particularly in the behaviour of $\omega_o$. Yu reports a sharp decrease from the value for water then a plateau interrupted only by a peak at around 0.5 mg m$^{-1}$. Whilst the data reported by Richards et al. also exhibit a decrease, then plateau out, there is no subsequent peak.

In comparing the surface visco-elastic parameters reported by the two groups the problem of the different data analyses is encountered. Since the spectral fitting and the existence of $\gamma'$ are now well established the discussion will be mainly confined to the results reported by Richards et al. Surface parameters were determined at a fixed $q$ value as the polymer surface concentration was varied. Over the whole region of $\Gamma_s$ studied, the SQELS surface tension was larger than the zero frequency value. This implies that a frequency dependent relaxation occurs to give the lower measured value at $\omega_o = 0$. The variation in $\gamma'$ reflects the behaviour of $\Gamma$, although it is notable that the values were up to an order of magnitude larger than those reported for PMMA. The SQELS dilational modulus qualitatively followed the Gibbs elasticity, as determined from the surface pressure isotherm, although there was deviation at the highest surface concentration studied, 1.2 mg m$^{-2}$. The dilational viscosity was reported to exhibit some quite remarkable behaviour, and as a consequence is not as simple to interpret as $\gamma'$. As $\Gamma_s$ increased to 0.6 mg m$^{-2}$ $\varepsilon'$ increased, then at 0.6 mg m$^{-2}$ exhibited a sudden fall to negative values which, although subsequently approaching a value of zero, remained negative. In contrast whilst Yu's values of $\varepsilon_o$ agree with the static values the values of $\varepsilon'$ do not show any of the features reported by Richards et al. The discrepancy is undoubtedly due to the assumptions made by Yu in his analysis.

The implications of negative dilational viscosity are not understood fully. Whilst there had been reports of such behaviour for low molecular weight surfactants, this work is notable as being the first observation of such behaviour in polymers. One interpretation of negative dilational viscosities is that for this system the surface wave evolution, particularly the dilational waves, is complex. At the simplest level a negative value of $\varepsilon'$ implies that rather than being damped the dilational waves are sustained at the surface by some unidentified source of energy. In the case of low molecular weight surfactant solutions candidates for this additional energy source include complex adsorption and desorption phenomena. Translation of this to spread polymer films is rather difficult, however it is noted that the onset of negative $\varepsilon'$ for spread PEO films is coincident with the point at which subphase penetration is observed from NR studies. It seems likely that the origin of negative values of $\varepsilon'$ lies in some kind of subphase interaction, although a satisfactory molecular explanation is not forthcoming. Whilst the data indicate that additional forces are influencing surface wave evolution, $\varepsilon'$ is perhaps best viewed as an effective parameter indicating that modification of current theories is required.

The frequency dependence of the surface parameters may aid understanding of the processes occurring on the surface. The frequency dependence of a 0.6 mg m$^{-2}$ film of PEO was studied and all the surface parameters were found to exhibit some variation. The surface tension exhibited a weak increase with $\omega_o$. The variation in $\gamma_o$ with $\omega_o$ for a higher concentration film, $\Gamma_s = 1.0$ mg m$^{-2}$, was fitted using a Maxwell model for the relaxation. By fitting the data at fixed $q$ the relaxation time, $\tau$, was determined to fall in the range $0.1-1 \times 10^{-4}$ s. These values are at least an order of magnitude longer than those exhibited by PMMA films, suggesting different relaxation mechanisms. As expected $\gamma'$ falls with increasing frequency, otherwise the loss term will become prohibitively large. The dilational modulus also falls as the capillary wave frequency increases from 7 mN m$^{-1}$ at $0.3 \times 10^5$ s$^{-1}$ to circa 1 mN m$^{-1}$ at $2.5 \times 10^5$ s$^{-1}$. The dilational viscosity remained negative over the entire frequency range studied but less so at the highest frequencies.

The phenomenon of mode mixing has already been mentioned and given the observation of negative $\varepsilon'$ deserves more consideration here. For mode mixing to take place, not only must the real frequencies of the dilational and capillary waves coincide but, in addition, the damping constants must converge. Generally dilational waves are more heavily damped than capillary waves. However, the effect of $\varepsilon' < 0$ is to reduce the dilational wave damping and in certain cases this has been shown to be sufficient to cause mode mixing. Identification of mode mixing is not easy and it is most informative to consider the measured wave frequency and damping normalised by the first order approximations, $\left(\frac{\gamma q^3}{\rho}\right)^{1/2}$ and $\left(\frac{2\eta q^2}{\rho}\right)$, respectively. In an attempt to investigate whether $\varepsilon'$ is sufficiently negative to cause mode mixing so called complex plane plots, i.e. normalised frequency against normalised damping, were presented. Although the variation in the normalised damping with $q$ showed a marked downturn symptomatic of negative $\varepsilon'$, i.e. the result is not an artefact of the fitting process, the identification of mode mixing from the complex plane plots is rather ambiguous. In fact now that appreciation of mode mixing and its origins have developed further it seems unlikely that at this concentration the system exhibits mode mixing.

It is clear from the discussion above that spread films of PEO exhibit behaviour unique from other polymers more akin to that of low molecular weight surfactants. This feature of behaviour, especially the onset of negative $\varepsilon'$ at 0.6 mg m$^{-2}$ is believed to be due to co-operative subphase penetration by the PEO (see neutron reflectometry data discussed above). Given the frequency dependence exhibited by $\varepsilon'$ it is expected that the origin of the negative values is dynamic in nature.

Having established the behaviour of spread films of PEO attention was turned to PEO solutions. Sauer and Yu had compared spread and adsorbed films of PEO in 1989 and concluded that behaviour was identical. These conclusions were based purely on light scattering data analysed in terms of the capillary wave frequency and damping only. Whilst there was undoubtedly good agreement in Sauer and Yu's data there was concern that subtle differences in behaviour may only be evident in the surface visco-elastic parameters. In addition the solution concentrations studied were low, at $10^{-4} - 10^{-3}\%$ $w/w$ much more dilute than the solution studied by NR.

The question over the surface visco-elastic parameter and the differences in the concentration between NR and SQELS experiments has been addressed recently by Richards and Taylor. Three different molecular weight PEO solutions, 0.1% $w/w$, were studied over a range of frequencies. A wide molecular weight range was studied, 15,000–828,000 g mol$^{-1}$, but no discernible molecular weight dependence was exhibited in the capillary wave frequency. Likewise the capillary wave damping was found to be essentially the same, once the molecular weight dependence of solution viscosities was accounted for.

This molecular weight independence was also evident in the surface parameters. The SQELS surface tension values were essentially coincident at all frequencies studied. The absence of any increase in $\gamma_o$ with $\gamma_o$ is associated with the absence of relaxation processes in the surface layer, this is in contrast to spread films of PEO. The dilational modulus exhibited an increase from low values, less than 1 mN m$^{-1}$, at low frequencies to approximately 3 mN m$^{-1}$ at a wave frequency of $0.2 \times 10^5$ s$^{-1}$. Thereafter the modulus decreased implying at least one relaxation process. The dilational viscosities were negative over the entire frequency range despite exhibiting a systematic increase with $\omega_o$. The data are presented in Figure 14 and, as will be discussed later, exhibit a frequency variation that has become synonymous with negative dilational viscosities. In an attempt to interpret the variation in $\varepsilon_o$ and $\varepsilon'$ with frequency a diffusion model developed by Luccassen was adopted. The model considers adsorption/desorption processes at the surface acting to even out the in plane surface concentration variation associated with the passage of a surface wave. The model includes a layer just below the surface such that molecules absorb and desorb to the surface from there. In this model the relative timescales of the measurement of the dilational parameters and changes in concentration lead to the frequency dependence of $\varepsilon_o$ and $\varepsilon'$. The experimental data were fitted well by this approach, negative $\varepsilon'$ can be 'explained' if $dc/d\Gamma_s$, the inverse slope of the adsorption isotherm, is negative. However, although this treatment was successful in this instance a diffusion model cannot be applied to insoluble polymers, (see below) where similar variations of $\varepsilon'$ and $\varepsilon_o$ are observed.

Whilst the authors do not make any comparisons between their data for spread and adsorbed PEO there seems to be little difference; certainly the behaviour of $\gamma_o$ and $\gamma'$ for the two systems is essentially the same. The only differences in the behaviour appear to be in the dilational parameters, although these are subtle. The frequency dependence of both $\varepsilon_o$ and $\varepsilon'$ whilst demonstrating similar features to spread PEO at higher frequencies, diverge at low frequencies. This observation is consistent, at least qualitatively, with the fact that it is the dilational parameters that exhibit unusual behaviour. One might anticipate the greatest differences in the surface visco-elasticities of spread and adsorbed PEO would be observed for spread film concentrations below 0.2 mg m$^{-2}$. For such dilute films subphase penetration

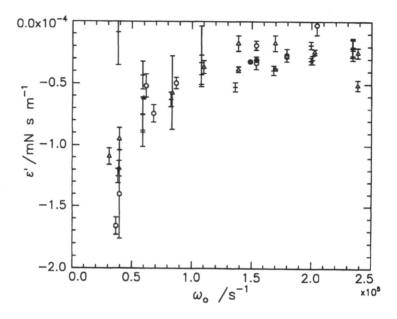

FIGURE 14. Dependence of dilational viscosity on capillary wave frequency for polyethylene oxides of different molecular weights dissolved in water at a concentration of 0.1 wt %.

is at a minimum and correspondingly, certainly at low $q$, $\varepsilon'$ is positive. Comparison of the frequency dependence of the surface visco-elastic parameters for fixed film concentrations of 0.3 and 1 mg m$^{-2}$, as well as an adsorbed film would be anticipated to demonstrate the largest divergence in properties if there is any. Such data are not available in the current literature. To summarise, it seems that any differences in the surface visco-elastic behaviour or adsorbed and spread films of PEO are confined to the dilational parameters. However, it should be acknowledged that other more subtle differences in the visco-elastic parameters of spread and adsorbed films may be masked by the current sensitivity of the experiment.

### 6.6.3 More Complex Cases

Having established the characteristics of the behaviour of PMMA and PEO films more complex systems, closer to real applications, have been considered. Two systems fall into this general subject area: a PEO-PMMA diblock copolymer and a graft copolymer of PMMA and PEO.

Richards et al. (Richards, Rochford et al., 1996) performed an extensive study of a PMMA-PEO diblock copolymer where both NR and SQELS were applied. The diblock had an equal mole ratio of MMA to EO, was insoluble in water but when spread formed a stable film. The full range of SQELS experiments were performed and data were collected at both constant $q$ for varying $\Gamma_s$ and constant $\Gamma_s$ for varying $q$.

Whilst the wave frequency decreased steadily with increasing $\Gamma_s$, reflecting to a large extent the static surface tension, there is some evidence of a maximum at low surface concentrations, circa $0.4\,\text{mg m}^{-2}$. This behaviour is confirmed by the SQELS surface tension where the values are above that for pure water; this reflects the dependence of $\omega_o$ on $\gamma$, since $\omega_o \propto \gamma^2$. Such behaviour is not as physically unrealistic as it may appear at first glance. One possible explanation lies in the frequency dependence of $\gamma_o$, alternatively the behaviour may be due to resonance between the capillary and dilational waves. After consideration of the variation in $\Gamma$ over the low surface concentration range the latter explanation seems most likely. At the same surface concentration associated with the maximum in $\omega_o$ the wave damping exhibits a much more marked, although broad, maximum. These data are consistent with some kind of resonance phenomenon occurring in the presence of a very dilute film.

The transverse shear viscosity, like $\Gamma$, exhibits a broad maximum spanning $0.4$–$2\,\text{mg m}^{-2}$ reaching values of $2 \times 10^{-5}\,\text{mN s m}^{-1}$. Outside this range the values remain essentially constant at $1 \times 10^{-5}\,\text{mN s m}^{-1}$. These values are of the same magnitude as those for PMMA and are much smaller than those observed for PEO. This similarity with the behaviour of PMMA is also evident in the dilational terms as the dilational modulus takes high values, $10$–$70\,\text{mN m}^{-1}$. The implication of such values of $\varepsilon_o$ is that the sensitivity of the capillary waves to both $\varepsilon_o$ and $\varepsilon'$ is much reduced. The scatter on the reported values of $\varepsilon_o$ and $\varepsilon'$ at both constant $q$ and $\Gamma_s$ reflects this point.

Given the considerations above, as for PMMA, we concentrate on the reported values of $\gamma_o$ and $\gamma'$ and their variation with frequency at $\Gamma_s = 3\,\text{mg m}^{-2}$. The SQELS surface tension, $\gamma_o$, increases with wave frequency behaviour symptomatic of visco-elastic relaxation processes. The transverse shear viscosity retains a constant value up to wave frequencies of $1 \times 10^5\,\text{s}^{-1}$ then falls to just less than half its initial value. This behaviour was interpreted in terms of a visco-elastic relaxation using a Maxwell model. Fitting data for fixed $q$, variable $\Gamma_s$, gave relaxation times in the range $1$–$10 \times 10^{-4}\,\text{s}$, two orders of magnitude slower than for PMMA but very similar to spread films of PEO which also exhibit a Maxwell relaxation. This observation was noted by the authors who attributed it to the dominant influence of PEO on the surface visco-elastic parameters. However, given the magnitude of both $\gamma'$ and $\varepsilon_o$ it is clear that the PMMA component of the diblock does have a significant influence on the surface dynamic behaviour. Recent observations of the behaviour of both PEO and other PEO containing copolymers implies that the dilational parameters measured by SQELS may be influenced most by the PEO component. However, unfortunately due to the high values of $\varepsilon_o$ in this system their determination was not possible. Despite this limitation the dynamic properties of the diblock quite clearly demonstrate features associated with the individual homopolymers. A composition dependent study of this system, as yet to be performed, would offer further correlations between molecular architecture and surface visco-elasticity with the ability to enable us to tailor dynamic properties.

The final copolymeric system considered here is a graft copolymer consisting of a PMMA backbone with pendant PEO side chains 54 units long. The graft copolymer was prepared by free radical polymerisation of MMA and a methoxy PEO methacrylate and by varying the ratios of these monomers three different composition copolymers were prepared. All three graft copolymers were studied as spread films using SQELS (Peace, 1996) at constant $q$ i.e, essentially tracking along the surface pressure isotherm.

The variation in the capillary wave propagation with $\Gamma_s$ exhibited the same features regardless of composition; $\omega_o$ increased from the value for clean water, reached a maximum at a surface concentration denoted $\Gamma_s^*$, then fell sharply thereafter maintaining a constant value. In contrast the capillary wave damping remained at the same value for water unit $\Gamma_s = \Gamma_s^*$ whereupon the values increased to approximately 2500 s$^{-1}$, almost three times the initial value. Thereafter the values decreased slightly as $\Gamma_s$ increased, this decrease in $\Gamma$ being directly proportional to the PEO content. The surface concentration associated with the transition in both $\omega_o$ and $\Gamma$, $\Gamma_s^*$, was found to be highly composition dependent. The higher the PEO content of the copolymer the lower $\Gamma_s^*$ and in all systems the PEO surface concentration at $\Gamma_s^*$ was found to be $0.3 \pm 0.1$ mg m$^{-2}$.

The maximum exhibited in both $\omega_o$ and $\Gamma$ would initially seem to indicate resonance between the dilational and capillary waves. However, the value of $\varepsilon_o/\gamma_o$ was far below the resonance condition in all the systems. By substituting the fitted surface parameters back into the dispersion equation the capillary and dilational wave frequencies can be obtained. This calculation, whilst reproducing the measured capillary wave frequencies, showed that rather than converging the frequencies of the two wave modes diverge at $\Gamma_s^*$ and exhibit the largest difference. This behaviour can be explained by considering a classical coupled oscillator (Pippard, 1985; Pippard, 1989). If two oscillators are coupled, as the capillary and dilational waves are, the oscillations or modes of the coupled system will depend on the frequency of the free oscillations and the coupling strength. Using this simplistic model it can be demonstrated that when the strength of the coupling is highest the frequency of the low frequency mode exhibits a minimum whereas the high frequency mode exhibits a maximum. From this analogy it is concluded that at $\Gamma_s^*$ the coupling between the capillary and dilational waves is at a maximum implying that $\Gamma_s^*$ represents some kind of transition in the system. From a combination of NR data and considerations of brush type behaviour, it is believed that $\Gamma_s^*$ corresponds to the surface concentration at which the PEO side chains penetrate the subphase co-operatively.

The transverse surface visco-elastic parameters, $\gamma_o$ and $\gamma'$, demonstrate the same qualitative dependence of $\gamma_s$ as $\omega_o$ and $\Gamma$ respectively. The importance of side chain-subphase interactions in this system has already been alluded to and it can be no coincidence that once again $\varepsilon'$ is negative. In fact the dilational parameters exhibit remarkable behaviour, $\varepsilon_o$ is much lower than the static values and $\varepsilon'$ is negative over the entire surface concentration range. The variation of $\varepsilon_o$ and $\varepsilon'$ with $\Gamma_s$ is unique (Figure 15a and 15b), $\varepsilon_o$ falls to a minimum at $\Gamma_s^*$, $\varepsilon'$ exhibits a divergence in its dependence on the surface concentration. On first sight these data may seem extraordinary but may be explained by considering the physics of forced oscillators and transfer functions. In simple terms a transfer function is the proportionality term relating the output of a driven system to the input. If a sinusoidal force is considered to act on this system and the 'velocity' or time derivative of some parameter of the dilational waves is the output the transfer function, $\xi(\omega)$, is complex. Splitting $\xi(\omega)$ into imaginary and real parts gives functions which reproduce the variation of $\varepsilon'$ and $\varepsilon_o$ with $\Gamma_s$, (see Figure 15). Although $\xi(\omega)$ is given as a function of $\omega$ and our experimental variable $\Gamma_s$ can be rationalised by remembering that as $\Gamma_s$ changes so too does the wave propagation with the consequence that a range of wave frequencies are being sampled. This result implies that some kind of sinusoidal force is acting on the dilational waves and that at $\Gamma_s^*$ resonance is established. Clearly, the dilational parameters represent the transfer function of the system.

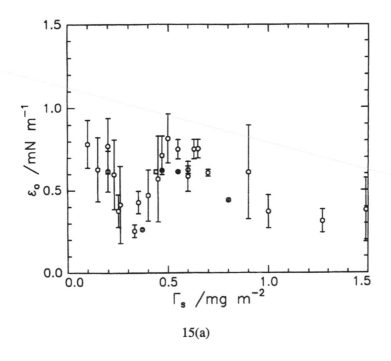

15(a)

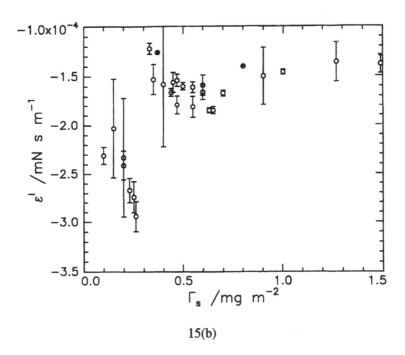

15(b)

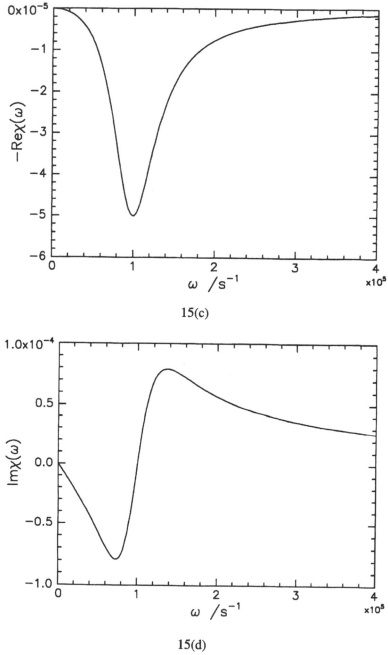

15(c)

15(d)

FIGURE 15.  Parameters obtained by SQELS for the graft copolymer of polymethyl methacrylate and polyethylene oxide spread at the air water interface. (a) Dilational modulus ; (b) Dilational viscosity; (c) Real part of the velocity transfer function (multiplied by −1 for an adsorption), (d) Imaginary part of the velocity transfer function.

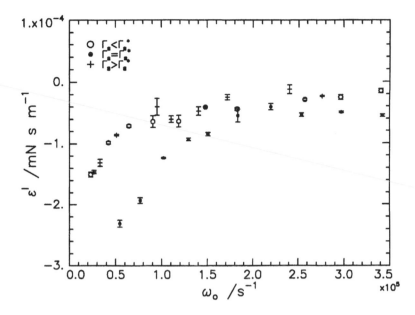

FIGURE 16. Dilational viscosity as a function of frequency for the graft copolymer with the highest content of polyethylene oxide at three different surface concentrations.

The origin of the sinusoidal force has not been identified, although the capillary waves are obvious candidates the difference in the computed wave frequencies prohibits this. Other considerations include molecular motion, for example Rouse type breathing modes, however as yet a satisfactory molecular level explanation is not forth coming.

Whilst the treatment outlined above reproduces the variation in $\varepsilon'$ with $\Gamma_s$ this does not explain the observation of negative values. As for PEO the origin of the behaviour is believed to be interactions between the side chains and the subphase influencing dilational wave behaviour. This is consistent with the behaviour of both $\varepsilon_o$ and $\varepsilon'$ which indicate that additional forces are influencing the dilational wave propagation.

A frequency study was also performed at fixed film concentrations for the two extreme copolymer compositions, 30% and 70% by mass PEO. Surface concentrations above, below and at the transition surface concentration, $\Gamma_s^*$, were studied. Generally the frequency dependence of $\gamma_o$ and $\gamma'$ were much the same; $\gamma_o$ fell with increasing wave frequency as did $\gamma'$. Clearly, this behaviour cannot be explained by phenomenological models as a 'negative' relaxation is implied. The values of $\varepsilon_o$ increased with $\omega_o$ and then decreased at the highest frequencies, rather similar to the behaviour of adsorbed PEO.

The frequency dependence of $\varepsilon'$ is noteworthy. Just as for PEO the values become less negative with increasing $\omega_o$ but never become positive. Comparison of the data for the highest PEO content copolymer for concentrations which span $\Gamma^*$ demonstrates this behaviour, see Figure 16. In particular notice that the values of $\varepsilon'$ are essentially the same for concentrations above and below $\Gamma_s^*$. However, at $\Gamma_s^*$ the values of $\varepsilon'$ whilst exhibiting the same frequency dependence are always lower. This is consistent with the frequency variation with $\Gamma_s$ exhibited by the capillary and dilational waves. Whilst the capillary wave frequency exhibits a maximum at $\Gamma_s^*$ its magnitude is much less than the minimum observed in the dilational wave frequency. It follows that the dilational wave frequency associated with $\Gamma_s^*$ is significantly lower than at other surface concentrations. As Figure 16 demonstrates the lower the wave frequency the more negative $\varepsilon'$, i.e. the more destabilised the wave is by energy transfer from another source. Thus the variation in $\varepsilon'$ with frequency may be rationalised.

Clearly, this particular system exhibits some complex behaviour that is not fully understood. This may be due to an inappropriate dispersion equation in which not all surface phenomena are parameterised. Whatever the short comings of current theories this study demonstrates how important composition is in determining dynamic behaviour. These data support the suggestion that subphase — polymer interactions are also important in influencing dynamic properties. The marked differences in behaviour of the graft copolymer and diblock, even for similar molar compositions, demonstrates that molecular architecture is also of vital importance.

## 6.7 SUMMARY

Evidently, provided that all the options for deuterium labelling are accessible, neutron reflectometry can provide a highly detailed description of the composition and dimensions of polymer containing surface regions. These data can be used to test current theories of polymers attached to surfaces to provide the necessary information which correlates structural changes with changes in such properties as surface pressure and surface viscosity. There will always be a question concerning the uniqueness of the analysis of neutron reflectivity data due to the loss of phase information but use of data from other sources can often help to 'pin down' the organisation. We have pointed out that accurate information necessitates that each of the partially deuterium labelled polymers have the same composition for copolymers. Nonetheless sufficiently close compositions can be obtained to give a useful description of the polymer organisation. The use of deuterated polymers in neutron reflectometry is paramount and the availability of such polymers is vital to the success of obtaining the maximum information the data and quite often it is only the absence of the necessary deuterated polymers which prevents full application being accessible.

From the short review given above, it is clear that variations in the organisation of the polymer can be closely correlated with the observed changes in dynamic properties of the surface layer as obtained by surface quasi-elastic light scattering. What is also evident is that the range of dynamic behaviour of polymers appears to be wide. Furthermore, although the phenomenology of the variations in frequency and damping can be explained using basic continuum physics, a molecular basis for the behaviour is not yet forthcoming. It is perhaps the combination of neutron reflectometry and surface quasi-elastic light scattering together with custom synthesis of specific polymer architectures which will reveal generic features of the dynamic behaviour of polymers at liquid surfaces. Clearly, the next area of polymers at fluid interfaces which needs investigation is at liquid-liquid interfaces. Some work is already in progress but there are quite severe difficulties with neutron reflectometry experiments, they are being approached and hopefully will be overcome.

## Acknowledgements

We would like to thank Dr M.R. Taylor for all his efforts in improving the SQELS apparatus at Durham, Drs J. Penfold, J.R.P. Webster and D.G. Bucknall at the Rutherford Appleton Laboratory for their continuing efforts to improve the quality of neutron reflectometry data.

## References

Alexander, S. (1977) Polymer Adsorption on Small Spheres. A Scaling Approach, *Journal de Physique (Paris)*, **38**, 977.

Bijsterbosch, H.D., de Haan, V.O. *et al.* (1995) Tethered Adsorbing Chains:Neutron Reflectivity and Surface Pressure of Spread Diblock Copolymer Monolayers, *Langmuir*, **11**, 4467.

Crowley, T.L., Lee, E.M. *et al.* (1991) The Application of Neutron Reflection to the Study of Layers Adsorbed At Liquid Interfaces, *Colloids and Surfaces*, **52**, 85–106.

de Gennes, P.G. (1979) *Scaling Concepts in Polymer Physics*. Ithaca, NY, Cornell University Press.

de Gennes, P.G. (1980) Conformations of Polymers Attached to an Interface, *Macromolecules*, **13**, 1069–1075.

Earnshaw, J.C., McGivern, R.C. *et al.* (1990) Light-Scattering-Studies of Surface Viscoelasticity – Direct Data-Analysis, *Langmuir*, **6**(3), 649–660.

Earnshaw, J.C., McGivern, R.C. *et al.* (1988) Viscoelastic Relaxation of Insoluble Monomolecular Films, *Journal De Physique*, **49**(7), 1271–1293.

Earnshaw, J.C. and McLaughlin, A.C. (1991) Waves At Liquid Surfaces – Coupled Oscillators and Mode Mixing. Proceedings of the Royal Society of London Series-a, *Mathematical and Physical Sciences*, **433**(1889), 663–678.

Earnshaw, J.C. and McLaughlin, A.C. (1993) Waves At Liquid Surfaces .2. Surfactant Action and Coupled Oscillators Proceedings of the Royal Society of London Series a, *Mathematical and Physical Sciences*, **440**(1910), 519–536.

Ferry, J.D. (1980) *Viscoelastic Properties of Polymers*. New York, Wiley.

Fleer, G.J., Cohen-Stuart, M.A. *et al.* (1993) *Polymers at Interfaces*. London, Chapman and Hall.

Granick, S. (1985) Surface Pressure of Linear and cyclic Polydimethylsiloxanes in the Transition Region, *Macromolecules*, **18**, 1597-1602.

Hard, S. and Neumann, R.D. (1981) Laser Light Scattering Measurements of Viscoelastic Monomolecular Films, *Journal of Colloid and Interface Science*, **83**(2), 313–334.

Henderson, J.A., Richards, R.W. *et al.* (1991) Neutron Reflectometry From Stereotactic Isomers of Poly (Methyl Methacrylate) Monolayers Spread At the Air-Water-Interface, *Polymer*, **32**(18), 3284–3294.

Henderson, J.A., Richards, R.W. *et al.* (1993) Neutron Reflectometry Using the Kinematic Approximation and Surface Quasi-Elastic Light-Scattering From Spread Films of Poly (Methyl Methacrylate), *Macromolecules*, **26**(1), 65–75.

Henderson, J.A., Richards, R.W. *et al.* (1993) Partial Structure Factor Analysis of Neutron Reflectmetry from Spread Monolayers of Isotactic Polymethyl Methacrylate. *Acta Polymerica*, **44**, 184–191.

Henderson, J.A., Richards, R.W. *et al.* (1993) Organization of Poly (Ethylene Oxide) Monolayers At the Air-Water-Interface, *Macromolecules*, **26**(17), 4591–4600.

Israelachvili, J. (1994) Self-Assembly in 2 Dimensions Surface Micelles and Domain Formation in Monolayers, *Langmuir*, **10**, 3774.

Kawaguchi, M., Sauer, B.B. *et al.* (1989) Polymeric Monolayer Dynamics at the Air-Water Interface by Surface Light Scattering, *Macromolecules*, **22**, 1735–1743.

Kent, M.S., Lee, L.-T. *et al.* (1995) Tethered Chains in Good Solvent Conditions: An Experimental Study Involving Langmuir Diblock Copolymer Monolayers, *Journal of Chemical Physics*, **103**(6), 2320–2342.

Kent, M.S., Lee, L.-T. *et al.* (1992) Characterization of Diblock Copolymer monolayers at the Liquid-Air Interface by Neutron Reflectivity and Surface Tension Measurements, *Macromolecules*, **25**, 6240–6247.

Langevin, D. (1992) *Light Scattering by Liquid Surfaces and Complementary Techniques*. New York, Marcel Decker.

Lu, J.R., Su, T.J. *et al.* (1996) The Determination of Segment Density Profiles of Polyethylene Oxide Layers Adsorbed At the Air-Water-Interface, *Polymer*, **37**(1), 109–114.

Lucassen-Reynders, E.H. and Lucassen, J. (1969) Properties of Capillary Waves, *Advances in Colloid and Interfacial Science*, **2**, 347–395.

Milner, S.T., Witten, T.A. *et al.* (1988) Theory of the Grafted Polymer brush, *Macromolecules*, **21**, 2610–2619.

Peace, S.K. (1996) PhD Thesis, University of Durham.

Pippard, A.B. (1985) *Response and Stability*. Cambridge, Cambridge University Press.

Pippard, A.B. (1989) *The Physics of Vibration*, Omnibus Edition. Cambridge, Cambridge University Press.

Poupinet, D., Vilanove, R. *et al.* (1989) *Macromolecules*, **22**, 2491–2496.

Reynolds, I., Richards, R.W. *et al.* (1995) Organization of Spread Monolayers of Poly (Lauryl Methacrylate) At the Air-Water-Interface From Neutron Reflectometry On Partially Labeled Isomers, *Macromolecules*, **28**(23), 7845–7854.

Richards, R.W., Rochford, B.R. *et al.* (1996) Surface Quasi-Elastic Light-Scattering From Spread Monolayers of a Poly (Methyl Methacrylate) Poly (Ethylene Oxide) Block-Copolymer, *Macromolecules*, **29**(6), 1980–1991.

Richards, R.W. and Taylor, M.R. (1996) Long-Wavelength Dynamics of Spread Films of Poly (Methyl Methacrylate) and Poly (Ethylene Oxide) At the Air/Water Interface, *Journal of the Chemical Society-Faraday Transactions*, **92**(4), 601–610.

Russell, T.P. (1990) X-ray and Neutron Reflectivity for the Investigation of Polymers, *Materials Science Reports*, **5**(4,5), 171–271.

Sauer, B.B., Yu, H. *et al.* (1987) Surface Light Scattering Study of a Poly (Ethylene oxide) Polystyrene block Copolymer at the Air–Water and Heptane–Water Interfaces, *Macromolecules*, **20**, 393–400.

Thomas, R.K. (1995) Neutron Reflection from Polymer Bearing Surfaces. in *Scattering Methods in Polymer Science*. edited by R.W. Richards. London, Ellis Horwood.

Vilanove, R. and Rondelez, F. (1980) Scaling Description of Two Dimensional Chain Conformations in Polymer Monolayers. *Physical Review Letters*, **45**(18), 1502–1505.

# 7. The Structure of Surfactant Monolayers at the Air-Water Interface Studied by Neutron Reflection

JIAN R. LU[1,*] and ROBERT K. THOMAS[2]

[1]*Department of Chemistry, University of Surrey, Guildford GU2 5XH, UK*
[2]*Physical and Theoretical Chemistry Laboratory, South Parks Road, Oxford OX1 3QZ, UK*

## 7.1 INTRODUCTION

Surfactants are the molecules with hydrophobic chains and hydrophilic heads which display a wide range of amphiphilic properties, e.g. a strong tendency to congregate at interfacial regions and solubility in both oil and water. Surfactants are widely used in many industrial systems, e.g. in detergency, cosmetics, shampoos, paints, pharmaceutical preparations and fabric softening (Porter, 1994), and their performance mostly relies on properties associated with either interfacial adsorption or aggregation in solution. Although much has been published on their phase and aggregation behaviour, and their interactions with polymers, little is known about the structure of their monolayers at planar interfaces (Schick, 1987). There have been few direct measurements of the structure of soluble surfactant monolayers at the planar interface because, until recently, there have been almost no suitable techniques. Although surface tension measurement in conjunction with the Gibbs equation can provide estimates of surface excess or the area per surfactant molecule it does not provide any information about the distribution of the adsorbed surfactant layer. X-ray reflection and ellipsometry are potentially useful techniques for studying surfactant adsorption at the air-water interface, but their inability in distinguishing the surfactant layer from water

---

*Correspondence: Dr. J.R. Lu, Department of Chemistry, University of Surrey, Guildford GU2 5XH, UK.

makes it difficult for them to determine the distribution of the disorganised surfactant layer (Thomas, 1995). Neutron reflection is the newly developed technique capable of probing the distributions of surfactant layers at flat interfaces. When combined with isotopic labelling it is extremely effective at revealing the distribution of molecular fragments in a layer with a resolution at the Å level.

The detailed molecular structure of a surfactant is bound to influence its interfacial and aggregation behaviour and the relation between chemical architecture and surface activity is vital for understanding surfactant function [Israelachicili et al., 1976]. Neutron reflection is the first experimental technique capable of in situ identification of the effects of specific structural characteristics of different surfactants on their structural differences at interfaces. This opens up the possibility of addressing some of the important issues concerning differences in behaviour of different surfactants. For example, it is well known that nonionic surfactants form microemulsions more readily than ionic surfactants. This difference must stem from some detail of the structural arrangement within the monolayers at the interface, detail which may now be accessible.

Over the last few years we have developed and applied the technique of neutron reflection with the aim of understanding the main structural characteristics of surfactant monolayers. In this article we summarise our recent work on the structure of the nonionic surfactants $C_{12}H_{25}(OC_2H_4)_mOH$ ($C_{12}EO_m$) and the cationic surfactants $C_NH_{2N+1}N(CH_3)_3^+Br^-$ ($C_NTAB$) at the air-water interface. These are all water soluble materials and the layers formed are equilibrium monolayers, to be distinguished from those formed by spreading insoluble amphiphiles on water, which are not at equilibrium with the sub-phase. Partial deuterium labelling was used explicitly to aid the determination of the thickness of a given fragment in the layer and its relative location to other composition. The conformational distributions, chain tilting, and roughness of the monolayers of the two series of surfactants are quite different and the relevance of these differences to our understanding of surfactant adsorption will be discussed.

## 7.2  THEORY

The specular reflection of neutrons at interfaces can be described by almost the same equations as light polarised perpendicular to the plane of reflection. These equations depend only on the refractive index profile across the interface, which is directly convertible to the coherent scattering length density. The scattering length density profile is related to the chemical composition and number density of the scattering species. Thus, detailed structural information across the interface can be determined by following the changes in reflectivity profiles.

Neutron reflectivity profiles can be analysed using the optical matrix method or the kinematic approximation. A detailed description of the optical matrix method has been given by Born & Wolf (Born et al., 1970) and Lekner (Lekner, 1987). The procedure for extracting structural information from experimental profiles is straightforward. A structural model is assumed and the exact reflectivity calculated using the optical matrix formula. The calculated reflectivity is then compared with the measured one and the structural parameters modified in a least-squares iteration (Thomas, 1995). The parameters used in the calculation are mainly the thickness of the layers, $\tau$, and the corresponding scattering length densities, $\rho$, which depend on the number densities of each atom species, $n_i$, and their known scattering lengths, $b_i$ :

$$\rho = \Sigma n_i b_i \tag{1}$$

Because $b_i$ varies from isotope to isotope changes in the isotopic labelling can be used to vary the scattering length density of a given layer and hence provide a set of different reflectivity profiles for a given chemical structure. The fitting of a set of isotopic compositions to a single structural model greatly reduces any ambiguity in the interpretation of the data, although it may add to the complexity of the fitting procedure. Deuterium/hydrogen substitution is particularly useful for surfactants and polymers. Since the scattering lengths of D and H are of opposite sign, the scattering length density of water can be varied over a wide range. The simplest example is that of water made by mixing one mole of $D_2O$ with nine moles of $H_2O$, which has exactly the same scattering length density as air. Such a solution does not reflect neutrons at all and is referred to as null reflecting water (NRW). When a deuterated surfactant is adsorbed at the air-water interface of NRW the surfactant is the only species which contributes to the specular reflectivity. Under this condition, if the adsorbed layer is assumed to be uniform, then the area per molecule, $A$, is given by

$$A = \frac{\Sigma m_i b_i}{\rho \tau} \tag{2}$$

where $m_i$ is the number of the $i$-th atom. In fitting the reflectivity profiles with a uniform layer model it is usually found that although $\tau$ and $\rho$ can be varied over a limited range, these variations cancel in their contribution to $A$ making the determination of the surface coverage independent of the assumption that the layer is uniform.

The use of the kinematic approach gives the opportunity for a more realistic description of the structure of the different components constituting the interface (Crowley, 1993; Simister et al., 1992; Lu et al., 1996a). In this approximation the reflectivity, $R(Q)$, is given by

$$R(Q) = \frac{16\pi^2}{Q^4} \mid \widehat{\rho}'(Q) \mid^2 \tag{3}$$

where $Q$ is momentum transfer ($Q = \frac{4\pi \sin \theta}{\lambda}$, $\theta$ is the incidence angle and $\lambda$ is the wavelength), $\widehat{\rho}'(Q) = Q^2 \widehat{\rho}(Q)$ and $\widehat{\rho}(Q)$ is the one dimensional Fourier transform of $\rho(z)$, the average scattering length density profile in the direction normal to the surface

$$\widehat{\rho}'(Q) = Q^2 \int_{-\infty}^{\infty} \exp(-iQz)\rho(z)dz \tag{4}$$

In terms of the distributions of the simplest fragments that can be regarded as the constituent parts of the surfactant monolayer, i.e. the chain ($c$), the head ($h$) and water ($w$), the scattering length density profile can be written as

$$\rho(z) = b_c n_c(z) + b_h n_h(z) + b_w n_w(z) \tag{5}$$

Substituting (5) into (4) and (3) gives

$$R(Q) = \frac{16\pi^2}{Q^2}[b_c^2 h_{cc}(Q) + b_h^2 h_{hh}(Q) + b_w^2 h_{ww}(Q) + 2b_c b_h h_{ch}(Q)$$
$$+ 2b_c b_w h_{cw}(Q) + 2b_h b_w h_{hw}(Q)] \tag{6}$$

where $h_{ii}(Q)$ and $h_{ij}(Q)$ are respectively called the self and cross partial structure factors and are defined in terms of the fragment number density distributions by

$$h_{ii}(Q) = |\widehat{n}_i(Q)|^2 \tag{7}$$

and

$$h_{ij}(Q) = \mathrm{Re}\,|\widehat{n}_i(Q)\widehat{n}_j(Q)| \tag{8}$$

where $\widehat{n}_i(Q)$ is the one-dimensional Fourier transform of the appropriate number density and can be obtained from an equation similar to (4). If the number density distributions corresponding to the two self structure factors $h_{ii}$ and $h_{jj}$ are symmetrical then the distance between the centres of the two distributions can be determined through the cross term in the structure factor

$$h_{ij}(Q) = \pm\sqrt{h_{ii}h_{jj}}\cos(Q\delta_{ij}) \tag{9}$$

where $\delta_{ij}$ is the distance between the centres of the two distributions. If one distribution is even and the other is odd, then

$$h_{ij}(Q) = \pm\sqrt{h_{ii}h_{jj}}\sin(Q\delta_{ij}) \tag{10}$$

The distributions of surfactants and their fragments across the interface will always approximate closely to these two limits. Thus equations (9) and (10) offer a route for determining the separation between pairs of distributions without either Fourier transformation or specific assumptions about the type of distribution apart from its being even or odd.

The choice of the optimum route for deriving number density distributions from self partial structure factors is less straightforward. In principle number density distributions can be obtained directly by Fourier transformation, but in practice the errors are too large because of the limited range of momentum transfer over which the data are measured. At present, reflectivities can only be measured up to relatively low momentum transfers because of the significant background from incoherent scattering. The use of suitable analytic expressions for the partial structure factors is therefore a more convenient approach. A simple and appropriate model to represent a soluble surfactant monolayer is a Gaussian distribution

$$n_i(z) = n_o \exp(-4z^2/\sigma^2) \tag{11}$$

where $n_i$ denotes the number density distribution, $n_o$ is the number density at the centre of the distribution, and $\sigma$ is the full width at $n_o/e$. Substitution of (10) into (4) and (7) gives

$$h_{ii}(Q) = \Gamma^2 \exp(-\sigma^2 Q^2/8) \tag{12}$$

where $\Gamma$ is the surface excess and is equal to $\sigma n_o \sqrt{\pi}/2$. The distributions of the alkyl chain, the head and the whole molecule are each fairly symmetrical and are expected to be well described by equation (11). For the water distribution across the interfacial region a tanh function has been found to be suitable for which

$$h_{ww}(Q) = n_o^2 (\pi \xi/2)^2 csch^2 (\pi \xi Q/2) \tag{13}$$

where $\xi$ is the width parameter of the tanh function.

## 7.3 THE STRUCTURE OF SURFACTANT MONOLAYERS

The major difference between a soluble monolayer and an insoluble monolayer is that a soluble monolayer has a larger area per molecule, which leads to a larger degree of disorder within the latter. The most suitable way of defining the layer structure is then in terms of the thicknesses of key fragments and their relative locations within the layer. The structure of a $C_{16}TAB$ monolayer at the air-water interface has been determined in great detail (Lu *et al.*, 1994a, 1995) and the results will be used here to illustrate some of the principles of neutron reflection measurements.

Ideally, one would like to determine the distribution of each individual methylene group in the hexadecyl chain, but this is limited by the level of the attainable signal to background ratio. A compromise between what is required and what is accessible is to divide the hexadecyl chain into four fragments, each containing four carbons. The thickness of the butylene group closest to the head group, for example, can be approximately determined by measuring the reflectivity from a layer of $C_{12}H_{25}C_4D_8N(CH_3)_3Br$ ($hC_{12}dC_4hTAB$), the basic principle being to restrict any reflected signal to just the fragment of interest. In this labelled species the head group is null reflecting because the positive scattering lengths of the C, N and Br nuclei cancel out the negative contribution from the protons to make the overall scattering length of this group zero. A similar cancellation also makes the $C_{12}H_{25}$ group approximately null reflecting. For more precise work exact matching of $C_{12}H_{25}$ can be achieved by adding a given amount of $dC_{16}hTAB$ so that the total contribution from $C_{12}H_{25}$ is exactly zero ($0C_{12}dC_4hTAB$). This can be done by using the equation of $b_1 n_1 + b_2 n_2 = 0$, where $b_1$ and $b_2$ denote the scattering lengths for the hydrogenated and deuterated dodecyl chains, and $n_1$ and $n_2$ are their corresponding molar numbers. Since the scattering length of $C_{12}H_{25}$ is $-13.7 \times 10^{-5}$ Å and that of $C_{12}D_{25}$ is $246.6 \times 10^{-5}$ Å, $\frac{b_1}{b_2}$ is $-0.0556$. The exact molar ratio of the hydrogenated to the deuterated dodecyl chains for zero scattering length is the reverse of 0.0556, that is, 18. Thus, if the solution uses null reflecting water the reflectivity from the $0C_{12}dC_4hTAB$ layer is only from the labelled butylene group and equation (6) reduces to

$$R(Q) = \frac{16\pi^2}{\kappa^2} b_c^2 h_{cc}(Q) \tag{14}$$

where $b_c$ is the scattering length for the labelled butylene group. The corresponding number density along the direction of the surface normal could be obtained through Fourier transformation using equations (4) and (7). However, a Gaussian distribution is generally found to be a good representation of the surfactant fragments and hence the area per molecule and the thickness of the fragment are most conveniently determined using an equation similar to equation (12)

$$h_{cc}(Q) = \frac{1}{A^2} \exp(-\sigma_c^2 Q^2/8) \tag{15}$$

where $A = 1/(\Gamma N_a)$, $N_a$ is Avogadro's constant and $\sigma_c$ is the thickness of the fragment. Although it might seem premature to assume a model of a Gaussian distribution, the exact nature of the distribution is not important for most of the following discussion. The assumption will be examined at a later stage.

Reflectivity measurements have been made for labelled $C_4$ fragments at different positions in the hexadecyl chain and for the head group. The resulting partial structural factors $h_{cc}(Q)$ were all found to be the same within error, i.e. the distributions of each of these labelled groups are identical. Figure 1 shows the measured $h_{cc}(Q)$ plotted against Q. Because the distributions are identical and the signal from each labelled group is relatively weak and hence has fairly large statistical errors, we determine the mean thickness by using the average of the four $C_4$ and one head group measurements. The continuous line is calculated using equation (15) with $A = 45 \pm 3$ Å$^2$ and $\sigma_c = 14 \pm 2$ Å. The total thickness is greater than the fully extended length of the fragments, suggesting that there is a substantial contribution from the roughness of the layer. To disentangle any structural features of the layer from the roughness requires a more subtle experiment which is described below.

The surfactant molecule can be labelled with deuterium such that the fragment detected by the neutrons is progressively lengthened. Then the thickness of the labelled part is determined by measuring the reflectivities of molecules in the series $0C_{16-N}dC_NdTAB$ in null reflecting water as a function of $N$ where $N = 0, 4, 8, 12, 16$. The measured thickness $\sigma$ contains contributions from the roughness and the width of the intrinsic distribution of the labelled fragment along the surface normal. The latter will begin to be significant when it becomes comparable with or larger than the roughness. The increase of the measured thickness is plotted as a function of $N$ in Figure 2a. The plot is not a straight line, but clearly reaches a limit of 14 Å as $N$ is extrapolated to zero. Thus the two compounds $0C_{14}dC_2dTAB$ and $0C_{16}dTAB$ have almost identical thicknesses. At large $N$ the plot flattens off, which suggests that the segments in the outer part of the chain are tilted further from the surface normal than the inner segments. For Gaussian distributions the intrinsic thickness of the chain fragment and the roughness approximately add in quadrature

$$\sigma_N^2 = l_N^2 + w^2 \tag{16}$$

where $l_N$ is the intrinsic thickness of the fragment along the surface normal, and $w$ is the roughness. This equation suggests that if the intrinsic thickness is proportional to the number $N$ of labelled carbon atoms a plot of $\sigma_N^2$ against $N^2$ will be linear. The results are plotted in this form in Figure 2b. The limiting intercept of the plot as $N \longrightarrow 0$ is about 195, again giving the result that $w$ is 14 Å. The slope of the plot decreases as $N$ increases, indicating that the carbons closest to the head group are oriented closer to the surface normal than the outer ones. This alternatively suggests that the carbon chain is bent further away from the surface normal the further it is away from the head group. However, Figure 2 does not show the most direct way of determining the chain orientation. It is better to use a direct determination of the separation between any pair of the fragments in the monolayer along the surface normal.

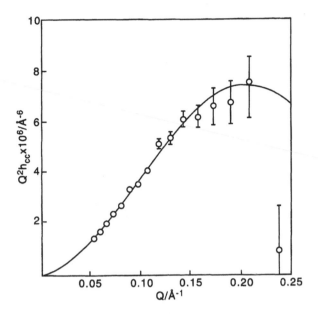

FIGURE 1. The combined partial structure factor of the $C_4$ fragments and the head of a monolayer of $C_{16}$TAB. The continuous line was calculated using equation (15) with $\sigma = 14$ Å and $A = 45$ Å$^2$.

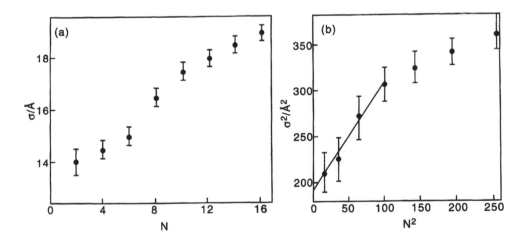

FIGURE 2. The relationship between the total apparent width ($\sigma$) and the number of deuterium labelled carbons (N) in the series of labelled compounds $0C_{16-N}dC_NdTAB$: (a) $\sigma$ plotted against N and (b) $\sigma^2$ plotted against $Q^2$.

For a surfactant monolayer in null reflecting water where the molecules contain a labelled head group and one labelled $C_4$ fragment in the chain, the reflectivity is determined by three partial structure factors and the following relation can be derived from equation (6):

$$R(Q) = \frac{16\pi^2}{Q^2}[b_c^2 h_{cc}(Q) + b_h^2 h_{hh}(Q) + 2b_c b_h h_{ch}(Q)] \tag{17}$$

where $h_{ch}(Q)$ is the cross partial structure factor between the $C_4$ and the head. In the present situation the distributions of $h_{hh}(Q)$ and $h_{cc}(Q)$ are identical within error and can be obtained using equation (14) applied to reflectivity profiles from either head deuterated or $C_4$ deuterated molecules in null reflecting water. Hence $h_{ch}(Q)$ can be determined from the additional measurement on the molecules with both $C_4$ and head group labelled. The relationship between the cross partial structure factor and the self partial structure factor has been given in equation (9) with $\delta_{ch}$ as the unknown separation between the centre of the labelled $C_4$ and that of the labelled head. Because all the terms in equation (9) except $\delta_{ch}$ are known, $\delta_{ch}$ can be obtained directly from the data without any need for modelling or Fourier transformation.

Measurements have been made for the distances between the centres of each of the chain fragments and the head using $C_{16}TAB$ samples where the chain has been divided into eight $C_2$ or four $C_4$ fragments. Separate measurements have also been done to determine the distances between different fragments within the hexadecyl chain and of those between the chain fragments and water (Lu et al., 1995). The fits of equation (9) to the cross partial structure factors between the $C_4$ fragments and the head are shown in Figure 3. The values for $\delta_{ch}$ are found to be $2.5 \pm 1$, $7.5 \pm 1$, $8 \pm 1$ and $12 \pm 1$ Å for the distances of each of the four $C_4$ groups from the head group. Similar results for $\delta_{ch}$ based on the $C_2$ labelling scheme were also obtained.

A model-independent shape for the whole distribution of the monolayer can now be constructed by superimposing the distributions of each $C_2$ or $C_4$ fragment at the appropriate distance. In the present case the eight $C_2$ or four $C_4$ units have identical thicknesses because their thicknesses are dominated by the roughness of the interface. If it is assumed that each $C_2$ fragment is best described by a Gaussian distribution we obtain the result shown in Figure 4a and, if we assume the individual distributions to be uniform layers, then the result shown in Figure 4b is obtained. The two distributions are very similar showing that the shape of the total distribution depends much more on the distances between fragments than on the shape of their individual distributions, and therefore that a Gaussian distribution is the most accurate description of the overall distribution of the chain. If a Gaussian is the best description of the whole chain then it must be an even better description of the individual fragment distributions. This is what would be expected for smaller fragments where surface roughness dominates the reflected signal and roughness will certainly follow a Gaussian distribution.

Figure 4 showed the average distribution of the whole molecule along the direction normal to the interface. A clearer model of the chain can be given if the assumption is made that all chain fragments are tilted at the same angle with respect to the surface normal. This is unlikely to be correct because each fragment almost certainly has a range of orientations, but it may give a useful guide to the chain configuration. Figure 5 shows the mean projection

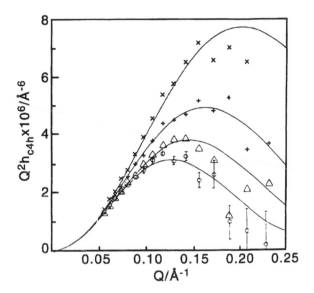

FIGURE 3. The cross partial structure factors between the $C_4$ fragments and the head group. The respective labelled compounds are $dC_4OC_{12}dTAB$ ($\times$), $OC_4dC_4OC_8dTAB$ (+), $OC_8dC_4OC_4dTAB$ ($\triangle$) and $OC_{12}dC_4dTAB$ ($\bigcirc$). The continuous lines are calculated using equation (9) with $\delta_{ch} = 12.0, 8.0, 7.5, 2.5 \pm 1$ Å, respectively.

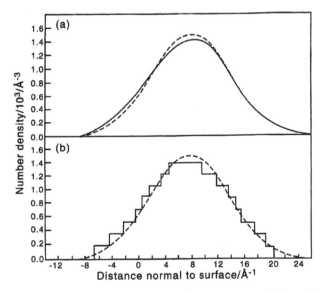

FIGURE 4. The distribution of the $C_{16}TAB$ monolayer constructed from individual $C_2$ and $C_4$ fragments assuming (a) a Gaussian distribution and (b) a uniform layer distribution for the individual fragments. The dashed line is from the direct measurement using $dC_{16}$ hTAB in NRW with $\sigma = 16.5$ Å and $A = 45 \pm 2$ Å$^2$.

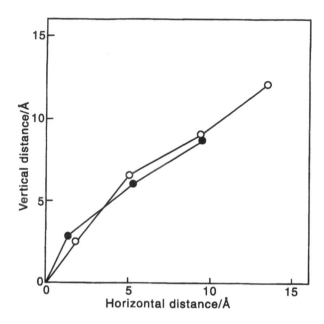

FIGURE 5. Comparison of the projection of $C_4$ fragments in $C_{16}$ TAB (○) and $C_{12}$ TAB (●) onto perpendicular and horizonal planes. The projection onto the horizonal plane was calculated assuming that all the $C_4$ fragments have their fully extended length.

of the hydrocarbon chain in terms of the four $C_4$ units where the known lengths of the fully extended chain fragments have been used together with the experimental values of $\delta_{ch}$ to determine the orientation of each fragment. It is clear that the chain is on average strongly tilted away from the surface normal and that this tilt increases for the outer half of the chain. Even the first $C_4$ fragment, next to the head group, is tilted away from the surface normal and this is possibly determined by the balance between the need for the head group to maximise its hydration by orienting the CN bond in the vertical direction, and the tendency for the alkyl chain to retain its all-*trans* configuration at an angle tilted away from the surface normal. The greater tilt of the second $C_4$ unit suggests that there is a significant chance of *gauche* defects within the first few segments. A final interesting observation is that the total projection of the molecule along the surface normal is 12.5 Å, which is actually smaller than the roughness of 14 Å.

A similar labelling scheme has been applied to the short chain cationic surfactant $C_{12}$TAB (Lu *et al.*, 1996b) and the corresponding chain configuration is also shown in Figure 5 for comparison. What is interesting is that the $C_{12}$TAB chain and the first 12 methylene groups in $C_{16}$TAB are identical within experimental error, i.e. the $C_{12}$ chain is also on average tilted away from the surface normal. This tilt progressively increases with increasing distance from the head group, indicating a significant chance of *gauche* defects occurring in the first few segments.

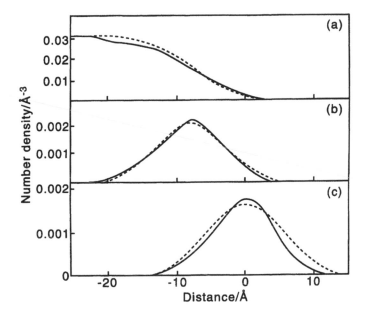

FIGURE 6. Comparison of the fragment distributions within a $C_{16}$ TAB monolayer from experiment (solid lines) and computer simulation (dashed lines): (a) the solvent, (b) the head group and (c) the hexadecyl chain.

Bocker et al. (Bocker et al., 1992) have recently used computer simulation to calculate the structure of a monolayer of $C_{16}$TACl at the air-water interface. Although the counterion is $Cl^-$ rather than $Br^-$ as used in the actual experimental, the calculation was done at exactly the same area per molecule of 45 $Å^2$ and the comparison is therefore a very direct one. Figure 6 compares simulation and experiment for water, the head and the whole hexadecyl chain respectively in terms of number density along the distance normal to the interface. There is excellent agreement in the relative positions of the three distributions. The simulated water profile has virtually the same width and distribution as the measured one, indicating that, although the resolution of the experiment is fairly insensitive to the shape of water profile, the measurement is nevertheless reasonably reliable for the approximate determination of the solvent distribution across the interface. It also suggests that the assumption of a tanh-shaped profile is appropriate for the water distribution. The simulated width of the head group distribution also agrees well with experiment, which, because the intrinsic dimensions of the head group are so small, confirms that roughness makes a major contribution to the thickness. The less satisfactory agreement is in the width of the alkyl chain, the experimental distribution being wider than the simulated one. Furthermore, the experimental distribution is more skewed than the simulated one, which results from the chain distribution shown in Figure 5 which is wider on the water side because of its more vertical orientation. In another simulation, but on the slightly shorter $C_{14}$TAB monolayer, Tarek et al. (Tarek et al., 1995) obtained a better agreement between calculated and simulated alkyl chain distributions.

Although roughness is an important parameter for surfactant monolayers adsorbed at the air-water interface, this is the first direct measurement of its contribution to the total thickness of the interfacial layer of a surfactant. In most data analysis roughness is often incorporated as a fitting parameter in modelling neutron reflectivity profiles. For pure liquids, on the other hand, off-specular scattering can be used to determine the surface roughness. For example, the roughness of pure water has been found to be 3 Å, close to the value predicted from the capillary wave model (Schwartz *et al.*, 1990). Since the capillary wave roughness varies inversely with the square root of the surface tension the capillary wave roughness can be estimated for the surface of $C_{16}TAB$ solutions as follows. The surface tension of the $C_{16}TAB$ solution at its critical micellar concentration is about 40 mNm$^{-1}$ and that of pure water is 72 mNm$^{-1}$. The square root of the ratio of the two tensions is 1.34. Allowing for a factor of 2.3 due to the difference in the definition of the roughness, the roughness for the surfactant solution surface is estimated to be 9.2 Å as compared with our measured value of 14 Å. Thus, the capillary wave roughness appears to be significantly smaller than that observed. Taking the roughness and the intrinsic thickness to add in quadrature, it is found that there is an additional 10 Å roughness not accounted for by the simple theory. Tarek *et al.* (Tarek *et al.*, 1995) have attributed this additional roughness to structural disorder within the layer.

## 7.4 THE MONOLAYER STRUCTURE OF A SERIES OF CATIONIC SURFACTANTS

The monolayer structure of a series of cationic surfactants ($C_N TAB$) with alkyl chain lengths of 10, 12, 14, 16 and 18 carbons at the air-water interface have been studied (Lu *et al.*, 1993a, 1993b, 1994a, 1994b, 1996b; Lyttle *et al.*, 1995; Simister *et al.*, 1992). The area per molecule for each chain length surfactant in the series falls within the range from 43 to 48 Å$^2$ except for $C_{10}TAB$ whose area per molecule at the critical micellar concentration (CMC) was found to be $56 \pm 3$ Å$^2$. The total thickness of the monolayer for different $C_N TABs$ can be determined by plotting equation (15) in the form $\ln(h_{aa})$ against $Q^2$, which should give a straight line with slope proportional to the square of the layer thickness and intercept proportional to the coverage. Figure 7 shows these plots for $C_{12}TAB$ and $C_{18}TAB$ at a similar coverage corresponding to an area per molecule of $45 \pm 3$ Å$^2$. The plots are almost identical in slope and intercept showing that not only is the area per molecule the same but so is the thickness of the monolayer. This is a somewhat unexpected result because the fully extended length of the two molecules differs by about 7.5 Å. The similarity in thickness is found for the whole series as shown by the data given in Table 1. The total thickness only changes by 3 Å while the fully extended length increases by 9 Å over the whole range. In general, layer thickness is found to increase with decrease in the area per molecule. This is illustrated by the slight increase of the layer thickness of 16.5 to 17 Å for $C_{12}TAB$ on addition of NaBr when the area per molecule decreases from 48 Å$^2$ to 44 Å$^2$.

The factors that contribute to this relatively constant thickness with chain length are rather subtle. When a full structural analysis is made it is found that the alkyl chain thicknesses (as opposed to the whole surfactant) are all between 16 to 17 Å: a variation well within experimental error. It is then clear from the thicknesses of the head group regions, as listed

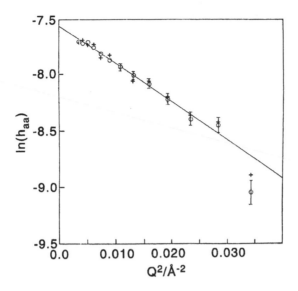

FIGURE 7. Plots of $\ln[h_{cc}]$ versus $Q^2$ for the alkyl chains of $C_{12}TAB$ ($\bigcirc$) and $C_{18}TAB$ (+), both having the thicknesses of $16.5 \pm 1$ Å and $A = 45 \pm 3$ Å$^2$.

in Table 1, that most of the increase in thickness of the whole monolayer with chain length occurs in parallel with a change in the width of the head groups. Since for $C_{16}TAB$ the width of the head group region was found to be determined almost entirely by the roughness, it can be concluded that the increase in the thickness of the whole monolayers with chain length is caused by the change in roughness. Since the surface tension at the CMC decreases by a few $mNm^{-1}$ as the chains lengthen, a small part of this change may be attributed to the increase in thermal capillary waves. When the contribution of the roughness to the overall width is taken into account, the intrinsic thickness of the alkyl chain region of the layer actually decreases as the length of the fully extended chain increases. This trend can be shown to continue through to $C_{10}TAB$. The value for $C_{10}TAB$ in Table 1 was obtained at a larger area per molecule of about 56 Å$^2$ than for the other members of the series. However, the thickness of the chain region is always observed to decrease as $A$ increases so the value of 16 Å for $\sigma$ of $C_{10}TAB$ is less than it would be if the monolayer were compressed to the same area per molecule as the other members of the series. The unexpected decrease in chain thickness with increasing chain length must result from the chains on average becoming more tilted away from the surface normal. This is in apparent conflict with the observation that the values of $\delta_{ch}$ increase with chain length. However, $\delta_{ch}$ is determined by the separation of the head group from the *lower* part of the chain. Thus the increase in $\delta_{ch}$ suggests that the tilt angle of the part of the alkyl chain closest to the head group is the same for all members of the series while the tilt angle of the outer part of the alkyl chain increases substantially with chain length. Both these observations agree with the detailed structural analysis of $C_{16}TAB$ shown in Figure 5.

TABLE 1. Structural parameters for the cationic surfactant series $C_N$TAB.

| parameters / Å | $C_{10}^*$ | $C_{12}$ | $C_{14}$ | $C_{16}$ | $C_{18}$ |
|---|---|---|---|---|---|
| $\sigma_a \pm 1$ | $16.5 \pm 1.5$ | 17.0 | $17.5 \pm 2$ | 18.5 | 19.0 |
| $\sigma_c \pm 1$ | 16.0 | 16.0 | 16.0 | 16.5 | 17.0 |
| $\sigma_h \pm 2$ | 11.0 | 12.5 | 13.0 | 14.0 | 14.0 |
| $l_c$ | 14.2 | 16.7 | 19.2 | 21.7 | 24.2 |
| $\delta_{ch} \pm 1$ | 6.0 | 6.5 | $6 \pm 2$ | 8.5 | 9.0 |
| $2\delta_{ch}$ | 12.0 | 13.0 | 12 | 17.0 | 18.0 |

*parameters were obtained at $A = 56 \pm 3$ Å$^2$, all others at $A = 45 \pm 3$ Å$^2$.

A large roughness seems to be an important feature of soluble surfactant monolayers. It might be speculated that repulsion between charged head groups is a factor forcing surfactant molecules to form a roughened layer. If this were the case, the addition of salt, which will screen the charges more effectively, would be expected to reduce the roughness. In Figure 8 we compare the head group and whole molecule partial structure factors for $C_{10}$TAB in the presence and absence of NaBr in the form of the linear plot of equation (15). The almost identical slopes for the head groups demonstrate that addition of salt has little effect on the roughness, although it causes a slight decrease in the area per molecule. The almost identical slopes for the whole monolayer also indicates that salt has not affected the total layer thickness.

## 7.5 THE MONOLAYER STRUCTURE OF THE NONIONIC SURFACTANT SERIES $C_{12}E_m$

The monoalkyl ethers of the oligoethylene glycols ($C_N H_{2N+1}(OC_2H_4)_m OH$) are a widely studied group of nonionic surfactants (Aveyard *et al.*, 1990; Gradzielski *et al.*, 1995; Glatter *et al.*, 1996). A major difference between the nonionic $C_n E_m$ and the cationic $C_N$TAB is that the former can form microemulsions without the need to add cosurfactant. For both series of surfactants, a key factor in the various structures adopted by their concentrated phases is the bending elasticity of the surfactant monolayers. Curved monolayers around aggregates such as spherical and cylindrical micelles are in thermodynamic equilibrium with the flat monolayer at the air-water interface, and the structural and thermodynamic properties of the latter are more easily investigated.

We have determined the structure of monolayers of the nonionic series $C_{12}H_{25}(OC_2H_4)_m$ OH ($C_{12}E_m$) for $m = 2$–6 (Lu *et al.*, 1993c, 1993d, 1993e, 1993f, 1998), 8 and 12 (Lu *et al.*, 1994c, 1997) at the air-water interface. Just as in the measurements described for $C_N$TAB the thickness of the whole monolayer can be obtained by determining the reflectivity using the fully deuterated surfactant in null reflecting water. Figure 9 shows plots of $\ln(h_{aa})$ versus $Q^2$ for the four fully deuterated nonionic surfactants with $m = 2, 4, 6$ and 8. The area per molecule was found to be $33 \pm 3$ Å$^2$ for $C_{12}E_2$, $44 \pm 3$ Å$^2$ for $C_{12}E_4$, $55 \pm 3$ Å$^2$ for $C_{12}E_6$ and $62 \pm 3$ Å$^2$ for $C_{12}E_8$, all at their CMCs, with corresponding thicknesses of 19.5, 20.5, 21.0 and $24.0 \pm 1$ Å. The limiting coverage of the monolayer at the CMC decreases substantially

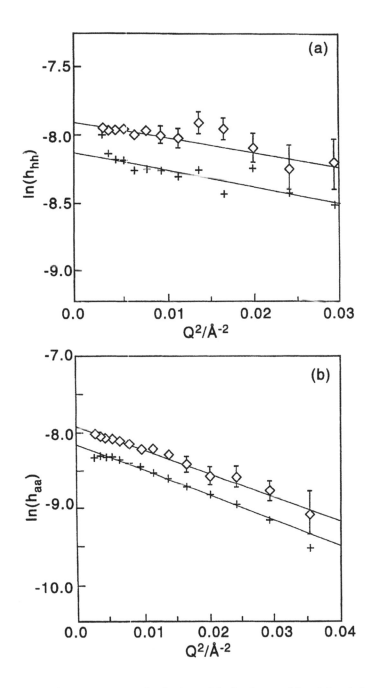

FIGURE 8. The effect of salt on the distributions of the head group (a) and the whole monolayer (b) of $C_{10}$ TAB in NRW. The presence of 0.3 M NaBr ($\diamond$) has decreased $A$ from 57 to $53 \pm 3$ Å$^2$ but the thicknesses remain constant at $11 \pm 2$ Å for the head and $16.5 \pm 2$ Å for the whole monolayer.

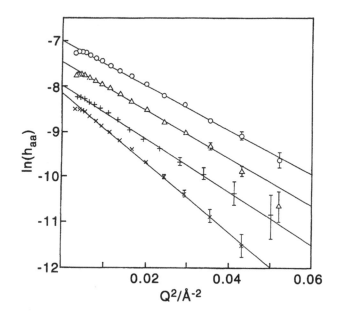

**FIGURE 9.** Comparison of the distributions for the whole monolayer of $C_{12}E_m$ with $m = 2$ (○), 4 (△), 6 (+) and 8 (×). The areas per molecule are 32, 44, 55 to $62 \pm 3$ Å$^2$ and the total thicknesses ($\sigma_T$) are 17.5, 18.5, 19.0 to 21.5 Å, respectively.

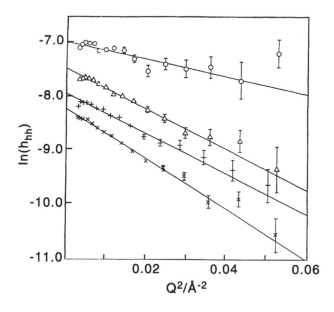

**FIGURE 10.** Variation of the thickness of the ethoxylate head groups in $C_{12}E_m$ with $m = 2$ (○), 4 (△), 6 (+) and 8 (×). The thicknesses are 11, 17.5, 16.5 to 19 $\pm$ 2 Å, respectively.

TABLE 2. Structural parameters for the nonionic surfactant series $C_{12}E_m$.

| parameters | $E_2$ | $E_4$ | $E_6$ | $E_8$ | $E_{12}^*$ |
|---|---|---|---|---|---|
| $A \pm 3/\text{Å}^2$ | 32 | 44 | 55 | 62 | 73 |
| $\sigma_a \pm 1/\text{Å}$ | 17.5 | 18.5 | 19.0 | 21.5 | $28 \pm 4$ |
| $\sigma_c \pm 1/\text{Å}$ | 17.0 | 16.5 | 16.0 | 15.5 | $15.0 \pm 3$ |
| $\sigma_h \pm 1/\text{Å}$ | 11.0 | 17.5 | 16.5 | 19.0 | $22.0 \pm 4$ |
| $l_h/\text{Å}$ | 7.2 | 14.4 | 21.6 | 28.8 | 43.2 |
| $\delta_{ch} \pm 1/\text{Å}$ | 8.5 | 7.5 | 9.0 | 10.5 | 12.0 |
| $2\delta_{ch}/\text{Å}$ | 17.0 | 15.0 | 18.0 | 21.0 | 24.0 |

*the parameters were obtained from an indirect labelling scheme (Lu *et al.*, 1997).

with increase of $m$, but the width of the layer only increases by a relatively small fraction. In the case of small head groups, the area per molecule is dominated by the steric requirements of the alkyl chain. However, as the size of the ethylene oxide group increases, it will be less likely to adopt a linear conformation and instead, will prefer to coil or bend and this will result in a bulky head group, which will lead to a larger area per molecule. The monolayer structure of $C_{12}E_{12}$ has also been determined very recently using a less direct, but simpler deuterium labelling of alkyl chain deuterated and fully hydrogenated surfactants. The area per molecule was found to be $75 \pm 3$ Å$^2$ and the total thickness ($\sigma$) $28 \pm 4$ Å.

The thickness of fragments within the monolayer, in particular, the dodecyl chain and the ethoxylate head group, can be obtained using the same method as described for $C_{16}$TAB using appropriately labelled fragments. This enables us to compare the widths of the chain and the head group distributions as a function of the size of the ethoxylated head group. Figure 10 shows the plot of $\ln(h_{ee})$ versus $Q^2$ for the head groups of the four $C_{12}E_m$, and Table 2 summarises the derived width parameters for the chain, head and whole molecule for $m = 2, 4, 6, 8$ and 12. Whilst the area per molecule increases with the size of the head group the thickness of the dodecyl chain shows a gradual decrease. Since the surface tension at the CMC for different $m$ steadily increases with head group size, the intrinsic thicknesses of the alkyl chain after removal of the capillary wave roughness will be even smaller and the trend of decreasing thickness with $m$ is more pronounced. As is apparent from inspection of Table 2, the intrinsic thicknesses are significantly smaller than the fully extended chain length (16.7 Å), suggesting that the alkyl chains are on average tilted away from the surface normal. The extent of tilting is more or less comparable for $m = 2$ to 6, but increases for $m = 8$ and 12. The explanation for this is that with the increase of the head group size the area per molecule becomes larger and there is more space available for the alkyl chains to tilt. The extent of tilting may also be coupled with increasing intermixing of the alkyl chain and the head group. The thicknesses of the head groups at the CMC show a steady increase with $m$, although a closer examination suggests that there is a small minimum in thickness at $m = 6$, which may be associated with a transition in the structural arrangement of the head group. For small $m$, the head groups appear to be fully extended and quite closely aligned to the surface normal, but this changes to something more like an anchored polymer chain at large $m$. Thus, the fully extended length for one unit of ethylene oxide is 3.6 Å and the maximum thickness for the head group in the absence of roughness

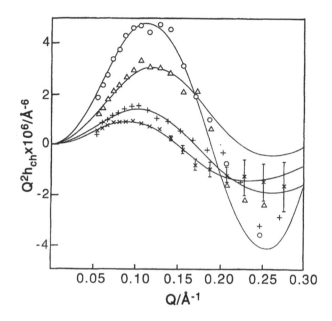

FIGURE 11. Cross partial structure factors for $C_{12}E_m$ with $m = 2$, ($\bigcirc$), 4 ($\triangle$), 6 ($+$) and 8 ($\times$), where $\delta_{ch} = 8.5, 7.5, 9.0$ and $10.5 \pm 1$ Å.

is 7.2 and 14.5 Å for $m = 2$ and 4 respectively. For longer ethoxylate groups the width of the head group distribution increases more slowly with $m$. Sarmoria et al. (Sarmoria et al., 1992) have used the rotational isomeric state model to analyse how polymer-like behaviour might develop in anchored ethoxylate groups and concluded that polymer-like behaviour may develop at around $m = 10$ for anchored chains.

The separation between the centres of the dodecyl chain and the head group can be obtained from equation (9) and the resulting values of $\delta_{ch}$ are given in Table 2. Figure 11 shows the measured and calculated cross partial structure factors for compounds with four different head groups and the difference in $\delta_{ch}$ can clearly be seen from the shape of the curves. For the first three head groups the increase in head group size has little effect on the separations. Since the values of $\delta_{ch}$ are independent of roughness this observation suggests that there is an increasing tendency for chain and head group to mix as $m$ increases. This is supported by comparison of the intrinsic thicknesses of the monolayers with $2\delta_{ch}$. An increasing difference in these two values has to be caused either by increased tilting of the outer part of the chain or head or increased intermingling of chain and head. The detailed structure measurement using the partially labelled surfactants suggests the latter.

Although the structural distributions within the nonionic surfactant monolayers have not been studied in as much detail as for cationic surfactants, we have made some coarse resolution studies of the distributions of hexaethylene glycol groups by labelling them in two halves. The width of the $E_3$ group closest to the dodecyl chain was found to be greater than that of the outer $E_3$, which should be deeper in the solution. The reconstructed distribution for the whole head group is therefore skewed towards the alkyl chain. This skewness must be caused by the attachment of the head group to the alkyl chain, which constrains the first half of the head group to be more vertical. Similar labelling of the dodecyl chain in $C_{12}E_3$ into two halves showed that the inner half of the dodecyl chain closest to the head group was more tilted away from the surface normal than the outer half of the chain, which results in a wider distribution for the outer part of the chain. The combination of these two results suggests that differences in tilting cannot account for the difference between the intrinsic thicknesses and $2\delta_{ch}$. Intermixing of the two chains within the monolayer is probably the major factor.

## 7.6  CONCLUDING REMARKS

The potential in applying neutron reflection to the study of disordered interfacial structures should be clear from the results presented. What now needs to be developed are relationships between the detailed structural information being obtained from neutron reflection to surfactant phase behaviour and microemulsion properties. This kind of information, which was not available a few years ago, should make a significant contribution to the development of sound theory in this expanding area. A further observation is that the same kind of isotopic labelling used here opens up a relatively easy route to the study of surfactant or surfactant/polymer mixtures, another important area of technological application, somewhat neglected because of a lack of experimental techniques.

### Acknowledgements

We thank the Engineering and Physical Science Research Council (EPSRC) for support.

### References

Aveyard, R., Binks, B.P., Clark, S. and Fletcher, P.D.I. (1990) Effects of Temperature on the Partioning and Adsorption of $C_{12}E_5$ in Heptane-Water Mixtures, *J. Chem. Soc., Faraday Trans.*, **86**, 3111.

Bocker, J., Shlenkrich, M., Bopp, P. and Brickmann, J. (1992) Molecular Dynamics Simulation of a N-Hexadecyltrimethylammonium Chloride Monolayer, *J. Phys. Chem.*, **96**, 9915.

Born, M. and Wolf, E. (1970) *Principles of Optics*, Pergamon, Oxford.

Crowley, T.L. (1993) A Uniform Kinematic Approaximation for Specular Reflectivity, *Physica*, **A195**, 354.

Glatter, O., Strey, R., Schubert, K.-V. and Kaler, E.W. (1996) Small Angle Scattering Applied to Microemulsions, *Ber. Bunsenges. Phys. Chem.*, **100**, 323.

Gradzielski, M., Langevin, D., Magid, L. and Strey, R. (1995) Small Angle Neutron Scattering from Diffuse Interfaces 2: Polydisperse Shells in Water-in-Alkane-$C_{10}E_4$ Microemulsions, *J. Phys. Chem.*, **99**, 13232.

Israelachivili, J., Mitchell, D.J. and Ninham, B.W. (1976) Theory of Self-Assembly of Hydrocarbon Amphiphiles into Micelles and Bilayers, *J. Chem. Soc., Faraday Trans. II*, **72**, 1525.

Lekner, J. (1987) *Theory of Reflection*, Dordrecht, Nijhoff, 1987.

Lu, J.R., Simister, E., Thomas, R.K. and Penfold, J. (1993a) Adsorption of Alkyltrimethyl Ammonium Bromide at the Air-Water Interface, *Progr. Colloid Polymer Sci.*, **93**, 92.

Lu, J.R., Simister, E., Thomas, R.K. and Penfold, J. (1993b) Structure of an Octadecyltrimethylammonium Bromide Layer at the Air-Water Interface Determined by Neutron Reflection: Systematic Errors in Reflectivity Measurements, *J. Phys. Chem.*, **97**, 6024.

Lu, J.R., Li, Z.X., Su, T.J., Thomas, R.K. and Penfold, J. (1993c) Structure of Adsorbed Layers of Ethylene Glycol Monododecylether Surfactants with 1, 2 and 4 Ethylene Oxide Groups, as Determined by Neutron Reflection, *Langmuir*, **9**, 2408.

Lu, J.R., Lee, E.M., Thomas, R.K., Penfold, J. and Flitsch, S.L. (1993d) Direct Determination by Neutron Reflection of the Structure of Triethylene Glycol Monododecyl Ether Layers at the Air-Water Interface, *Langmuir*, (1993d), **9**, 1352.

Lu, J.R., Hromadova, M., Thomas, R.K. and Penfold, J. (1993e) Neutron Reflection from Triethylene Glycol Monododecyl Ether Adsorbed at the Air-Liquid Interface: the Variation of the Hydrocarbon Chain Distribution with Surface Concentration, *Langmuir*, **9**, 2417.

Lu, J.R., Li, Z.X., Thomas, R.K., Staples, E., Tucker, I. and Penfold, J. (1993f) Neutron Reflection from a Layer of Monododecyl Hexaethylene Glycol Adsorbed at the Air-Liquid Interface: the Configuration of the Ethylene Glycol Chain, *J. Phys. Chem.*, **97**, 8012.

Lu, J.R., Hromadova, M., Simister, E., Thomas, R.K. and Penfold, J. (1994a) Neutron Reflection from Hexadecyltrimethylammonium Bromide Adsorbed at the Air-Liquid Interface: the Variation of the Hydrocarbon Chain Distribution with Surface Concentration, *J. Phys. Chem.*, **98**, 11519.

Lu, J.R., Hromadova, M., Simister, E., Thomas, R.K. and Penfold, J. (1994ba) Determination of the Structure of the Monolayer of Hexadecyltrimethyl Ammonium Bromide Adsorbed at the Air-Water Interface, *Physica B*, **198**, 120.

Lu, J.R., Li, Z.X., Thomas, R.K., Staples, E., Thompson, L., Tucker, I. and Penfold, J. (1994c) Neutron Reflection from a Layer of Monododecyl Octaethylene Glycol Adsorbed at the Air-Liquid Interface: the Structure of the Layer and the Effects of Temperature, *J. Phys. Chem.*, **98**, 6559.

Lu, J.R., Li, Z.X., Smallwood, J., Thomas, R.K. and Penfold, J. (1995) Detailed Structure of the Hydrocarbon Chain in a Surfactant Monolayer at the Air-Water Interface: Neutron Reflection from Hexadecyltrimethylammonium Bromide, *J. Phys. Chem.*, **99**, 8233.

Lu, J.R., Lee, E.M. and Thomas, R.K. (1996a) The Analysis and Interpretation of Neutron and X-ray Specular Reflection, *Acta Cryst.*, *A52*, 11.

Lu, J.R., Li, Z.X., Thomas, R.K. and Penfold, J. (1996b) Structure of Hydrocarbon Chains in Surfactant Monolayers at the Air-Water Interface: Neutron Reflection from Dodecyl Trimethylammonium Bromide, *J. Chem. Soc., Faraday Trans.*, **92**, 403.

Lu, J.R., Li, Z.X., Thomas, R.K., Binks, B.P., Crichton, D., Fletcher, P.D.I. and McNab, J.R. (1998) The Structure of Monododecyl Pentaethylene Glycol Monolayers with and without Added Dodecane at the Air/Solution Interface: a Neutron Reflection Study, *J. Phys. Chem. B*, **102**, 5785.

Lu, J.R., Su, T.J., Li, Z.X., Thomas, R.K., Staples, E., Tucker, I. and Penfold, J. (1997) Structure of Monolayers of Monododecyl Dodecaethylene Glycol at the Air-Water Interface Studied by Neutron Reflection, *J. Phys. Chem. B*, **101**, 10332.

Lyttle, D.J., Lu, J.R., Su, T.J., Thomas, R.K. and Penfold, J. (1995) Structure of a Dodecyltrimethylammonium Bromide Layer at the Air-Water Interface Determined by Neutron Reflection: Comparison of the Monolayer Structure of Cationic Surfactants with Different Chain Lengths, *Langmuir*, *11*, 1001.

Porter, M.R. (1994) *Handbook of Surfactants*, 2nd ed., Chapman & Hall, 1994.

Sarmoria, C. and Blankschtein, D. (1992) Conformational Characteristics of Short Poly(ethylene oxide) Chains Terminally Attached to a Wall and Free in Aqueous Solution, *J. Phys. Chem.*, **96**, 1978.

Schick, M.J. (1987) Nonionic Surfactants, in *Surf. Sci. Ser.*, **27**, Dekker.

Schwartz, D.K., Schlossman, M.L., Kawamoto, E.H., Kellogg, G.J., Pershan, P.S. and Ocko, B.M. (1990) Thermal Diffuse X-Ray Scattering Studies of the Water-Vapour Interface, *Phys. Rev.*, **A41**, 5687.

Simister, E., Lee, E.M., Thomas, R.K. and Penfold, J. (1992) Structure of a Tetradecyltrimethylammonium Bromide Layer at the Air-Water Interface Determined by Neutron Reflection, *J. Phys. Chem.*, **96**, 1373.

Simister, E., Thomas, R.K., Penfold, J., Aveyard, R., Binks, B.P., Cooper, P., Fletcher, P.D.I., Lu, J.R. and Sokolowsyi, A. (1992) Comparison of Neutron Reflection and Surface Tension Measurements of the Surface Excess of Tetradecyltrimethylammonium Bromide Layers at the Air-Water Interface, *J. Phys. Chem.*, **96**, 1383.

Tarek, M., Tobias, D.J. and Klein, M.L. (1995) Molecular Dynamics Simulation of Tetradecyltrimethylammonium Bromide Monolayer at the Air-Water Interface, *J. Phys. Chem.*, **99**, 1393.

Thomas, R.K. (1995) Neutron Scattering from Polymer-Bearing Interfaces, Chapter 4 in *Scattering Method in Poly. Sci.*, ed. Richards, R.W., Ellis Horwood.

# 8. Polymer Blends

V. ARRIGHI[1] and J.S. HIGGINS[2]

[1] *Chemistry Department, Heriot-Watt University, Edinburgh EH14 4AS, UK*
[2] *Chemical Engineering Department, Imperial College, London SW7 2BY, UK*

## 8.1 INTRODUCTION

The technological importance of polymer blends has long been recognized by polymer scientists and blending is now a well established method to design new materials with specifically tailored properties possessed by neither of the two constituent polymers separately. Miscibility at a molecular level is not required for a blend to be of practical interest although a certain extent of compatibility between the constituents is necessary. In many cases one therefore deals with a multiphase system whose properties depend on the amount and composition of the phases, their morphology as well as the characteristic of the interphases.

The characterization and design of polymer blends requires that good understanding of the thermodynamics of mixing (Utracki, 1989; Olabisi *et al.*, 1979) in the melt state is achieved. Because of the complex nature of polymer molecules their miscibility behaviour is a function of various parameters, not only composition and temperature, but also type of intermolecular interactions, molecular weight, chain architecture and tacticity.

Small angle neutron scattering (SANS) has now become a very powerful technique to determine polymer-polymer interaction parameters via measurements of the concentration fluctuations close to the phase boundary. This is achieved using the same method that enabled the first measurement of polymer conformation in the bulk (Cotton *et al.*, 1974). The natural contrast in the mixture is enhanced by partially or completely deuterating one of the constituent polymers. Thus for a blend in the one phase region it is possible to determine the interaction parameter and to extract information about molecular dimensions.

Because the spatial range available with SANS varies between a few and thousands of Angstroms this technique is not only suitable to studies in the one-phase region but has been used to investigate the kinetics of the phase separation process, the morphology of the two-phase system which are formed as well as domain sizes. The information obtained by SANS are similar to those from light scattering (LS) and small angle X-ray scattering (SAXS) (Higgins, 1985; Lindner *et al.*, 1991) but while LS and SAXS studies rely on the difference between the refractive indices or electron densities of the two components, respectively SANS investigations can be performed on any system where one of the components can be fully or partially deuterated. So if two polymers have similar refractive indices and little or no contrast for SAXS, deuteration of one of the two polymer species provides a suitable and advantageous method to investigate the polymer blend, due to the large difference between the scattering length of hydrogen ($b_H = -3.739 \times 10^{-13}$ cm) and deuterium ($b_D = 6.671 \times 10^{-13}$ cm). Many recent studies on polyolefin blends have only been possible because of the large contrast obtained using the labelling method (Schipp *et al.*, 1996).

Although labelling was initially regarded as a non-intrusive method to investigate a binary system more recent studies have concentrated on the effect of deuteration on miscibility. It has been shown that mixtures of hydrogenous and deuterated polymer of the same chemical species are characterized by a small positive interaction parameter, $10^{-4} < \chi < 10^{-3}$ and exhibit an upper critical solution temperature behaviour (UCST) (Bates, 1988). The isotopic effect indicates that particular care must be taken in interpreting the interaction parameters obtained by SANS for a blend of two dissimilar polymers. For this reason mixtures of hydrogenous and deuterated polymers are also discussed in this chapter.

Other neutron scattering techniques are becoming increasingly important in the study of polymeric systems. Over the past ten years, neutron specular reflection (NSR) has emerged as one of the most powerful techniques for the study of interfaces in a variety of systems from semiconductors to biological membranes or surface magnetism. Various features of NSR renders the technique attractive to polymer scientists. Because of the large penetration depth of neutrons NSR can be employed to investigate "buried" interfaces and the labelling method used to highlight specific layers in a specimen but most importantly, these features are combined with a resolution of the order of less than a nanometer. For polymer blends valuable information can be obtained using NSR on the interfacial shape and size between the components.

Neutron scattering is not only a probe used in structural studies but can also be employed to study molecular motion which occurs in the frequency range from $10^7$ to $10^{14}$ Hz. This dynamic range includes vibrational motion, rotation of side groups and main chain motion above the glass transition temperature. Although the neutron scattering techniques such as inelastic neutron scattering, quasielastic neutron scattering and neutron spin echo have proved to be successful in investigating the dynamics of simple polymeric materials (Higgins *et al.*, 1993) there is a lack of similar studies in blends. The capabilities of these techniques in relation to polymer blends will be discussed in section 8.5. A brief introduction to polymer blends is given in the following section followed by a description of the applications of neutron scattering to studies of polymer blends.

Scattering techniques have largely contribute to our understanding of the conformation and morphology in copolymer systems. This topic has been recently reviewed in detail (Hamley, 1998) and will not be treated in this chapter.

## 8.2  THERMODYNAMICS AND PHASE DIAGRAMS

The mechanism of miscibility is governed by the Gibbs free energy of mixing $\Delta G_m$

$$\Delta G_m = \Delta H_m - T \Delta S_m \tag{1}$$

which depends upon the enthalpy and entropy of mixing, $\Delta H_m$ and $\Delta S_m$, respectively. For low-molecular weight materials the large entropic change is the driving force for mixing and as temperature increases $\Delta G_m$ becomes more negative and miscibility is enhanced. It follows that low-molecular weight materials generally mix upon heating thus exhibiting an upper critical solution temperature behaviour (UCST) as shown in Figure 1(a).

The situation changes dramatically when the two species involved in the mixing process have high molar mass. In this case the entropic contribution to mixing is very small and any small interaction in the mixture which is unfavorable to mixing renders $\Delta G_m > 0$. This observation lead to the conclusion in the early days of polymer blending that polymeric materials are immiscible.

Despite the unfavorable entropic term, a number of polymer pairs have been found to be miscible (Olabisi et al., 1979; Paul et al., 1978). Miscibility is usually due to the existence of specific interactions such as dipole-dipole or hydrogen bonding among the two components in the polymer mixture. In these systems $\Delta H_m \leq 0$ and $\Delta G_m$ is therefore also negative. As temperature increases the specific interactions between the two components may weaken due to molecular motion and the mixture phase separates. Most polymer blends show the behaviour represented in Figure 1(b) which is described as a lower critical solution temperature behaviour (LCST).

A UCST behaviour may still be observed in some polymeric binary systems when the two polymers are very similar and $\Delta H_m \approx 0$. In this case, even if the entropic term is small it is still larger than $\Delta H_m$ and $\Delta S_m$ governs miscibility. A small enthalpic term is typical of systems with very similar chemical structures and isotopic polymer blends i.e mixtures of hydrogenous and deuterated materials with the same chemical structure.

For a binary blend the Gibbs free energy of mixing $\Delta G_m$ may be calculated from the mean-field rigid lattice model developed by Flory and Huggins (Flory, 1953):

$$\frac{\Delta G_m}{k_B T} = \frac{\phi_1}{N_1} \ln \phi_1 + \frac{\phi_2}{N_2} \ln \phi_2 + \phi_1 \phi_2 \chi \tag{2}$$

where $N_i$ and $\phi_i$ are the degree of polymerization and the volume fraction of component $i$, respectively, $k_B$ is Boltzmann's constant and $\chi$ is the polymer-polymer interaction parameter. The first two terms in equation (2) represent the combinatorial entropy of mixing which decreases with increasing $N_i$ i.e. when the molecular weight of component $i$ is large. As $N_i$ increases the last term in the above equation becomes increasingly important in governing polymer miscibility.

As originally defined by Flory-Huggins theory the interaction parameter is independent of concentration and inversely dependent on temperature. In practice systematic studies of polymer blends have shown that $\chi$ varies with blend composition, molecular weight, chain architecture, tacticity as well as temperature (and this temperature dependence may not be as predicted by Flory-Huggins theory). Hence $\chi$ which is determined experimentally by interpreting data in terms of equation (2) is a complex function of various parameters and in this respect differs considerably from the original Flory-Huggins parameter.

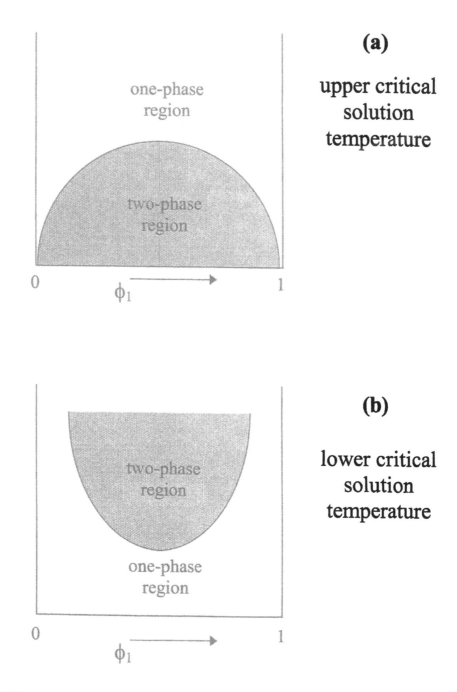

FIGURE 1. Typical phase diagrams for low molecular weight (a) and high molecular weight components: (a) the system show UCST behaviour and mixing is observed upon heating and (b) the system shows LCST behaviour and the two components phase separate on heating.

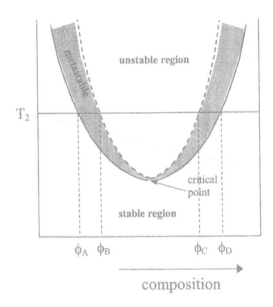

FIGURE 2. Phase diagram for a partially miscible blend exhibiting lower critical solution temperature behaviour. The spinodal separates the unstable and metastable regions while the binodal represents
the boundary between metastability and stability.

## 8.3 MECHANISMS OF PHASE SEPARATION

The phase diagram for a partially miscible polymer system exhibiting LCST behaviour is
illustrated in Figure 2. The solid line is called the *binodal* and it separates the stable from the
metastable regions of the phase diagram. The dashed line is the *spinodal* which separates
the metastable and unstable regions. The spinodal touches the binodal at the critical point,
$T_C$.

A typical dependence of $\Delta G_m$ as a function of composition and temperature is shown in
Figure 3. At a temperature $T_1$ which is below $T_C$ the quantity $\Delta G_m$ is always negative and the
system is completely miscible whereas at $T_3$ one has $\Delta G_m > 0$ and the blend is immiscible.
$\Delta G_m$ presents a more complex behaviour at a temperature $T_2$ as shown in Figure 3. Here
the binodal is defined by the points of common tangent, $\phi_A$ and $\phi_D$ whereas the spinodal
corresponds to the inflection points $\phi_B$ and $\phi_C$ where the relationship $\partial^2 \Delta G_m / \partial \phi^2 = 0$
holds.

For compositions between $\phi_B$ and $\phi_C$, $\partial^2 \Delta G_m / \partial \phi^2 < 0$ and because any small
concentration fluctuations lower the free energy the system tends to phase separate into
two phases of composition $\phi_A$ and $\phi_D$. Phase separation will take place spontaneously by
a process called spinodal decomposition. Between $\phi_A$ and $\phi_B$ and $\phi_C$ and $\phi_D$ separation
proceeds by nucleation and growth and the system phase separates if there is a fluctuation

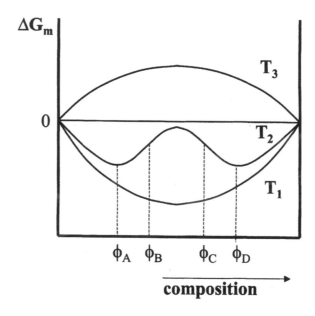

FIGURE 3. Temperature and composition dependence of the Gibbs free energy of mixing. At $T_3$ the system is immiscible whereas it is miscible at the temperature $T_1$. At $T_2$ partial miscibility is observed (see Figure 2).

in concentration large enough to overcome the potential barrier to demixing. The system shown in Figures 2 and 3 implies symmetry in terms of molecular weight and interactions. Very frequently this will not be the case and both $\Delta G_m$ and the phase diagrams show very asymmetrical dependence on composition.

SANS can be applied to studies of polymer blends both in the one and two phase regions of the phase diagram providing information on the polymer-polymer interaction parameters as well as the mechanism and kinetics of the phase separation process.

## 8.4 SMALL ANGLE NEUTRON SCATTERING

### 8.4.1 One-phase Region

The coherent elastic scattering from a homogeneous polymer blend of components 1 and 2 can be expressed in terms of the differential scattering cross section, $d\sigma(Q)/d\Omega$ which represents the number of neutrons scattered per second into a small solid angle $d\Omega$ (neutron s$^{-1}$) with respect to the incident neutron flux (neutrons cm$^{-2}$ s$^{-1}$) in units of cm$^{-2}$. This is related to the quantity measured experimentally, the normalized differential scattering cross section per unit volume, $d\Sigma(Q)/d\Omega$, expressed in units of cm$^{-1}$. $d\Sigma(Q)/d\Omega$ is a function of the scattering vector or momentum transfer $Q(= (4\pi/\lambda)\sin(\theta/2)$, where $\theta$ is the scattering angle and $\lambda$ the neutron wavelength) and it is related to the structure factor $S(Q)$ which is the Fourier transform of the density correlation function relative to the scattering units:

$$\left(\frac{d\Sigma(Q)}{d\Omega}\right) = \left(\frac{b_1}{v_1} - \frac{b_2}{v_2}\right)^2 S(Q) \qquad (3)$$

The term within brackets in equation (3) is the contrast which is calculated from the scattering lengths of the monomer units $b_1$ and $b_2$ and the volumes per monomeric unit, $v_1$ and $v_2$.

It is the contrast that determines the magnitude of the observed scattered intensity via the difference between the ratios $b_i/v_i$ for the two monomer units in the binary polymer mixture. Although it may be possible that there is sufficient contrast between the hydrogenous polymers 1 and 2 it is usually found that the scattering lengths of the two components are very similar. In this case the large difference between the scattering lengths of hydrogen and deuterium ($b_H = -3.739 \times 10^{-13}$ cm and $b_D = 6.671 \times 10^{-13}$ cm) can be conveniently used to enhance the scattered intensity. For this reason SANS measurements on blends are usually performed by deuterating one of the two polymeric components. It has been pointed out in the introduction that isotopic substitution may affect the phase behaviour. This was first demonstrated for blends of poly(vinyl methylether) and hydrogenous or deuterated poly(styrene) where the LCST was found to increase $40°C$ upon deuteration (Yang et al., 1983). Following predictions of Buckingham and Hentschel (Buckingham et al., 1980) that mixtures of hydrogenous and deuterated polymers should exhibit a UCST behaviour due to small differences between segment volumes, various authors have measured experimentally interaction parameters for a number of binary mixtures of hydrogenous and deuterated polymers having the same chemical structure (Bates et al., 1988; Lapp et al., 1985). Although studies of the isotopic effect in simple mixtures of hydrogenous and deuterated materials may be at first regarded as being of purely academic interest they are extremely important for a correct interpretation of the interaction parameters measured by SANS. Because SANS measurements on blends rely on one of the components being deuterated the overall interaction parameter measured by the technique is due both isotopic and microstructural effects (Budkowski et al., 1993). These may combine in a complicated manner as indicated by the large differences in critical temperature which are obtained by exchange of the isotopic labelling. For example the critical temperatures of model polyolefin blends structurally analogous to copolymers of ethylene and butene-1 were found to differ by nearly $70°C$ when reversing the deuterium labels (Balsara et al., 1992). For these reasons, studies of isotopic blends are important and have been included in this chapter.

### 8.4.1.1   Isotopic Blends

Neutron scattering experiments performed in the early seventies to verify Flory's predictions that the polymer conformation in the bulk is the same as in a $\theta$-solvent (Flory, 1969) exploited the labelling method in order to achieve the necessary contrast for measurements of single chain scattering (Cotton et al., 1974; Kirste et al., 1975). These experiments were carried out at low concentrations of labelling (low concentration of either the hydrogenous or deuterated components) and treated the mixture of hydrogenous and deuterated polymers as being ideal.

It was later realized that higher concentrations of labelling could be used close to a 1:1 ratio although the effect of high concentrations combined with high molecular weights lead in some cases to immiscibility between the components in the isotopic mixture. This was first observed by SANS in hydrogenous deuterated mixtures of 1,4-polybutadiene (Bates et al., 1985) and polydimethylsiloxane (Lapp et al., 1985) as a small positive interaction parameter, $10^{-4} < \chi < 10^{-3}$. Similar findings were reported for other isotopic mixtures and it is now generally accepted that hydrogenous and deuterated materials with the same chemical structure do not form ideal solutions. This supports theoretical predictions of Buckingham and Hentschel (Buckingham et al., 1980) that isotopic mixtures should exhibit an upper critical solution temperature due to small differences in segment volume of the components. This approach has been extended by Bates et al. (Bates et al., 1988) who predicted interaction parameters of the order of those measured experimentally from the shorter C-D bond length compared to the C-H bond length and changes in polarizability of the bonds. The general treatment developed below considers that the polymer isotopes do not form an ideal binary mixture.

For a homogeneous mixture (single phase) of two chemically identical polymeric species the structure factor $S(Q)$ in equation (3) is given by the mean-field random phase approximation (RPA) derived by de Gennes (de Gennes, 1970) and Binder (Binder, 1983):

$$\frac{1}{S(Q)} = \frac{1}{\phi_1 v_1 N_1 P_1(Q)} + \frac{1}{\phi_2 v_2 N_2 P_2(Q)} - \frac{2\chi}{v_0} \tag{4}$$

where $N_i$ and $\phi_i$ are the degree of polymerization and the volume fraction of component $i$, respectively and $\chi$ is the segment-segment interaction parameter. The reference volume $v_o$ is usually taken as the geometric average:

$$v_0 = \sqrt{v_1 \cdot v_2} \tag{5}$$

Since the difference between the volumes of hydrogenous and deuterated species is small (Bates et al., 1988) $v_1$ is approximately equal to $v_2$. By combining equations (3) and (4), the differential scattering cross section can be written in the form:

$$\frac{d\Sigma(Q)}{d\Omega} = \frac{(1/v_0)(b_H - b_D)^2}{[\phi_D N_D P_D(Q)]^{-1} + [\phi_H N_H P_H(Q)]^{-1} - 2\chi} \tag{6}$$

where the subscripts $H$ and $D$ indicate the hydrogenous and deuterated components, respectively. For monodisperse Gaussian chains the variation of the scattered intensity with scattering vector $Q$ represented by $P_i(Q)$ in equations (4) and (6) is described by the Debye equation (deGennes, 1979):

$$P_i(Q) = \left(\frac{2}{Q^4 R_{g,i}^4}\right) [Q^2 R_{g,i}^2 - 1 + \exp(-Q^2 R_{g,i}^2)] \tag{7}$$

where $R_{g,i}$ is the radius of gyration of polymer $i$ which is related to the degree of polymerization $N_i$ and the statistical segment length $a$:

$$R_{g,i}^2 = \left(\frac{N_i a_i^2}{6}\right) \tag{8}$$

The validity of using the Debye equation to model the $Q$ dependence can be verified by determining the slope of the scattered intensity in a log(intensity)-log($Q$) plot. At intermediate $Q$ values a power law decay of slope $-2$ should be observed if equation (7) is valid.

Various methods have been used to extract the relevant parameters from equation (6) i.e the interaction parameter and the radii of gyration of the components. A schematic representation of different approaches to the data analysis is given in Figure 4 where scattering data have been generated for a 50:50 poly(dimethylsiloxane) (PDMS) isotopic blend using the RPA approximation (a random noise has been added to the calculated data). The molecular weights of the hydrogenous and deuterated PDMS have been assumed to be equal to 70,000 g mol$^{-1}$. The interaction parameter between the hydrogenous and deuterated components and the statistical length have been taken from literature data (Beaucage *et al.*, 1996). The differential scattering cross section has been plotted versus the scattering vector $Q$ in Figures 4(a) and 4(b) on a logaritmic and linear scale, respectively. Although equation (6) can be employed to fit the scattering data in the whole $Q$ range, approximate expressions are oftem used to extract information from either the low $Q$ (Figure 4(c)) or the high $Q$ (Figure 4(d)) regions. A detailed discussion of various procedures of data analysis is given below.

In the limit of small $Q$ values when $Q \leq R_{gi}^{-1}$ equation (7) can be approximated to

$$P_i(Q) \approx \left[ 1 - \frac{Q^2 R_{g,i}^2}{3} \right] \tag{9}$$

and the form factor becomes independent of particle shape.

A correlation length $\xi$ for the composition fluctuation in the mixture is usually defined (Bates *et al.*, 1988) as

$$\xi = \frac{a}{6} [\phi_D \phi_H (\chi_s - \chi)]^{-1/2} \tag{10}$$

where $\chi_s$ is the spinodal boundary which represents the limit of thermodynamic stability for a single phase system :

$$\chi_s = \frac{1}{2} \left[ \frac{1}{\phi_D N_D} + \frac{1}{\phi_H N_H} \right] \tag{11}$$

By substituting equations (9) and (10), equation (6) can be written in the Ornstein-Zernike form:

$$\frac{d\Sigma(Q)}{d\Omega} = \frac{(1/v_o)(b_H - b_D)^2}{2(\chi_s - \chi)[1 + Q^2 \xi^2 \phi]} \tag{12}$$

and the composition fluctuation length can be determined from the slope of a plot of $1/(d\Sigma(Q)/d\Omega)$ versus $Q^2$ (Figure 4(c)). The zero angle scattered intensity $(d\Sigma(0)/d\Omega)$ of an Ornstein-Zernike plot is related to the second concentration derivative of the Gibbs free energy of mixing, $\partial^2 \Delta G_m / \partial \phi^2$, via the relationship:

$$\frac{d\Sigma(0)}{d\Omega} = (1/v_0)(b_H - b_D)^2 k_B T \left[ \frac{\partial^2 \Delta G_m}{\partial \phi^2} \right]^{-1}$$

$$= (1/v_0)(b_H - b_D)^2 \left[ \frac{1}{\phi_D N_D} + \frac{1}{\phi_H N_H} - 2\chi \right]^{-1} \tag{13}$$

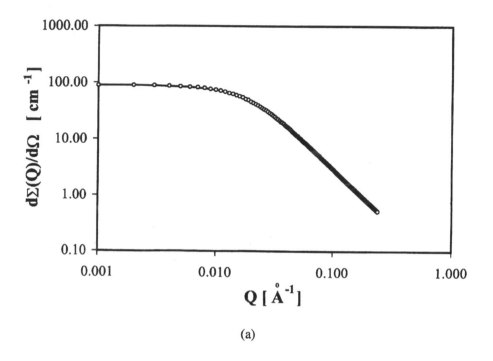

(a)

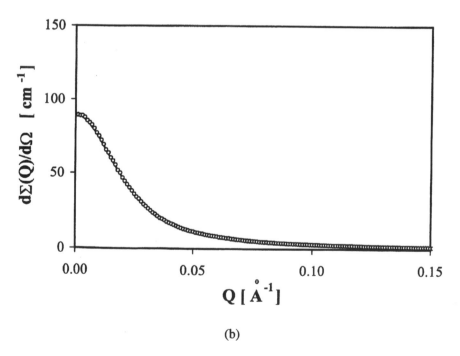

(b)

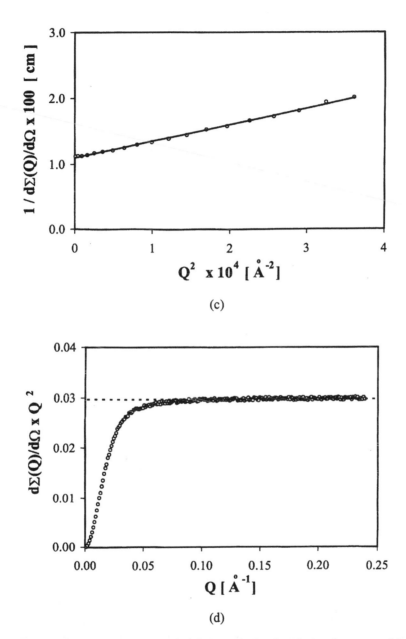

FIGURE 4. Schematic representation of various methods of analysing the scattered intensity for a blend of hydrogenous and deuterated polymers. Data points were generated using the RPA approximation for a poly(dimethylsiloxane) isotopic blend: $\phi_H = \phi_D = 0.5$, $M_w(H) = M_w(D) = 70{,}000$ g mol$^{-1}$, $a = 5.61$ Å and $\chi = 0.00062$ (Beaucage et al., 1996). (a) and (b) represent the differential scattering cross section versus scattering vector plots in a logaritmic (a) and linear scale (b). The initial data have also been plotted according to the Oernstein-Zernike or Zimm equation (c). (d) is a Kratky plot for the determination of the statistical length.

and therefore provides a measure of the interaction parameter if known values of degree of polymerization and composition are substituted in equation (13).

It has been implicitly assumed in the treatment discussed so far that the hydrogenous and deuterated polymers are monodisperse. Polydispersity affects the scattering curves but equations (12) and (13) can still be used if the degrees of polymerization are replaced by their weight averages and the $z$-average radii of gyration are used (Higgins *et al.*, 1993). A detailed account of the effect of polydispersity and differences between the distribution of molecular weights of the two polymers has been given by Lapp *et al.* (Lapp *et al.*, 1985).

In the intermediate $Q$ range the approximate expression of the form factor $P(Q)$ is no longer valid and for $Q \geq R_g^{-1}$ the complete Debye function should be used. This is given by equation (7) for a monodisperse system (Figures 4(a) and (b)). For a sample with polydispersity index $u = M_w/M_n - 1$, the Debye equation has been modified assuming a Schulz-Flory distribution of molecular mass (Greschner, 1973):

$$P_i(Q) = \frac{2}{(u+1)v^2}[(1+uv)^{-1/u} - 1 + v] \tag{14}$$

where $v = \langle R^2 \rangle_z Q^2/(1+2u)$ with $\langle R^2 \rangle_z$ being the mean-square $z$-average radius of gyration. After substituting the appropriate expression for the form factor, equation (6) can be used to fit the experimental data using non-linear least squares regression analysis. Since $M_w$ and $M_w/M_n$ are usually known from independent GPC measurements this leaves only three unknown parameters to be determined: the radii of gyration of the two components and the interaction parameter.

As pointed out by various authors (O'Connor *et al.*, 1991; Shibayama *et al.*, 1985) $P_1(Q)$ and $P_2(Q)$ in equation (4) are correlated and therefore the two $R_{g,i}$ values cannot be determined independently. Usually equation (8) is used with an average statistical length which is the same for the deuterated and hydrogenous polymers. This assumption is generally justified for isotopic blends of the same chemical species except for short chains where deviations from Gaussian statistics are expected (Higgins *et al.*, 1993; O'Connor *et al.*, 1991).

The statistical length $a$ can also be determined in the high $Q$ range $a^{-1} \ll Q \ll R_{gi}^{-1}$ from a Kratky plot, $Q^2 d\Sigma(Q)/d\Omega$ versus $Q^2$ (Figure 4(d)). In this $Q$ range equation (4) reduces to

$$S(Q) = \frac{12\phi_D\phi_H}{a^2 Q^2} \tag{15}$$

The choice of the method to be used to analyze the scattering data depends on the experimental $Q$ range compared to the particle size and while the use of Ornstein-Zernike plots relies on sufficiently low $Q$ values so that $Q \leq R_{gi}^{-1}$, evaluation of the statistical length from equation (15) requires the Kratky plateau to be well defined (Figure 4(d)). This depends on the polymeric system under study and on range of validity of Gaussian statistics. The use of the Debye equation also rests on the validity of Gaussian scaling and while for highly flexible polymers such as poly(dimethylsiloxane) the RPA function was found to fit the experimental data over three decades in size (Beaucage *et al.*, 1996) for other polymers such as stereoregular poly(methyl methacrylate) (PMMA) deviations from Gaussian statistics have been observed (O'Reilly *et al.*, 1985; Bates *et al.*, 1988;

Hopkinson *et al.*, 1994). It has been noted that for syndiotactic PMMA there is a tendency for appearance of particular stereochemical sequences along the chain (Yoon *et al.*, 1976) and this gives rise to deviations from the scattering predicted for a Gaussian distribution of segments (Hopkinson *et al.*, 1994).

The temperature, composition and molecular weight dependence of the interaction parameters in isotopic polymer blends have been investigated by various authors. The results differ from system to system investigated and while for example composition dependent interaction parameters have been reported for isotopic blends of syndiotactic PMMA (Hopkinson *et al.*, 1994) no composition dependence has been observed for mixtures of hydrogenous and deuterated poly(styrene) (PS) (O'Connor *et al.*, 1991; Londono *et al.*, 1994). The composition dependence has been interpreted in terms of a theory developed for the thermodynamics of polymer solutions (Muthukumar *et al.*, 1986).

Measured interaction parameters have been usually found to depend upon the molecular weight of the components (O'Connor *et al.*, 1991; Beaucage *et al.*, 1996; Bates *et al.*, 1988). A pronounced molecular weight dependence has been observed for bimodal poly(styrene) blends where a high $M_w$ deuterated component ($M_w \approx 100,000$) was mixed with low molecular weight hydrogenous PS ($M_w \approx 1,000$–$3,000$) (O'Connor *et al.*, 1991). A $\chi$ value of $6 \times 10^{-4}$ was measured at $M_w = 2,950$ and 0.02 at $M_w = 1,050$. The unusually high interaction parameters measured for this blend were attributed to non-ideal mixing between the low and high molecular weight components as a result of chain end effects. These were also found to be crucial to account for the molecular weight dependence in isotopic PDMS blends (Beaucage *et al.*, 1996).

As for generic mixtures of two polymeric species, isotopic blends present $\chi$ values which follow a linear temperature dependence of the form (Beaucage *et al.*, 1996; Bates *et al.*, 1988):

$$\chi = A + \frac{B}{T} \tag{16}$$

where $A$ is an entropic component which is not accounted for in the original Flory-Huggins theory (Flory, 1969). The interaction parameters increase with decreasing temperature indicating the existence of a upper critical solution temperature for these mixtures. Thus increasing temperature leads to a reduction of the scattering intensity (Figure 5) which is consistent with the UCST behaviour. Phase separation is expected to occur when the interaction parameter equals a critical value $\chi_c$

$$\chi_c = \frac{1}{2} \frac{(\sqrt{N_H} + \sqrt{N_D})^2}{N_H N_D} \tag{17}$$

which for a symmetric blend with $N_H = N_D = N$ is given by $(N\chi_c) = 2$. The parameter $\chi_c$ decreases with increasing degree of polymerisation and therefore phase separation is more likely to occur for high molecular weight blends.

Only a limited number of phase separated systems have been reported in the literature. In many cases the critical temperature is well below the polymer glass transition and therefore phase separation cannot be observed experimentally. For PDMS, a polymer with low $T_g$, SANS studies were carried out in a wide temperature range. Phase separation was observed in symmetric blends with $M_w = 300,000$ g/mol at room temperature and for lower

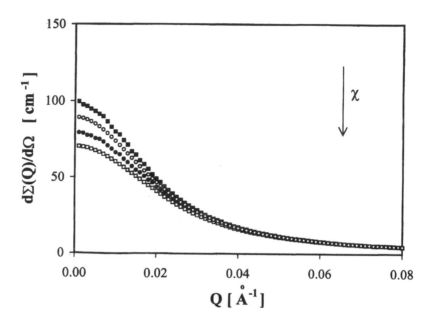

FIGURE 5. Effect of temperature on the scattered intensity for an isotopic blend of hydrogenous and deuterated PDMS (50:50 blend, $M_w(H) = M_w(D) = 70,000$ g mol$^{-1}$, $a = 5.61$ Å). Data points were generated using the RPA approximation with values of interaction parameters calculated from data reported in the literature (Beaucage *et al.*, 1996): (■) $\chi = 0.00062$ ($T = 25$ C), (○) $\chi = 0.00047$ ($T = 50$ C), (●) $\chi = 0.00022$ ($T = 100$ C) and (□) $\chi = 0$.

molecular weight samples ($M_w < 100,000$) below room temperature using a cryostat (Beaucage *et al.*, 1996).

### 8.4.1.2   Blends of Chemically Dissimilar Polymers

SANS can be applied to studies of polymer blends both in the one and two phase regions of the phase diagram. Figure 6 provides an example of experimental data for a blend of fully deuterated PMMA and solution chlorinated polyethylene (SCPE) as a function of annealing. The normalized coherent cross section, $d\Sigma(Q)/d\Omega$, of a one-phase polymer blend extrapolated to $q = 0$ is inversely proportional to the second derivative of the Gibbs free energy of mixing $\partial^2 \Delta G_m/\partial\phi^2$ and therefore to $\chi$.

The scattered intensity increases with increasing temperature. This is consistent with the reduction of $\partial^2 \Delta G_m/\partial\phi^2$ towards zero as the spinodal is approached and the lower critical solution temperature behaviour of this blend (Clark *et al.*, 1993). The top two curves in Figure 6 correspond to data from the same blend at two temperatures in the two-phase region. The scattering patterns differs considerably from those lower temperatures and this provides the basis for discriminating between one-phase and two-phase behaviour. Phase separation will be discussed in the following section while here the focus is on the data analysis of mixed blends.

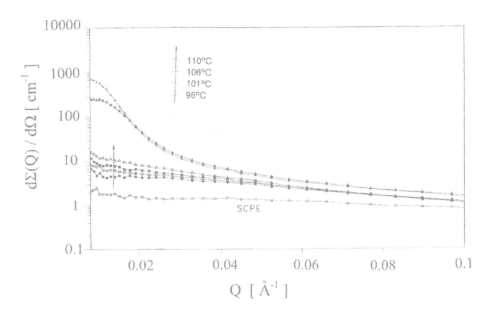

FIGURE 6. Small angle neutron scattering data for (○) an SCPE sample and for a $d_8$-PMMA/SCPE blend at four annealing temperatures inside the one-phase region from 96 to 110°C and at two temperatures inside the two-phase region (▲, ∗). The solid lines connecting the experimental points have been drawn for clarity. Reprinted with permission from Clark *et al.* (1993). Copyright 1993, American Chemical Society.

The discussion on isotopic blends and the scattering equations given in section 8.4.1.1 can be used as the starting point in the analysis of the scattering of any binary mixture. Obviously, the difference between the segmental volumes of components 1 and 2 needs to be explicitly considered and therefore equations (3) and (4) cannot be simplified. Provided there is some contrast between the two components SANS studies could be performed on mixtures of unlabelled polymers. This requires that the two monomeric units contain different carbon to hydrogen ratios, a condition which has been rarely achieved for partially miscible polymers. For this reason most of the SANS studies reported in the literature have been performed by deuterating one of the two components.

Warner *et al.* have extended the RPA expression given by equation (4) to include blends of two polymers 1 and 2 where a portion c of polymer 1 is deuterated (Warner *et al.*, 1983). The scattered intensity for such a mixture is given by:

$$\frac{d\Sigma(Q)}{d\Omega} = \frac{[cb_{1,D} + (1-c)b_{1,H} - \beta b_2)^2]/v_1}{[\phi_1 N_1 P_1(Q)]^{-1} + \beta[\phi_2 N_2 P_2(Q)]^{-1} - 2\chi}$$
$$+ \frac{c(1-c)(b_{1,D} - b_{1,H})^2 \phi_1 N_1 P_1(Q)}{v_1} \qquad (18)$$

where $b_{1,H}$, $b_{1,D}$ and $b_2$ are the scattering lengths per monomer unit of the hydrogenous and deuterated polymer 1 and of polymer 2, respectively. $\phi_i$ and $N_i$ are the volume fraction and degree of polymerisation of component $i$ and $\beta = v_1/v_2$. $\chi$ is the interaction parameter per segment mole of component 1.

Equation (18) consists of two terms: the first term represents the scattering due to concentration fluctuations which contains information on the polymer-polymer interactions and the second term is the form factor of component 1. A weighted subtraction of the scattering of two samples with the same composition of polymers 1 and 2 but differing in the hydrogenous and deuterated content of polymer 1 should yield the single chain scattering of component 1, thus providing a measure of the radius of gyration of 1 in the presence of polymer 2. Alternatively, the same information could be obtained if the contribution to the scattering of the structure factor (first term in equation (18)) is eliminated by choosing a "contrast matched" concentration c which fulfills the condition $[cb_{1,D} + (1 - c)b_{1,H} - \beta b_2 = 0]$ so that the first term in equation (18) equals zero (Maconnachie et al., 1989).

It must be noted that in view of the existence of a more or less pronounced isotopic effect the procedure described can only be applied after careful thermodynamic considerations about the system under study. This is clearly evidenced by the large shifts (as large as 40°C) of the cloud point which have been observed upon deuterating one of the components in a binary mixture (Yang et al., 1983). It follows that two samples differing in the composition of hydrogenous and deuterated component 1 will present different cloud points even if the composition of the 1 and 2 components is identical in the two samples.

The procedure which is now preferred makes use of the original RPA expression developed by deGennes. The interaction parameters $\chi$ are extracted from the neutron scattering data using equation (4) but different approaches may be used. As discussed for isotopic blends if the condition $QR_g < 1$ is not satisfied then the Debye equation is used instead to represent the form factor $P(Q)$. Although the degrees of polymerisation of polymers 1 and 2 need to be known from independent measurements to calculate the interaction parameter, different assumptions can be made to extract information on polymer dimensions. For example the radii of gyration of the two components can either be assumed equal to their unperturbed dimensions and used as fixed parameters (Hopkinson et al., 1995) or an average molecular dimension, a radius of gyration or a statistical segment length, is obtained from the fitting. The latter procedure has been used to analyse the scattering of deuterated polystyrene/polyvinylmethylether blends with an average statistical segment length $a$ given by (Han et al., 1988):

$$R_{gi}^2 = \frac{N_i v_i}{6}\left(\frac{a^2}{v_o}\right)$$

(19)

Alternatively, an Ornstein-Zernike plot (equation (12)) can be used to determine the interaction parameter from the intercept at $Q = 0$ and the correlation length $\xi$ from the slope. This is related to the average statistical length as described by equation (10) with $a$ now given by (Han et al., 1988):

$$\frac{\overline{a^2}}{v_0} = \phi_1\phi_2\left(\frac{a_1^2}{\phi_1 v_1} + \frac{a_2^2}{\phi_2 v_2}\right)$$

(20)

where $a_1$ and $a_2$ are the segment statistical lengths of polymer 1 and 2, respectively. Equation (20) can also be modified to account for molecular weight distributions (Beaucage et al., 1993). Kratky plots of the high $Q$ data could be used to determine the average statistical segment length of the blend and these values could be compared with those determined from the Ornstein-Zernike plots or the fitting using the RPA equation.

A procedure largely adopted to extract the interaction parameters in model polyolefin blends makes use of mixtures of hydrogenous and deuterated polymers having the same molecular weight and molecular architecture to determine the statistical lengths $a_1$ and $a_2$ as well as the interaction parameters between hydrogenous and deuterated materials (Balsara et al., 1992). These parameters are then used to calculate the structure factor for ideal mixing:

$$\left(\frac{1}{S(Q)}\right)_{\text{ideal}} = \frac{1}{\phi_1 v_1 N_1 P_1(Q)} + \frac{1}{\phi_2 v_2 N_2 P_2(Q)} \tag{21}$$

where $P_1(Q)$ and $P_2(Q)$ are calculated from $a_1$ and $a_2$ obtained from two independent measurements. The interaction parameter for a hydrogenous and deuterated mixture of components 1 and 2 is then calculated from the difference function:

$$E(Q) = \frac{v_0}{2}\left[\frac{1}{\phi_1 v_1 N_1 P_1(Q)} + \frac{1}{\phi_2 v_2 N_2 P_2(Q)} - \frac{1}{S(Q)}\right] \tag{22}$$

where $S(Q)$ is the structure factor for the blend as defined by equation (4). According to RPA and Flory-Huggins theory $E(Q)$ should be $Q$-independent although a weak $Q$ dependence has been experimentally observed (Balsara et al., 1992; Graessley et al., 1993). The interaction parameter is therefore obtained by extrapolation to $Q = 0$.

According to Flory-Huggins theory, the interaction parameter is defined as a local parameter which arises from short-range, segment-segment interactions and should therefore be $Q$-independent. In practice a $Q$-dependent $\chi$ has been invoked to explain anomalous SANS results. For example it has been suggested that the non-linearity of $R_g$ values versus composition observed for deuterated PMMA/poly(styrene-co-acrylonitrile) blends could be interpreted in terms of a $Q$-dependent interaction parameter (Hahn et al., 1992). Similarly the anomalous temperature dependence of the scattered intensity for the system polytetramethyl carbonate/deuterated polystyrene was successfully interpreted using a short-range spatial model for $\chi(r)$ (Brereton et al., 1987).

By carrying out SANS measurements at different temperatures in the one-phase region of the phase diagram, it is possible to obtain the temperature dependence of $\partial^2 \Delta G_m/\partial \phi^2$ and therefore of $\chi$ and to calculate the spinodal temperature $T_s$. The interaction parameter is usually found to vary linearly with inverse temperature with either positive or negative slopes for UCST or LCST behaviour, respectively.

Interaction parameters measured by SANS are better described as effective parameters which are often not only temperature but also composition (Londono et al., 1994; Hopkinson et al. 1995) and molecular weight dependent (Shibayama et al., 1985; Han et al., 1988).

Upward and downward curvatures in $\chi$ with composition have been measured experimentally (Hopkinson et al., 1995) with a change between the two occurring as temperature increases. However it has been noted that upward curvatures are generally reported for polymer blends.

Various explanations have been proposed for the composition dependence of the interaction parameters. For example Kumar (Kumar, 1994) has attributed the composition dependence of the SANS interaction parameters to small excess volume changes on mixing which are not incorporated in the incompressible random phase approximation. Kumar's results on Monte Carlo simulations indicate that very small volume changes (as small as 0.05%) are sufficient to produce a significant composition dependence at the extremes of the composition range and explain some of the trends observed by SANS.

The effect of deuteration on phase behaviour has been extensively discussed in the literature. The phase diagrams have been found to depend upon deuteration with large shifts in LCST behaviour being observed when one of the components is deuterated compared to the unlabelled system (Yang et al., 1983). The effect of deuteration is also clearly evidenced by the difference in $\chi$ values which has been observed for a binary blend upon changing the component of the blend that carries the deuterium label (Krishnamoorti et al., 1994). Attempts have been made to explain the "label-switching" effect in terms of the changes on the interactions between the components which take place when replacing hydrogen with deuterium as a consequence of the reduction in length and polarisability of the bond (Hopkinson et al., 1995).

### 8.4.1.3  Polyolefins

The thermodynamic behaviour of polyolefin blends is of considerable technological importance and has been extensively investigated by SANS during the past few years. The small difference between the physical properties of the constituents in these blends makes SANS the main technique for thermodynamic studies and clearly shows the power of the labelling method.

Commercial polyolefin blends are multicomponent systems whose physical properties depend upon the extent of miscibility among the constituent polymers. Chain architecture and branching certainly play a crucial role in determining the thermodynamic behaviour.

Most of the SANS studies in polymer blends have been conducted using binary model polyolefin blends which were prepared by saturating the double bonds of a variety of anionically synthesized polydienes. Saturation with $H_2$ produces the hydrogenous polymer while introducing $D_2$ the partially deuterated analogue is obtained (Balsara et al., 1992; Graessley et al., 1993). This procedure ensures that the labeled and unlabelled components have closely matched molecular weights and microstructures.

A large number of model systems have been synthesized and SANS data have been reported for a variety of binary mixtures. Interaction parameters measured by SANS were generally found to be independent of molecular weight and insensitive to composition in the range $0.25 \leq \phi \leq 0.75$ (Balsara et al., 1992; Graessley et al. 1995). The temperature dependence is usually of the form given in equation (16) and the blends show an upper critical solution temperature.

As observed for other systems, the interaction parameters vary upon deuterium substitution and show the label-switching effect discussed above with differences in critical temperature between blends where the deuterium label is switched from one component to the other which are as high as 70°C (Graessley et al., 1993). The deuterium labelling effect appeared to exhibit a consistent pattern that could be explained considering the microstructure of the polymers: $\chi$ values were always found to be larger when the more branched component was deuterated (Graessley *et al.*, 1993). This is consistent with the observation that the C-H bond length and polarisability reduces upon deuterium substitution and therefore the cohesive energy density is also expected to decrease with labelling.

A scheme was developed to rationalize in a qualitative manner the interaction strengths of a number of model polyolefins using a solubility parameter formalism (Graessley *et al.*, 1993). The interaction parameters were expressed in terms of the Hildebrand equation (Hildebrand *et al.*, 1950):

$$\chi = \frac{v_0}{k_B T}(\delta_2 - \delta_1)^2 \tag{23}$$

where $v_o$ is a reference volume and $\delta_1$ and $\delta_2$ the solubility parameters of the components which are defined as the square roots of the cohesive energy density. By exploiting the label-switching effect to determine the sign of the term $(\delta_2 - \delta_1)$ and using an arbitrary reference it was possible to assign values of solubility parameters to many systems (Graessley *et al.*, 1993; Krishnamoorti *et al.*, 1994; Graessley *et al.*, 1995).

Given a blends of two components $\alpha$ and $\beta$, the results were tested using blends of the two components with a third component $\gamma$. For the solubility parameter formalism to be consistent the following condition has to be fulfilled:

$$(\delta_\alpha - \delta_\beta) = (\delta_\beta - \delta_\gamma) - (\delta_\alpha - \delta_\gamma) \tag{24}$$

meaning that the direct determination of the difference between the solubility parameters for the blend has to equal the value calculated from the right hand side of equation (24).

Blends which satisfy equation (24) are called *regular* blends whereas inconsistency with the above equation indicates that one or more blends behaves as *irregular* mixtures (Graessley *et al.*, 1995). Anomalous behaviour which could not be explained by the solubility parameter formalism was reported for polyisobutylene (Krishnamoorti *et al.*, 1995) and polypropylene (Reichart *et al.*, 1997) blends. Large negative values of interaction parameters and a LCST behaviour was observed for some of these hydrocarbon blends indicating a net attraction between the components. The origin of this attractive interaction is difficult to be understood for blends with non-polar components where positive $\chi$ values are expected with a UCST.

The departure from a solubility approach has been expressed in terms of an extra interaction strength (Graessley *et al.*, 1995; Reichart *et al.*, 1997):

$$X_E \phi_1 \phi_2 = \left[ \chi \frac{k_B T}{v_0} - (\delta_1 - \delta_2)^2 \right] \phi_1 \phi_2 \tag{25}$$

According to the definition above, $X_E$ equals zero if the measured interaction strength (the first term in equation (25)) is equal to the value calculated from the solubility parameters. As illustrated in Figure 7 for polypropylene blends, the miscibility behaviour largely depends on the second species with changes from regular to highly irregular mixing. The observed deviations from regularity can be either positive (favouring phase separation) or negative (favouring miscibility).

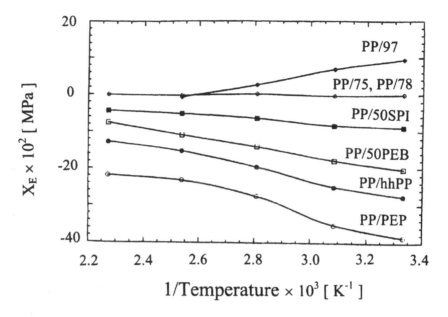

FIGURE 7. Temperature dependence of the extra interaction strength $X_E$ for blends of poly-(propylene) (PP) with head-to-head PP, saturated polyisoprene with a range of 3,4 contents (PEP, 50SPI and 75SPI), saturated polyethylbutadiene (50PEB) and saturated polybutadienes with 97% and 78% 1,2 content (97, 78). Reprinted with permission from Reichart *et al.* (1997). Copyright 1997, American Chemical Society.

### 8.4.1.3  Pressure Experiments

Recent interest in the pressure effect on miscibility in blends has stimulated a number of in-situ pressure experiments and theoretical developments of methods for interpretation of the SANS data in compressible systems.

The most investigated system is deuterated polystyrene/poly(vinyl methylether) (d-PS/PVME) whose thermodynamics at ambient pressure has been extensively investigated. For this as for other systems exhibiting LCST behaviour the scattered intensity increases with increasing temperature as the LCST is approached (Figure 6). A pressure increase has the opposite effect indicating an upward shift of the spinodal which can be as high as 25 to 30°C/ kbar. The pressure dependence of the intercept at $Q = 0$ is shown in Figure 8 for d-PS/PVME (Hammouda *et al.*, 1995).

The interaction parameters which have been determined from this and other systems (Janseen *et al.*, 1993) using the RPA formalism could be separated into an enthalpic and an entropic contribution. As expected, the enthalpic part was found to be independent of pressure while the entropic part decreases with increasing pressure.

Although the RPA equation developed for incompressible systems was used in some cases to derive $\chi$ parameters (Janseen *et al.*, 1993; Hammouda *et al.*, 1995) this cannot represent the correct procedure for an obviously compressible system.

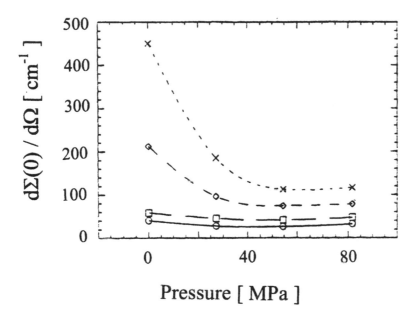

FIGURE 8.  Pressure dependence of the scattered intensity at $Q = 0$ for a 10/90 deuterated PS/PVME blend at (○) 60; (□) 80; (◇) 100 and (×) 110°C. Reprinted with permission from Hammouda *et al.* (1995). Copyright 1995, American Chemical Society.

A method of data analysis which makes use of the compressible RPA formulation and the Sanchez-Lacombe lattice fluid (LF) equation of state has been proposed (Bidkar *et al.*, 1995) and applied to explain the pressure dependence of the scattered intensity in d-PS/PVME data (Hammouda and Benmouna, 1995). This approach treats a compressible mixture as a three component incompressible blend where free volume is included as the third component. Free volume fractions and intermonomer interactions can be extracted from the experimental data by solving self-consistently the LF and RPA equations.

The procedure relies on a knowledge of the cohesive energy densities of the blend components which are taken from tabulated values. It has been noted that this may represent the main drawback of the methods since the LF-RPA approach appears to be very sensitive to slight changes in these values (Hammouda and Benmouna, 1995). A procedure for determining the internal pressure of the components from SANS transmission measurements has been proposed (Hammouda and Benmouna, 1995).

## 8.5 PHASE SEPARATION STUDIES

Phase separation in a polymer blend exhibiting a LCST behaviour as shown in Figure 2 takes place when the sample is heated above a critical temperature $T_C$.

The spinodal line is defined by the locus of the points with $\partial^2 \Delta G_m / \partial \phi^2 = 0$. By considering the relationship between the second derivative and the differential scattering cross section it is possible to determine experimentally the spinodal temperature from extrapolation of the inverse intercept $[d\Sigma(Q = 0)/d\Omega]^{-1}$ versus temperature (Han *et al.*, 1988).

As illustrated in Figure 6 the scattering patterns in the one and two-phase region of the phase diagram present distinct features thus providing a means to locate the phase boundary. The binodal is in fact determined as the temperature at which a strong forward scattering is first observed due to the appearance of a domain structure. As with light and X-ray scattering it is also possible by SANS to distinguish between the possible mechanisms of phase separation (Higgins, 1985): (a) nucleation and growth and (b) spinodal decomposition. The morphology formed differs in the two cases and so does the angular dependence of the scattered intensity which is measured experimentally.

### 8.5.1  *Spinodal Decomposition and Kinetics of Phase Separation*

The process of spinodal decomposition takes place by rapidly heating (for LCST) or cooling (for UCST) the sample from the one-phase stable region into the unstable region of the phase diagram where $\partial^2 \Delta G_m / \partial \phi^2 < 0$. As a result of the temperature jump concentration fluctuations will grow in amplitude and during the early stages of the phase separation process, a very regular co-continuous structure develops which gives rise to a diffraction maximum at $Q_m$ in the scattering pattern. The position of the maximum defines a characteristic size $\Lambda_m = 2\pi/Q_m$ which is interpreted in terms of a characteristic wavelength of the concentration fluctuations. For the system shown in Figure 9 the wavelength of the unstable concentration fluctuation is of the order of 1000 Å and the spinodal process cannot be investigated using light scattering: SANS or SAXS, techniques which probe smaller particles sizes, must be used.

The kinetics of the phase separation process can be studied by measuring the time dependence of the scattered intensity versus the scattering vector $Q$. For low molecular weight systems phase separation processes are fast and therefore difficult to investigate experimentally. Polymer phase separation occurs on a time-scale which is accessible by time-resolved scattering methods and for this reason polymeric materials have been used to test the validity of theoretical models developed for low molecular weight materials.

As shown in Figure 9 after a rapid temperature jump from a temperature $T_o$ in the one phase region to a temperature $T$ in the two phase region (the samples reached the final temperature in less than 20 s) the scattered intensity rises sharply and a maximum develops at $Q_m$ corresponding to the wavelength whose growth rate is fastest. These measurements were carried out at a lower $Q$ value than that covered in Figure 6 and therefore the complete spinodal peaks are clearly visible (see top two curves in Figure 6). The position of the maximum is initially time independent (Hill *et al.*, 1985; Higgins *et al.*, 1989; Higgins 1988) indicating that the difference between the composition of the two phases increases but the characteristic domain size remains constant.

During the early stages of spinodal decomposition (where $\partial^2 \Delta G_m / \partial \phi^2$ does not differ too strongly from its initial value) the time evolution of the scattered intensity at constant temperature is described by the Cahn-Hilliard-Cook theory of demixing (Binder, 1983;

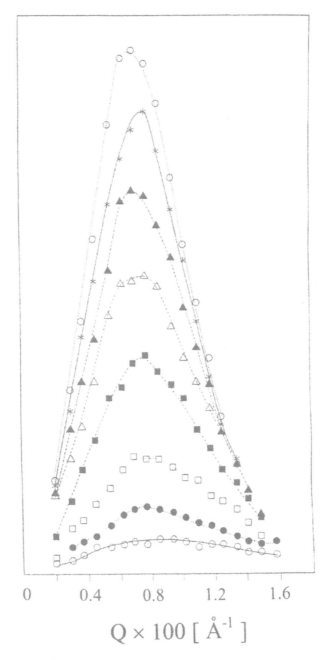

FIGURE 9. Time evolution of the scattered intensity as a function of scattering vector $Q$ for a $d_8$-PMMA/SCPE blend after a temperature jump from the one phase to the two-phase region. The data sets were taken at 16 s intervals except for the first set which was taken 26 s after the temperature jump.

Cahn *et al.*, 1959). The time-dependent structure factor $S(Q, t)$ is given by (Okada *et al.*, 1986):

$$S_T(Q, t) = S_T(Q) + (S_{T_o}(Q) - S_T(Q)) \cdot L^2(Q, t) \qquad (26)$$

where $S_{T_o}(Q)$ and $S_T(Q)$ are the equilibrium structure factors at temperatures $T_o$ and $T$:

$$S_T(Q) = S_T(Q, t \to \infty) \qquad (27)$$

The static structure factors $S_T(Q)$ are given by equations analogous to (12). Upon examination of equation (13), and remembering that $\partial^2 \Delta G_m / \partial \phi^2 < 0$ inside the spinodal it is apparent that $S_T(Q)$ would be negative at least over the lower $Q$ range. $S_T(Q)$ can only be obtained from extrapolation procedure (Müller *et al.*, 1996). It has been shown (Higgins *et al.*, 1989) that for temperatures away from the spinodal (i.e. $\Delta T > 2\,\text{K}$) $S_T(Q) \ll S_{T_o}(Q)$ and equation (26) reduces to a simpler form.

$L(Q, t)$ is the normalized dynamic structure factor which represents the relaxation function for a fluctuation with scattering vector $Q$. According to the Cahn-Hilliard-Cook theory during the early states of the spinodal process $L(Q, t)$ is expressed by an exponential function:

$$L(Q, t) = \exp[-R(Q)t] \qquad (28)$$

where $R(Q)$ is the growth rate of the concentration fluctuations. This is defined in terms of the collective diffusion constant $D_C$:

$$R(Q) = Q^2 D_C(Q) = \frac{Q^2 \Lambda(Q)}{S_T(Q)} \qquad (29)$$

which consists of a kinetic and a thermodynamic factor: the Onsager coefficient $\Lambda(Q)$ and the static structure factor at $T$, respectively.

The growth rate of the concentration fluctuations can be determined experimentally from plots of $\ln(L(Q, t))$ versus time at constant $Q$. Although linear $\ln(L(Q, t))$ versus time plots are expected according to theory, non-exponential time behaviour has been observed even after the $S_T(Q)$ forms have been accounted for (Müller *et al.*, 1996).

For polymers, the collective diffusion constant $D_C$ represents the centre of mass diffusion although this description is correct for $Q < R_g^{-1}$. The Onsager coefficient is then expected to be $Q$-dependent and proportional to the Debye function for Gaussian coils. Given that many systems phase separate with the peak in $S(Q)$ in the light scattering range, where the Debye curve is effectively constant and equal to unity, $Q$-dependent Onsager coefficients have been limited to neutron experiments. For $Q > R_g^{-1}$ a $Q^{-2}$ dependence is predicted and it has been observed experimentally (Schwahn *et al.*, 1992). However, the characteristic length defined by $\Lambda(Q)$ is always found to be larger than the molecular radius of gyration up to about five to seven times. This was interpreted as being due to elasticity effects of temporal networks which form in asymmetric entangled polymer mixtures (Onuki, 1994; Kawasaki *et al.*, 1993). Onsager interaction ranges larger than the radius of gyration have been also reported for symmetric isotopic blends (Jinnai *et al.*, 1993; Müller *et al.*, 1996).

In the intermediate and late stages of the spinodal decomposition process, the Cahn-Hilliard theory fails and coarsening of the concentration fluctuations yields to aripening domain structure. Due to the growth of the dimensions of the domains of the maximum at $Q_m$ shifts towards smaller $Q$ values with increasing time (Schwahn et al., 1987). At the end of the process the morphology may be indistinguishable from that developed during nucleation and growth.

### 8.5.2 Domain Structure

In the metastable region between the binodal and spinodal curves (Figure 1), nucleation processes are responsible for phase separation but the process requires an activation energy and is at least an order of magnitude lower than spinodal decomposition. No regular structure develops and the scattering does not show a maximum as observed for spinodal processes. The phase separation process is characterised by the appearance of strong forward scattering. A "neutron cloud point" can be defined in analogy to "cloud point" determined by light scattering (Olabisi et al., 1979) from the temperature at which a considerable increase of the scattered intensity is observed. Generally, conventional and neutron cloud points will differ, this difference being related to size of the domains probed by the two techniques which is of the order of several thousand Ångstroms for light and could be as small as hundredths of Ångstroms for SANS.

For a non-homogeneous blend constituted of two separate phases having random shape and size (Figure 10(a)) and characterized by sharp boundaries between them, the scattering can be described by the Debye-Bueche model (Wignall et al., 1982; Debye et al., 1957; Debye et al., 1949). The fluctuation of the scattering power of the system which for neutron scattering is the scattering length density (defined as the scattering length per unit molecular volume) is expressed in terms of the two-point correlation function:

$$\gamma(r) = \frac{\langle \rho_A \rho_B \rangle}{\langle \rho^2 \rangle} \tag{30}$$

where $\rho_A$ and $\rho_B$ represent deviations in scattering length density at points $A$ and $B$ from the average value $\langle \rho^2 \rangle$. The correlation function is expressed in terms of a Debye-Bueche correlation length $a_c$:

$$\gamma(r) = \exp\left(-\frac{r}{a_c}\right) \tag{31}$$

which can be regarded as a measure of domain size. For concentrated systems, the mean chord intercept of phases 1 and 2, $l_1$ and $l_2$ provide a better estimate of the domain sizes. These are defined as the average length between the segments intercepting phases 1 and 2 (Figure 10) and can be calculated from $a_c$ using the following relationships:

$$\overline{l_1} = \frac{a_c}{\phi_2}$$

$$\overline{l_2} = \frac{a_c}{\phi_1} \tag{32}$$

(a)

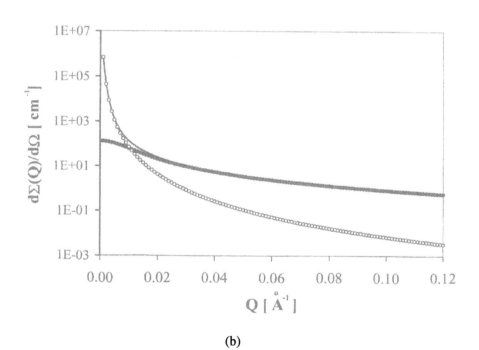

(b)

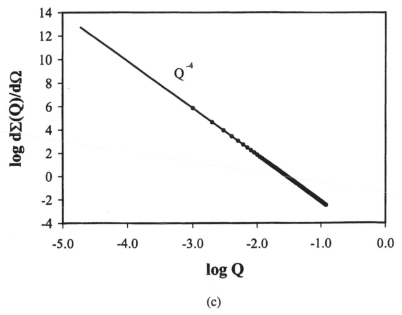

(c)

FIGURE 10. (a) Schematic illustration of a two-phase system showing the succession of chords $l_1$ and $l_2$ passing through the two phases. (b) Calculated scattered intensity for a phase separated blend. A simplified model has been used whereby the total scattered intensity (—) is considered as being due to two contributions: (a) the scattering within the domains calculated from the random phase approximation (•) and (b) the scattering from the two-phase structure modelled by the Debye-Bueche equation with a correlation length equal to 1 $\mu$m (o). For these large domain sizes the scattered intensity approaches Porod law as shown in (c) which is a log-log plot of the same (o) data from (b).

The neutron scattered intensity is then given by (Wignall *et al.*, 1982):

$$\frac{d\Sigma(Q)}{d\Omega} = \frac{8\pi a_c^3 \phi (1-\phi)\left(\frac{b_1}{V_1} - \frac{b_2}{V_2}\right)^2}{[1 + Q^2 a_c^2]^2} \tag{33}$$

and a plot of $[d\Sigma(Q)/d\Omega]^{-1/2}$ versus $Q^2$ should give a straight line. The correlation length and therefore domain dimensions can be calculated from the slope to intercept ratio via equation (32).

The Debye-Bueche model is only valid when $a_c \approx Q^{-1}$ and for SANS this translates to correlation lengths of approximately 1000 Å. When $Q \gg a_c^{-1}$ the Porod law and the Porod invariant (Higgins *et al.*, 1993) can be used to model the scattering. The application of these rules is valid for any two-phase structure with sharp boundaries for $Q \gg R^{-1}$, $R$ being a particle dimension. In these conditions the scattering varies as $Q^{-4}$ according to a formula

$$Q^4 \frac{d\Sigma(Q)}{d\Omega} = \left(\frac{b_1}{V_1} - \frac{b_2}{V_2}\right)^2 \frac{2\pi S}{V} \tag{34}$$

where $S$ and $V$ are the particle surface and volume, respectively.

Deviations from the Porod law have been observed. Positive contributions to the scattered intensity are attributed to void scattering and thermal density fluctuations while negative deviations might be caused by diffuse boundaries between the two phases. The existence of a diffuse interface has been incorporated in the Porod law and modifications have been developed for a number of morphologies such as spherical, cylindrical and lamellar domains (Richards *et al.*, 1983).

When the domain sizes are large compared to the experimental $Q$ range unambiguous interpretation of SANS data may not be possible. It is probably because of these difficulties that a unanimous consensus on the melt miscibility in linear and lightly branched polyethylene (PE) blends has not been reached.

Hill *et al.* have extensively investigated the possibility of liquid-liquid phase separation in polyethylene blends using indirect experiments in which rapidly quenched samples are investigated in the solid state by DSC and transmission electron microscopy (Schipp *et al.*, 1996). The existence of two distinct morphologies in these samples is then considered as evidence of phase separation in the melt prior to quenching. The domain sizes are of the order of several micrometers, too large to be detectable by SANS.

The difficulties encountered in interpreting the SANS data for these systems have been discussed recently (Schipp *et al.*, 1996) by analyzing the scattering from a blend of 10% linear deuterated PE and 90% branched polymer (50 short branched per 1000 backbone carbon atoms) which was expected from indirect experiments (where solid samples prepared by rapid quenching are analysed) to be phase separated. It was noted that due to the large domain sizes the scattering within the phases could not be neglected. The total scattering scattering was expressed by the sum of three contributions: the scattering within each of the two phases modeled by the RPA equation and the scattering due to the two-phase structure (Figure 10(b)). The latter was found to contribute negligibly to the scattered intensity within the experimental $Q$ range.

The effect of large domains on the scattered intensity is illustrated in Figure 10(b) where the scattering for a two-phase system has been calculated from equation (33) using a correlation length of $1 \mu m$. As shown in Figure 10(c) the scattered intensity approaches Porod law.

For $Q$ values larger than $0.02 \, \text{\AA}^{-1}$ the total intensity is well approximated by the scattering within the phases. It has been argued that the effect shown in Figure 10(b) could explain the difference in data interpretation reported in the literature (Schipp *et al.*, 1996). For example it was reported by other authors (Alamo *et al.*, 1994) who access rather lower Q values and would be sensitive to the low $Q$ upturn in Figure 10(b) that these blends are mixed at all composition when the existence of an isotopic effect between hydrogenous and deuterated species is taken into account.

Recent studies seem to indicate that a threshold value of branching is required for detecting liquid-liquid phase separation (Alamo *et al.*, 1997). According to these results homogeneous mixtures are obtained when the branch content is lower than 4 branches/100 backbone carbon atoms whereas phase separation occurs above 8 branches/100 backbone carbon atoms. Phase separation is also observed at low branch contents for high molecular weight species but it is attributed solely to isotope effects causing segregation between hydrogenous and deuterated chains. This is manifested in the SANS data as a change in the scattering law from $(\Sigma(Q)/d\Omega \propto Q^{-4})$ to $(d\Sigma(Q)/d\Omega \propto Q^{-2})$ i.e a change from Porod to Debye regime.

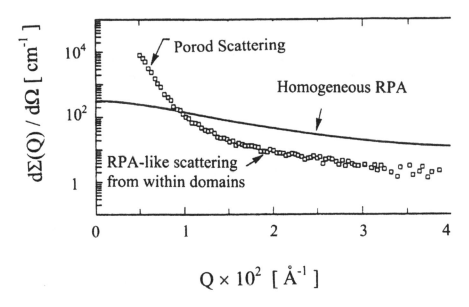

FIGURE 11. SANS data for a 75/25 blend of linear deuterated and hydrogenated polybutadiene (10.6 branches/100 backbone carbon atoms) compared to the RPA approximation at $T = 160°C$. The strong scattering at low $Q$ arises from the domain structure in the phase separated blend. Reprinted with permission from Alamo *et al.* (1997). Copyright 1997, American Chemical Society.

It has also been noted that although the scattering of a two-phase system will exhibit at high $Q$ the general shape $(d\Sigma(Q)/d\Omega \propto Q^{-2})$ of a single phase mixture, the RPA function cannot describe the scattering within the domains. At high $Q$ the scattered intensity of a two-phase mixture is a complicated function of volume fractions and compositions of the phases and it is expected to be lower than calculated for a homogeneous mixture as shown for polybutadiene in Figure 11.

## 8.6 INTERFACIAL STUDIES

Over the past ten years, neutron specular reflection has emerged as a powerful technique which can be used to study surfaces and interfaces in a variety of systems from semiconductors to biological membranes. For polymer blends the structure and composition profile of the interface between the two polymers is of considerable technological interest since the extent of mixing between the components plays a crucial role on the mechanical properties of the final product.

Neutron specular reflection provides information on the composition profile normal to a surface. This technique presents advantages compared to other experimental methods developed for investigating polymer-polymer interfaces such as transmission electron microscopy (TEM), infra-red densitometry, X-ray microanalysis (Goldstein *et al.*, 1981) or ion-beam techniques such as forward recoil spectrometry (Kramer et al., 1984) and secondary ion mass spectrometry (Benninghaven et al., 1987). The main advantage of neutron reflectivity is given by the high depth resolution of approximately 10 Å which is ideally suitable to study the detailed shape and size of the interface between two polymer species. As with SANS, the possibility of varying the scattering length via deuterium substitution can be exploited to highlight a specific layer in a multilayer structure. It is because of these features i.e. depth resolution and deuterium labelling that it has been possible to investigate the dynamics of the interdiffusion process between hydrogenous and deuterated polymers having the same chemical structure (Russell et al., 1988; Karim et al., 1994) or two different chemical species (Fernandez et al., 1988).

Neutron reflection can be treated using the same formalism developed to describe reflection and refraction phenomena of light or X-rays (Higgins et al., 1993). A neutron refractive index, $n$, is defined and, neglecting absorption, this is given by

$$n = 1 - \lambda^2 \left( \frac{\rho_b}{2\pi} \right) \tag{35}$$

where $\rho_b$ is the scattering length density of the medium defined as the average scattering length per unit volume. Neutron refractive indices are typically close to but smaller than unity and $(1 - n) \sim 10^{-6}$. From Snell's law an approximate expression for the critical glancing angle i.e. the angle of incidence for which total external reflection occurs is defined as

$$\frac{\theta_c}{\lambda} = \left( \frac{\rho_b}{2\pi} \right)^{1/2} \tag{36}$$

which is small ($\leq 1°$). The reflectivity is defined as the ratio of the reflected to the incident neutron intensities and it is unity for incident angles smaller than $\theta_c$. For $\theta > \theta_c$, the reflectivity falls off very sharply and the dependence on the angle of incidence can be described using Fresnel's law.

Data analysis is generally performed by calculating the reflectivity profiles using a multilayer matrix formalism (Lekner et al., 1987). This requires that assumptions are made on the interfacial shape which is then described as a sequence of thin homogeneous layers. Comparison between experimental and calculated reflectivity curves gives information on the refractive index as a function of depth normal to the surface of the sample. Because the refractive index is related to the scattering length density, the reflectivity data convey information on the composition profile normal to the surface.

Polymer-polymer interdiffusion is investigated using samples which consist of two layers of either the pure polymer components or two mixtures of different composition. The interfacial broadening is measured after annealing the samples at temperatures above the glass transition of the two components for a given length of time. The shape and size of the polymer-polymer interface depends on the thermodynamics of the system.

For a miscible blend or a partially miscible mixture annealed in the one-phase region, polymer-polymer interdiffusion occurs without any limit. Only the initial stages of the diffusion process can be investigated with neutron reflectivity (Sauer et al., 1991) since the technique has good sensitivity for interfaces up to a few hundred Angstroms (Stamm et al., 1992). By observing the development of the interfacial profile between hydrogenous and deuterated polymers it has been possible to test deGennes' theoretical predictions (Karim et al., 1994; deGennes, 1989) thus supporting the theory of reptation.

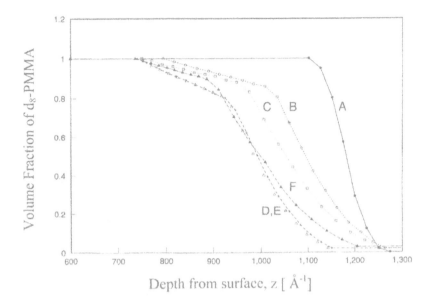

FIGURE 12. Interfacial profiles used to fit the experimental reflectivity data for a $d_8$-PMMA/SCPE bilayer: as made (A) and at different annealing stages at 120°C: 37 min (B), 141 min (C), 203 min (D), as D + 6 months at room temperature (E) and 243 min (F). Fernandez *et al.* (1991). Reproduced by Permission of the Royal Society of Chemistry.

For a partially miscible system annealed in the two-phase region an equilibrium interface is achieved at sufficiently long annealing times (Fernandez *et al.*, 1991) contrary to interdiffusion in the one-phase region. The time dependence of the interfacial broadening depends not only on the diffusion coefficients of the individual polymers but also on the interaction parameter $\chi$ and therefore on the thermodynamics of the system. Figure 12 illustrates the evolution of the interface for the blend of solution chlorinated polyethylene and deuterated PMMA annealed at 120°C inside the two-phase region as a function of annealing time. The concentration profiles represent the best fits to the reflectivity data after trying several model interfacial profiles. The interface presents a discontinuity on the PMMA side and a longer tail on the SCPE side. The shape and width of the interface do not change during the first annealing stages although the position of the interface shifts toward the PMMA side. After annealing (curve E), this process becomes slower and the bottom layer becomes enriched with PMMA as a result of its diffusion in the SCPE layer.

The interfacial shape is consistent with theoretical predictions of diffusion processes between two polymer species having different mobilities (Brochard-Wyart, 1990) where a net flux of matter across the boundary between the two polymers is expected towards the species of lower mobility (in this case SCPE).

Neutron reflectivity is particularly suited to investigate the narrow interface which is obtained between two immiscible polymers such as polystyrene and PMMA (Fernandez *et al.*, 1988). For two homopolymers with degree of polymerisation $N_1$ and $N_2$ the interaction parameter can be derived from the width of the interface between the two polymers, $l$ (Schubert *et al.*, 1995):

$$\chi = \left(\frac{b}{1\sqrt{6}}\right)^2 + \frac{\pi^2}{6}\left(\frac{1}{N_1} + \frac{1}{N_2}\right) \tag{36}$$

where $b$ is the average statistical segment length. As the molecular weight of the components increases the second term in equation (36) becomes negligible and the interaction parameter mainly depends on $b$ and $w$. Experimental $w$ values are generally found to be larger than expected. For example neutron reflectivity studies on the immiscible system polyethylene / polystyrene in the melt state give an interfacial width of 2.96 nm which is almost twice the theoretical interfacial width for this system (Hermes *et al.*, 1997). The discrepancy between calculated and experimental results is attributed to the measured interfacial width being a convolution of the true interfacial width $l_t$ and a secondary width $l_o$:

$$1 = (l_o^2 + l_t^2)^{1/2} \tag{37}$$

Different explanations have been given in the literature to account for the parameter $lo$ which has been attributed for example to a capillary interfacial width (Müller *et al.*, 1995) or to the interfacial width of the "as made" sample before annealing (Schubert, 1995). It has been shown that the experimental $\chi$ values depend on the specific definition of $l_o$ which is used with discrepancies between values as large as an order of magnitude (Hermes *et al.*, 1997).

By using deuterium substitution, neutron reflectivity can also be employed to study the shape of the concentration profile produced by a copolymer at the interface between an immiscible pair (Bucknall *et al.*, 1993, 1994). If the scattering length density profile is converted to copolymer volume fraction as a function of depth, the interfacial excess can be calculated and the results compared to self-consistent mean-field theory (SCMF).

The problem with the model fitting approach which is usually adopted to interpret the reflectivity data is that while a particular interfacial profile might fit the data, it is not certain whether the model describes the profile in a unique way. Other methods can be used to analyse the reflectivity data (Higgins *et al.*, 1993). For example the so-called kinematic approximation which provides analytical solutions for the reflectivity profiles and direct inversion of the experimental data are of limited applicability in polymer studies, although they may give unambiguous information on the depth profile (Higgins *et al.*, 1993). Bayesian analysis method (Sivia *et al.*, 1991) is now becoming an increasingly popular method to discriminate in an objective way between different possible interfacial profiles which describe the reflectivity data (Bucknall *et al.*, 1994).

The application of neutron reflectivity is not limited to interfacial studies. The technique can provide information on the near-surface composition which determines the properties of polymer coatings such as wetting and lubrication. Here ion-beam techniques have limited depth resolution (100 to 800 Å) and therefore neutron reflectivity provides a suitable probe of the detailed profile near the sample surface and a means to test theoretical predictions of surface enrichment (Genzer *et al.*, 1996; Schmidt *et al.*, 1985; Genzer *et al.*, 1994).

Current discussion focusses on the effect of surface and interfacial capillary waves in these polymeric systems (Sferrazza *et al.*, 1997). It has been recently suggested that the apparent extra broadening observed at the interface between two immiscible polymers is a result of thermally excited capillary waves and evidences have been provided which support this hypothesis.

## 8.7 DYNAMICS

Recently there has been increasing interest in the dynamics of polymer blends and a large number of studies have focused on the effect that blending produces on the motion of each of the components in a blend. One of the first criteria for blend compatibility is the existence of a single glass transition process, $T_g$ whereas an incompatible mixture will show two $T_g s$ which are located at the same temperature as those of the pure components. Numerous techniques can be employed in these studies, for example dynamic mechanical thermal analysis (DMTA), dielectric spectroscopy (DS) and NMR. These techniques give additional information compared to differential scanning calorimetry measurements (DSC) since the distribution of relaxation times associated to the glass transition can be evaluated from the width and shape of the mechanical or dielectric relaxation processes. The width of the distribution of relaxation times provides means to characterize the compatibility of a polymer mixture at the segmental level. Compatible blends that exhibit a single glass transition are often characterized by broader $T_g s$ compared to those of the single components. This is indicative of changes in the local environment due to the presence of the second polymeric component.

Incoherent inelastic neutron scattering (IINS), quasielastic neutron scattering (QENS) and neutron spin-echo (NSE) provide information on the dynamics of polymeric systems on a frequency range which varies from $10^7$ to $10^{14}$ Hz and on a distance scale up to approximately 100 Å (Higgins *et al.*, 1993). This is a suitable range to study vibrations and rotations of side groups in polymers below the glass transition temperature and segmental motion above the glass transition. The information obtained is complementary to that obtained from other techniques, from IR spectroscopy to NMR. Neutrons present the advantage that during the same experiment the frequency and the geometry of the motion are investigated simultaneously. This information can be used to distinguish between intra- and inter-molecular effects.

To date only the effect of blending on local motions below $T_g$ has been investigated in some detail (Floudas *et al.*, 1992; Arrighi *et al.*, 1995). Two systems have been studied: a blend of solution chlorinated polyethylene and selectively deuterated PMMA (SCPE/d$_5$-PMMA) and a blend of polyvinyl methyl ether and deuterated polystyrene (PS/PVME). QENS experiments focused on the effect of blending on the rotational motion of the ester methyl group in PMMA and the O-CH$_3$ in PVME and the results indicate that the technique is suitable to detect changes in the motion of methyl group occurring upon blending.

Figure 13 shows the QENS data for the SCPE/d$_5$-PMMA system at 140 K and $Q = 1.78$ Å$^{-1}$. Due to the absence of side groups, the scattering from SCPE is purely elastic while d$_5$-PMMA contributes to an elastic and a quasielastic component which is well described by rotational hopping of the CH3 groups over the three-fold potential barrier

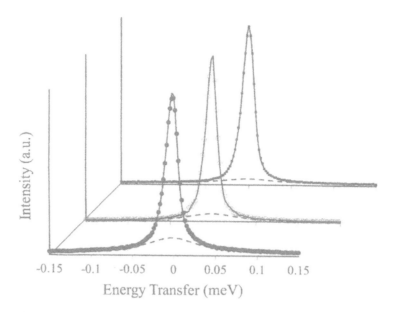

**FIGURE 13.** Quasielastic neutron scattering spectra at 140 K and $Q = 1.78\ \text{Å}^{-1}$ for (•) $d_5$-PMMA, (■) 50/50 SCPE/$d_5$-PMMA and (○) 30/70 SCPE/$d_5$-PMMA. The continuous line represent a fit to the experimental data and the dashed line the quasielastic component modeled by a Lorentzian line convoluted with the instrumental resolution. Reprinted with permission from Arrighi *et al.* (1996). Copyright 1996, American Chemical Society.

(Gabrys *et al.*, 1984; Floudas *et al.*, 1992; Arrighi *et al.*, 1995). The quasielastic broadening (simply modeled in Figure 13 by a Lorentzian line convoluted with the instrumental resolution) is considerably reduced in the blend compared to pure PMMA. The effect on the parameters characterizing the motion, the activation energy and the distribution of correlation times (modeled by a stretched exponential function) is illustrated in Figure 14(a) and (b). The activation energy increases with SCPE content indicating that the rotational motion is hindered by the presence of the second component. The $\beta$ parameter of the stretched exponential function decreases upon blending. In analogy to other studies reported in the literature for sub-$T_g$ motion (de los Santos Jones, 1994), blending does not always affect the dynamic properties of one of the constituents in binary mixture. The PS/PVME system for example did not show any effect due to the presence of PS on the rotational motion of the ether $CH_3$ group (Arrighi *et al.*, 1995).

## 8.8 SUMMARY

Neutron scattering is a well established probe of the structure and dynamics of condensed matter. In this chapter, the application of various neutron scattering techniques to polymer

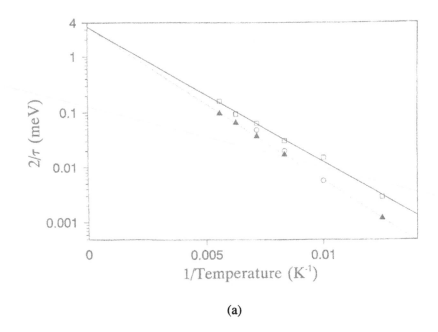

(a)

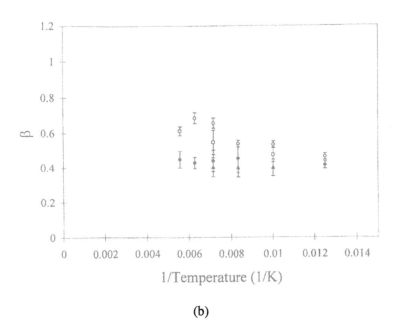

(b)

FIGURE 14. (a) Correlation times plotted as $2/\tau$ in meV, versus inverse temperature for ($\square$) $d_5$-PMMA, ($\circ$) 50/50 SCPE/$d_5$-PMMA and ($\blacktriangle$) 30/70 SCPE/$d_5$-PMMA. The lines indicate a fit to the PMMA and SCPE/PMMA data. (b) Parameter $\beta$ as a function of inverse temperature for ($\square$, $\circ$) $d_5$-PMMA, ($\blacktriangle$) 50/50 SCPE/$d_5$-PMMA and ($\bullet$) 30/70 SCPE/$d_5$-PMMA.

blend studies has been reviewed. Small angle neutron scattering provides a wealth of information on the miscibility behaviour and the phase separation mechanisms in polymer blends. Due to the large penetration depth of neutrons, it is possible to perform experiments using complex and bulky sample environments such as pressure and shear cells. In addition to SANS, neutron reflectivity has proved to be a powerful technique to investigate the interfacial profiles between miscible and immiscible polymer pairs. Dynamic studies using neutron scattering have been very limited so far. Over the past ten years, QENS and NSE have been successfully employed to investigate the molecular motion in simple polymeric systems. These studies will be certainly extended to more complex materials such as blends in the near future.

# References

Alamo, R.G., Londono, J.D., Mandlekern, L., Stehling, F.C. and Wignall, G.D. (1994) Phase behaviour of blends of linear and branched polyethylenes in the molten and solid states by small-angle neutron scattering. *Macromolecules*, **27**, 411.

Alamo, R.G., Graessley, W.W., Krishnamoorti, R., Lohse, D.J., Londono, J.D., Mandlekern, L., Stehling, F.C. and Wignall, G.D. (1997) Small angle neutron scattering investigations of melt miscibility and phase segregation in blends of linear and branched polyethylenes as a function of the branch content. *Macromolecules*, **30**, 561.

Arrighi, V., Higgins, J.S., Burgess, A.N. and Howells, W.S. (1995) Rotation of Methyl Side Groups in Polymers: A Fourier Transform Approach to Quasielastic Neutron Scattering. 1. Homopolymers. *Macromolecules*, **28**, 2745.

Arrighi, V., Higgins, J.S., Burgess, A.N. and Howells, W.S. (1995) Rotation of Methyl Side Groups in Polymers: A Fourier Transform Approach to Quasielastic Neutron Scattering. 2. Polymer Blends. *Macromolecules*, **28**, 4622.

Balsara, N.P., Fetters, L.J., Hadjichristidis, N. Lohse, D.J., Han, C.C., Graessley, W.W. and Krishnamoorti, R. (1992) Thermodynamic interactions in model polyolefin blends obtained by small angle neutron scattering. *Macromolecules*, **25**, 6137.

Bates, F.S., and Wignall, G.D., and Koehler, W.C. (1985) Critical behavior of binary-liquid mixtures of deuterated and protonated polymers. *Phys. Rev. Lett.*, **55**, 2425.

Bates, F.S., Fetters, L.J. and Wignall, G.D. (1988) Thermodynamics of isotopic polymer mixtures – poly(vinylethylene) and poly(ethylethylene). *Macromolecules*, **21**, 1086.

Beaucage, G. and Stein, R.S. (1993) Tacticity effects on polymer blend miscibility. 3. Neutron-scattering analysis. (1993) *Macromolecules*, **26**, 1617.

Beaucage, G., Sukumaran, S., Clarson, S.J., Kent, M.S., and Schaefer, D.W. (1996) Symmetric, isotopic blends of poly(dimethylsiloxane). *Macromolecules*, **29**, 8349.

Benninghaven, A., Rudenaur, F.G. and Werner, H.W. (1987) *SIMS – Chemical Analysis Series*, **86**, Wiley, New York.

Bidkar, U.R. and Sanchez, I.C. (1995) Neutron scattering from compressible polymer blends – A framework for experimental analysis and interpretation of interaction parameters. *Macromolecules*, **28**, 3963.

Binder, K. (1983) Collective diffusion, nucleation and spinodal decomposition in polymer mixtures. *J. Chem. Phys.*, **79**, 6387.

Brereton, M.G., Fischer, E.W., Herkt-Maetzky, Ch. and Mortensen, K. (1987) Neutron scattering from a series of polymer blends – Significance of the Flory Chi-F parameter. *J. Chem. Phys.*, **87**, 6144.

Brochard-Wyart, F. and deGennes, P.G. (1990) Hindered interdiffusion in asymmetric polymer-polymer junctions. *Makromol. Chem. Macromol. Symp.*, **40**, 167.

Buckingham, A.D. and Hentschel, H.G.E. (1980) Partial miscibility of liquid mixtures of protonated and deuterated high polymers. *J. Polym. Sci., Polym. Phys. Ed.*, **18**, 853.

Bucknall, D.G., Higgins, J.S., Penfold, J. and Rostami, S. (1993) Segregation behavior of deuterated poly(styrene-block-methyl methacrylate) diblock copolymer in the presence of poly(methyl methacrylate) homopolymer. *Polymer*, **34**, 451.

Bucknall, D.G., Fernandez, M.L. and Higgins, J.S. (1994) Neutron reflection studies of interfaces between partially miscible and compatibilised immiscible polymers. *Faraday Discuss*, **98**, 19.

Budkowski, A., Klein, J., Eiser, E., Steiner, U. and Fetters, L.J. (1993) Interference of microstructure and isotope labeling effects in polymer blend compatibility. *Macromolecules*, **26**, 3858.

Cahn, J.W. and Hilliard, J.E. (1959) Free energy of a non-uniform system. III. Nucleation in a two-component incompressible fluid. *J. Chem. Phys.*, **31**, 688.

Clark, J.N., Fernandez, M.L., Tomlins, P.E. and Higgins, J.S. (1993) Small-angle neutron scattering studies of phase-equilibria in blends of deuterated poly(methyl methacrylate) with solution chlorinated polyethylene. *Macromolecules*, **26**, 5897.

Cotton, J.P., Decker, D., Benoit, H., Farnoux, B., Higgins, J., Jannink, G., Ober, C., Picot, C. and des Cloizeaux, J. (1974) Conformation of polymer chain in the bulk. *Macromolecules*, **7**, 863.

Debye, P., and Bueche, M.A. (1949) Scattering by an inhomogeneous solid. *J. Appl. Phys.*, **20**, 518.

Debye, P., Anderson, H.R. and Brumberger, H. (1957) Scattering by an inhomogeneous solid. II. The correlation function and its applications. *J. Appl. Phys.*, **28**, 679.

de Gennes, P.G. (1979) *Scaling Concepts in Polymer Physics*. Cornell University Press, Ithaca, New York.

deGennes, P.G. (1989) Formation of a diffuse interface between two polymers. *C.R. Acad. Sci. Paris, Ser. B.*, **308**, 13.

de los Santos Jones, H., Liu, Y., Inglefield, P.T., Jones, A.A., Kim, C.K. and Paul, D.R. (1994) Coupled relaxations in a blend of PMMA and a polycarbonate. *Polymer*, **35**, 57.

Fernandez, M.L., Higgins, J.S., Penfold, J., R.C. Ward, Shackleton, J. and Walsh, D.J. (1988) Neutron reflection investigation of the interface between an immiscible polymer pair. *Polymer*, **29**, 1923.

Fernandez, M.L., Higgins, J.S., Penfold, J. and Shackleton, J. (1991) Interfacial structure for a compatible polymer blend for short and intermediate annealing times. *J. Chem. Soc. Faraday Trans.*, **87**, 2055.

Flory, P.J. (1968) *Statistical Mechanics of Chain Molecules*, John Wiley — Interscience, New York and London, p.34.

Flory, P.J. (1969) *Principles of Polymer Chemistry*. Cornell University Press, New York.

Floudas, G. and Higgins, J.S. (1992) Ester methyl-group rotation in poly(methyl methacrylate) and in the blend solution chlorinated polyethylene/ poly(methyl methacrylate) — A quasi-elastic neutron scattering study. *Polymer*, **33**, 4121.

Gabrys, B., Higgins, J.S., Ma, K.T. and Roots, I. (1984) Rotational motion of the ester methyl-group in stereoregular poly(methyl methacrylate). A neutron scattering study. *Macromolecules*, **17**, 560.

Genzer, J., Faldi, A. and Composto, R.J. (1994) Self-consistent mean-field calculation of surface segregation in a binary polymer blend. *Phys. Rev. E*, **50**, 2373.

Genzer, J., Faldi, A., Oslanec, R. and Composto, R.J. (1996) Surface enrichment in a miscible polymer blend: An experimental test of self-consistent field and long-wavelength approximation models. *Macromolecules*, **29**, 5438.

Gomez-Elvira, J.M., Halary, J.L., Monnerie, L. and Fetters, L.J. (1994) Isotope effects on the phase-separation in polystyrene poly(vinyl methyl-ether) blends. 2. Influence of the microstructure of linear and star block-copolymers. *Macromolecules*, **27**, 3370.

Goldstein, J.I., Newbury, D.E., Echlin, P.E., Coy, D.C., Fiori, C. and Lifshin, E. (1981) *Scanning Electron Microscopy and x-ray Microanalysis*, Plenum, New York.

Graessley, W.W., Krishnamoorti, R., Balsara, N.P., Fetters, L.J., Lohse, D.J., Schulz, D.N. and Sissano, J.A. (1993) Effect of deuterium substitution on thermodynamic interactions in polymer blends. *Macromolecules*, **26**, 1137.

Graessley, W.W., Krishnamoorti, R., Reichart, G.C., Balsara, N.P., Fetters, and L.J., Lohse (1995) Regular and irregular mixing in blends of saturated-hydrocarbon polymers. *Macromolecules*, **28**, 1260.

Hahn, K., Schmitt, B.J., Kirschey, M., Kirste, R.G., Salie, H. and Schitt-Strecker, S. (1992) Structure and thermodynamics in polymer blends — Neutron-scattering measurements on blends of poly(methyl methacrylate) and poly(styrene-co-acrylonitrile). *Polymer*, **33**, 5150.

Hamley, I.W. (1998) *The Physics of Block Copolymers*, Oxford University Press.

Hammouda, B. and Benmouna, M. (1995) Neutron-scattering from polymer blends under pressure. *J. Polym. Sci.: Polym. Phys.*, **33**, 2359.

Hammouda, B. and Bauer, B.J. (1995) Compressibility of two polymer blend mixtures. *Macromolecules*, **28**, 4505.

Han, C.C., Bauer, B.J., Clark, J.C., Muroga, Y., Matsushita, Y., Okada, M., Tran-cong, Q., Chang, T. and Sanchez, I.C. (1988) Temperature composition and molecular-weight dependence of the binary interaction parameter of polystyrene poly(vinyl methyl-ether) blends. *Polymer*, **29**, 2002.

Hermes, H.E., Higgins, J.S. and Bucknall, D.G. (1997) Investigation of the melt interface between polyethylene and polystyrene using neutron reflectivity. *Polymer*, **38**, 985.

Higgins, J.S. (1985) in *Polymer Blends and Mixtures*, Walsh, D.J., Higgins, J.S., and Maconnachie, A., (eds.) NATO ASI Series, Series E: Applied Sciences, **89**, Martinus Nijhoff Publishers, Dordrecht.

Higgins, J.S. (1988) Polymer Blends. *Makromol. Chem. Macromol. Symp.*, **15**, 201.

Higgins, J.S., Fruitwala, H.A. and Tomlins, P.E. (1989) Experimental evidence for slow theory of mutual diffusion-coefficients in phase separating polymer blends. *Br. Polym. J.*, **21**, 247.

Higgins, J.S. and Benoit, H.C. (1993) *Polymers and Neutron Scattering*. Oxford University Press, Oxford.

Hildebrand, J.H. and Scott, R.L. (1950) *The Solubility of Non-electrolytes*, Van Nostrand-Reinhold, Princeton, NJ.

Hill, R.G., Tomlins, P.E. and Higgins, J.S. (1985) Preliminary study of the kinetics of phase separation in high molecular weight poly(methyl methacrylate) solution chlorinated polyethylene blends. *Macromolecules*, **18**, 2555.

Hopkinson, I., Kiff, F.T., Richards, R.W., King, S.M. and Munro, H. (1994) Thermodynamics of isotopic mixtures of syndiotactic poly(methyl methacrylate) from small-angle neutron-scattering. *Polymer*, **35**, 1722.

Hopkinson, I., Kiff, F.T., Richards, R.W. and King, S.M. and Farren, T. (1995) Isotopic labeling and composition dependence of interaction parameters in polyethylene oxide poly(methyl methacrylate) blends. *Polymer*, **36**, 3523.

Jinnai, H., Hasegawa, H., Hashimoto, T. and Han, C.C. (1993) Time-resolved small-angle neutron-scattering study of spinodal decomposition in deuterated and protonated polybutadiene blends. 2. Q-dependence of Onsager kinetic coefficient. *J. Chem. Phys.*, **99**, 4845.

Janssen, S., Schwahn, D., Mortensen, K. and Springer, T. (1993) Pressure-dependence of the Flory-Huggins interaction parameter in polymer blends — A SANS study and a comparison to the Flory-Orwoll-Vrij equation of state. *Macromolecules*, **26**, 5587.

Karim, A., Felcher, G.P. and Russell, T.P. (1994) Interdiffusion of polymers at short times. *Macromolecules*, **27**, 6973.

Kawasaki, K. and Koga, T. (1993) Relaxation and growth of concentration fluctutations in binary fluids and polymer blends. *Physica A*, **201**, 115.

Kirste, R., Kruse, W.A. and Ibel, K. (1975) Determination of the conformation of polymers in the amorphous solid state and in concentrated solution by neutron diffraction. *Polymer*, **16**, 120.

Kramer, E.J., Green, P.F. and Palmstrom, C.J. (1984) Interdiffusion and marker movements in concentrated polymer polymer diffusion couples. *Polymer*, **25**, 473.

Krishnamoorti, R., Graessley, W.W., Balsara, N.P. and Lohse, D.J. (1994) Structural origin of thermodynamic interactions in blends of saturated-hydrocarbon polymers. *Macromolecules*, **27**, 3073.

Krishnamoorti, R., Graessley, W.W., Fetters, L.J., Garner, R.T. and Lohse, D.J. (1995) Anomalous mixing behavior of polyisobutylene with other polyolefins. *Macromolecules*, **28**, 1252.

Kumar, S.K. (1994) Chemical potentials of polymer blends from Monte-Carlo simulations — Consequences on SANS-determined Chi-parameters. *Macromolecules*, **27**, 260.

Lapp, A., Picot, C. and Benoit, H. (1985) Determination of the Flory interaction parameters in miscible polymer blends by measurement of the apparent radius of gyration. *Macromolecules*, **18**, 2437.

Lekner, J. (1987) *Theory of Reflection*, Nijhoff, Dordrecht.

Lindner, P. and Zemb, Th. (ed.) (1991) *Neutron, X-ray and light scattering: introduction to an investigative tool for colloidal and polymeric systems*, North Holland, Amsterdam..

Londono, J.D., Narten, A.H., Wignall, G.D., Honnell, K.G., Hsieh, E.T., Johnson, T.W. and Bates, F.S. (1994) Composition dependence of the interaction parameter in isotopic polymer blends. *Macromolecules*, **27**, 2864.

Maconnachie, A., Fried, J.R. and Tomlins, P.E. (1989) Neutron scattering studies of blends of poly(2,6-dimethyl-1,4-phenylene oxide) with poly(4-methylstyrene) and with polystyrene: concentration and temperature dependence. *Macromolecules*, **22**, 4606.

Müller, M., Binder, K. and Oed, W. (1995) Structural and thermodynamic properties of interfaces between coexisting phases in polymer blends — A Monte-Carlo simulation. *J. Chem. Soc. Faraday Trans.*, **91**, 2369.

Müller, G., Schwahn, D., Eckerlebe, H., Rieger, J. and Springer, T. (1996) Effect of the Onsager coefficient and internal relaxation modes on spinodal decomposition in the high molecular isotopic blend polystyrene deutero-polystyrene studied with small angle neutron scattering. *J. Chem. Phys.*, **104**, 5326.

Muthukumar, M. (1986) Thermodynamics of Polymer-Solutions *J. Chem. Phys.*, **85**, 4722.

O'Connor, K.M., Pochan, J.M. and Thiyagarajan, P. (1991) Small-angle neutron-scattering studies of interactions and chain dimensions in bimodal polystyrene blends. *Polymer*, **32**, 195.

Okada, M. and Han, C.C. (1986) Experimental study of thermal fluctuation in spinodal decomposition of a binary polymer mixture. *J. Chem. Phys.*, **85**, 5317.

Olabisi, O., Robison, L.M. and Shaw, N.T. (1979) *Polymer-Polymer Miscibility*. Academic Press, New York.

Onuki, A. (1994) Dynamic scattering and phase separation in viscoelastic 2-component fluids. *J. Non-Cryst. Solids*, **172–174**, 1151.

O'Reilly, J.M., Teegarden, D.M. and Wignall, G.D. (1985) Small-angle and intermediate angle neutron scattering from stereoregular poly(methyl methacrylate). *Macromolecules*, **18**, 2747.

Paul, D.R. and Newnman, S. (1978) *Polymer Blends*, Academic Press, New York.

Reichart, G.C., Graessley, W.W., Register, R.A., Krishnamoorti, R. and Lohse, D.J. (1997) Anomalous attractive interactions in polypropylene blends. *Macromolecules*, **30**, 3036.

Richards, R.W. and Thomason, J.L. (1983) Small-angle neutron scattering measurement of block co-polymer interphase structure. *Polymer*, **24**, 1089.

Russell, T.P. (1990) X-ray and neutron reflectivity for the investigation of polymers. *Mater. Sci. Rep.*, **5**, 171.

Russell, T.P., Karim, A., Mansour, A. and Felcher, G.P. (1988) Specular reflectivity of neutrons by thin polymer films. *Macromolecules*, **21**, 1890.

Sauer, B.B. and Walsh, D.J. (1991) Use of neutron reflection and spectroscopic ellipsometry for the study of the interface between miscible polymer films. *Macromolecules*, **24**, 5948.

Schipp, C., Hill, M.J., Barham, P.J., Cloke, V.M., Higgins, J.S. and Oiarzabal, L. (1996) Ambiguities in the interpretation of small-angle neutron scattering from blends of linear and branched polyethylene. *Polymer*, **37**, 2291.

Schmidt, I. and Binder, K. (1985) Model calculations for wetting transitions in polymer mixtures. *J. Phys. II (Paris)*, **46**, 1631.

Schubert, D.W., Abetz, V., Stamm, M., Hack, T. and Siol, W. (1995) Composition and temperature dependence of the segmental interaction parameter in statistical copolymer homopolymer blends. *Macromolecules*, **28**, 2519.

Schwahn, D. and Yee-Madeira, H. (1987) Spinodal decomposition of the polymer blend deutereous polystyrene (d-PS) and poly(vinyl methylether) (PVME) studied with high-resolution neutron small angle scattering. *Colloid Polym. Sci.*, **265**, 867.

Schwahn, D., Janssen, S. and Springer, T. (1992) Early state of spinodal decomposition with small angle neutron scattering in the blend deuteropolystyrene and polyvinylmethylether — A comparison with the Cahn-Hilliard-Cook theory. *J. Chem. Phys.*, **97**, 8775.

Sferrazza, M., Xiao, C., Jones, R.A.L., Bucknall, D.G. and Penfold, J. (1997) Evidence for capillary waves at immiscible polymer/polymer interfaces. *Phys. Rev. Lett.*, **78**, 3693.

Shibayama, M., Yang, H., Stein, R.S. and Han, C.C. (1985) Study of miscibility and critical phenomena of deuterated polystyrene and hydrogenated poly(vinyl methylether) by small angle neutron scattering. *Macromolecules*, **18**, 2179.

Sivia, D.S., Hamilton, W.A. and Smith, G.S. (1991) Analysis of neutron reflectivity data — Maximum entropy, Bayesian spectral analysis and speckle holography. *Physica B*, **173**, 121.

Stamm, M. (1992) Reflection of Neutrons for the Investigation of Polymer Interdiffusion at Interfaces. In *Physics of Polymer Surfaces and Interfaces*, Sanchez, I.C. (ed.), Butterworths-Heinemann, Boston, p.163.

Utracki, L.A. (1989) *Polymer Alloys and Blends. Thermodynamics and Rheology*. Hauser Publishers, Munich.

Walsh, D.J., Higgins, J.S. and Maconnachie, A. (1985) *Polymer blends and mixtures*, NATO ASI Series, Series E: Applied Sciences, **89**, Martinus Nijhoff Publ., Dordrecht.

Warner, M., Higgins, J.S. and Carter, A.J. (1983) Chain dimensions and interaction in neutron scattering from polymer blends with a labeled component. *Macromolecules*, **16**, 1931.

Wignall, G.D., Child, H.R. and Samuels, R.J. (1982) Structural characterization of semi-crystalline polymer blends by small angle neutron scattering. *Polymer*, **23**, 957.

Yang, H., Hadziioannou, G. and Stein, R.S. (1983) The effect of deuteration on the phase-equilibrium of the polystyrene poly(vinyl methyl ether) blend system. *J. Polym. Sci.: Polym. Phys. Ed.*, **21**, 159.

Yoon, D.Y. and Flory, P.J. (1976) Small angle neutron and X-ray scattering by poly(methyl methacrylate) chains. *Macromolecules*, **9**, 299.

# 9. Inelastic Neutron Scattering of Polymers

STEWART F. PARKER

*ISIS Facility, Rutherford Appleton Laboratory, Chilton, Didcot, Oxon OX11 0QX, UK*

## 9.1 INTRODUCTION

Vibrational spectroscopy is a commonly used technique in both industry and academia to provide both quantitative and qualitative information on molecular species and the functional groups present in them (Parker, 1995). It is equally applicable to the study of gases, liquids, crystals and amorphous solids. It is dominated by the optical techniques of infrared (photon adsorption or emission) and Raman (photon scattering) spectroscopies. The most familiar results of optical spectroscopy are the frequency *eigenvalues*, $\omega$, which it provides with great accuracy. Neutron spectroscopy can also be used to measure a vibrational spectrum, but, because of the unique properties of neutrons, these spectra are different from their optical counterparts. The most obviously exploitable neutron property is its atomic scattering potential, parameterised by the scattering cross-section. Because this potential is nuclear the ever present electrons are irrelevant, optical selection rules are irrelevant, optically black or reflective surfaces are irrelevant and photosensitivity is irrelevant. What is most relevant is the magnitude of the cross-section.

Hydrogen has the largest incoherent scattering cross-section and at 80 barns it is 20 times greater than the cross-sections of most other atoms. The response from hydrogen so completely dominates the spectra that to first order, we need consider only hydrogen dynamics when calculating the spectral intensities for hydrogenous systems. Taking advantage of the fact that the samples are molecular (i.e. high $\omega$ values) and that the spectra are obtained at low temperatures, the usual one-phonon expression for the scattering law

(Lovesey, 1987) reduces to:

$$S(Q, \omega) = \frac{Q^2 : B_\omega}{3} \cdot \exp(-Q^2 : \alpha_\omega) \tag{1}$$

where

$$\alpha_\omega = \frac{1}{5} \left\{ \frac{^{Tr}A}{3} + 2\frac{B_\omega : A}{^{Tr}B_\omega} \right\} \tag{2}$$

and

$$A = \sum_\omega B_\omega \tag{3}$$

The observed intensity, $S(Q, \omega)$, of a band at an energy, $\hbar\omega$, is a function of only two parameters; the momentum transferred, $Q$, during the scattering (which is determined by the spectrometer design) and the vibrational amplitude, $B_\omega$, of the hydrogen atoms in the mode at $\omega$ (which is determined by the forces in the molecule). Since the value of $\underline{Q}$ is controlled by the experimentalist the values of $B$ can be extracted. Thus the experimentalist is provided with direct access to the atomic displacements, the *eigenvectors*. This can be compared with the common practice of optical spectroscopy where the band intensities are usually ignored; largely because there is no entirely successful theory of optical intensities. In order to make full use of the information, a data-fitting programme CLIMAX (Kearley, 1986, 1995) has been developed to carry-out a normal co-ordinate analysis using both energy and intensity information as constraints.

In addition to the fundamentals (0 → 1 transitions), overtones (0 → 2, 3... transitions) and combination bands (simultaneous excitation of two vibrational modes) are allowed transitions in the harmonic approximation; this is a major difference between INS spectra and infrared and Raman spectra . It is important to realise that overtones and combinations are single quantum excitations and are distinct from multiple scattering where a neutron interacts in events, that are temporally and spatially separated, with two (or more) scattering centres. Combination bands (phonon wings) between internal modes and external (lattice) modes maybe particularly strong (Tomkinson, 1989). Their importance lies in the fact that the total spectral intensity is distributed between the internal mode and the phonon wing. The intensity of the wing relative to the band origin is dependent on $\underline{Q}^2$. CLIMAX can optionally calculate the spectrum with the inclusion of phonon wings, first overtones and binary combinations.

In the remainder of this chapter I will give a brief description of the spectrometer TFXA on which the work described below was carried out and illustrate the use of INS for the study of polymers. Two very different systems are considered: PMR15, an advanced composite at the leading edge of polymer technology and polyethylene, a commodity polymer that has been in use for almost 50 years, but still yields surprises.

## 9.2 THE INSTRUMENT: TFXA

The INS experiments were performed using the high resolution broadband spectrometer TFXA at the ISIS pulsed spallation neutron source at the Rutherford Appleton Laboratory,

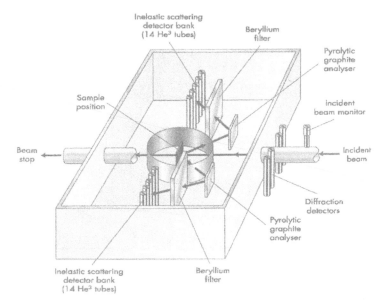

FIGURE 1. Schematic diagram of TFXA.

Chilton, UK (Penfold, 1986). A schematic diagram of the instrument is shown in Figure 1. This is an inverted geometry time-of-flight spectrometer where a pulsed, polychromatic beam of neutrons illuminates the sample at 12 m from the source. The backscattered neutrons are Bragg reflected by a pyrolytic graphite analyser and those with a final energy of ca. 32 cm$^{-1}$ (8.067 cm$^{-1}$ $\equiv$ 1 meV) are passed to the He$^3$ detector bank. In the data evaluation, the raw time-of-flight data are converted to energy transfer as a function of spectral intensity. TFXA measures the double differential cross-section, $d^2\sigma/d\Omega dE$; the number of neutrons of a given energy scattered into a given solid angle $\Omega$. This expression is related to the intermediate scattering function $S(Q, \omega)$ as follows:

$$S(Q, \omega) = \frac{k_o}{k_f} \frac{d^2\sigma}{d\Omega dE} \qquad (k \equiv 2\pi/\lambda) \qquad (4)$$

where $k_o$ and $k_f$ are the incident and final neutron wavevectors respectively and $\lambda$ is the wavelength of the neutron. $S(Q, \omega)$ is the generally preferred form for further data analysis since the low energy features are less emphasized and it also can be calculated via equation (1).

Figure 2a compares the resolution function of TFXA and that of previously employed instruments (Lynch, 1968; Jobic, 1982). The advantages are clear: TFXA is unique among INS spectrometers in that it offers both high resolution and wide spectral range.

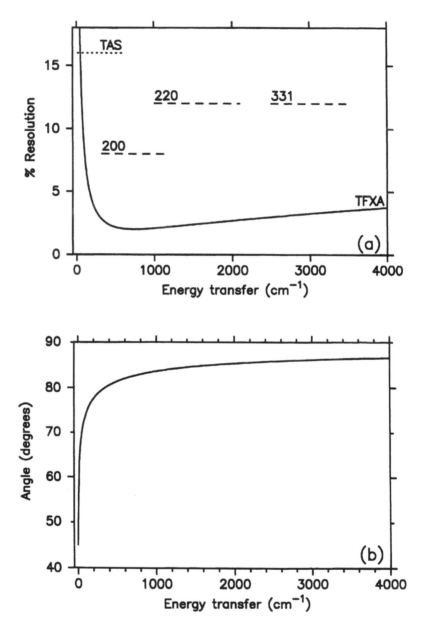

FIGURE 2. (a) Comparison of the spectral range and resolution functions of TFXA with those used in previous work. (TAS refers to the triple axis spectrometer used in the work of Lynch *et al.* (Lynch, 1968). 200, 220 and 331 are the planes of the copper single crystal monochromator in IN1BeF used by Jobic (Jobic, 1982). The value quoted is the average resolution for that plane. (b) The direction of the momentum transfer vector $\underline{Q}$ relative to the plane of the sample as a function of the energy transfer for TFXA.

The use of a small, fixed final energy has an important consequence. Neutrons transfer both energy and momentum when they are inelastically scattered. The momentum transfer vector $\underline{Q}$ is given by:

$$\underline{Q} = \underline{k}_o - \underline{k}_f \qquad (5)$$

Since for most energies the relation $k_o \gg k_f$ holds, it follows that both the magnitude and direction of $\underline{Q}$ are largely determined by $k_o$. Hence there is virtually no $\underline{Q}$ resolution on TFXA. This means that there is a unique value of $\underline{Q}$ for each energy transfer: $Q2(\text{Å}^{-2}) \approx E(\text{cm}^{-1})/16$. Further, $\underline{Q}$ is almost parallel to the incoming neutron beam and perpendicular to the sample plane across virtually the entire energy transfer range and this is illustrated in Figure 2b. The significance of this is that in order to observe an INS transition, the vibration must have a component of motion parallel to $\underline{Q}$. For a randomly oriented (i.e. polycrystalline) sample this condition will be satisfied for all the vibrations and all will be observed. However, this is not the case for an oriented sample and experiments directly analogous to optical polarization measurements are therefore possible.

Sample preparation is straightforward: a few grams of the material are wrapped in aluminium foil (if solid) or put in a thin-walled aluminium cell (if liquid) and cooled to 30 K or below using either a closed cycle refrigerator or a liquid helium cryostat. This is necessary in order to reduce the Debye-Waller factor (the exponential term in equation (1)) which has the effect of strongly damping the spectral intensity. Since the Debye-Waller factor is temperature dependent, its influence is drastically reduced by measuring the INS spectrum at low temperature. The measurement time varies from a few hours up to 24 hours depending on the hydrogen content of the sample.

## 9.3 ADVANCED COMPOSITES

Advanced composites are engineering materials that offer similar mechanical properties to metal alloys but are lighter than them. The materials consist of fibres embedded in a polymer matrix. Vibrational spectroscopy, particularly infrared spectroscopy, has been widely used for the characterisation of composites (Parker, 1992). However, there is a need for a spectroscopic technique that can examine the cured resins in the presence of the fibres to aid the understanding of the cure chemistry. INS has considerable potential in this regard since two common fibre types, glass and carbon, are effectively, invisible to neutrons because of the small incoherent cross-sections of silicon, oxygen and carbon.

The materials are commonly made by coating the fibres with the monomers to give a precursor fabric ("prepreg") that can be moulded to the desired shape. This is then heated to cure the resin and give the finished product. The chemistry of the cure processes is complex and frequently involves a number of stages. The ability to study the reaction(s) is greatly hampered by the nature of the products: they are often highly cross-linked and thus insoluble, and the presence of the fibre matrix makes them difficult to study spectroscopically.

A wide variety of polymers have been used including epoxies, bismaleimides and polyimides. One of the most common polyimides is PMR-15 (Wilson, 1988). The chemistry is complex, as shown in Figure 3, but consists essentially of two stages; polymerization to give a norbornene end-capped oligomer followed by reaction of the norbornene group to

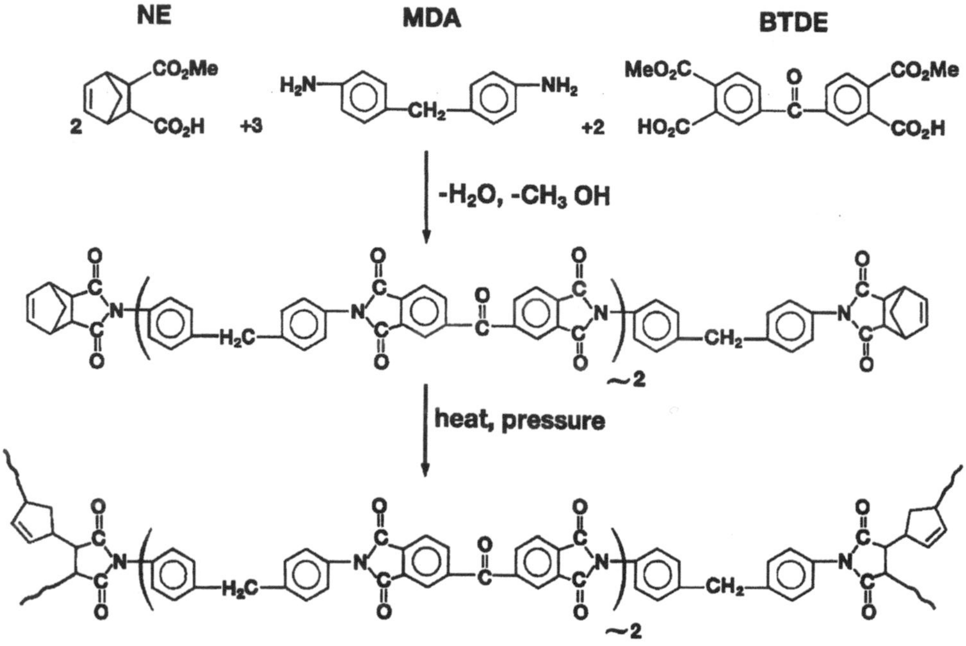

FIGURE 3. The chemistry of PMR-15.

give the cross-linked polymer. The temperature at which the cross-linking reaction is carried out has a major effect on the mechanical properties of the finished product, particularly its susceptibility to microcracking (Wilson, 1987), which worsens as the cure temperature is raised.

INS spectra of PMR-15 composites produced by compression moulding of prepreg and cured at 330 and 270°C are shown in Figure 4a and 4b respectively. There are clearly differences between the two spectra; bands at 1031 and 1114 cm$^{-1}$ have diminished in intensity at the higher cure temperature and there are also indications of changes in the region 200–400 cm$^{-1}$ and at 638, 720 and 1273 cm$^{-1}$.

The examples studied here are particularly difficult samples to investigate because by the cure temperatures employed here, a significant proportion of the end-groups have already reacted. In addition, the composites are only $\sim$30% by weight resin, of which only 2/7 of the molecules are the end-cap, hence the expected differences are likely to be small. Nonetheless, differences between the spectra characterising the two samples are apparent.

In order to understand these differences, it is necessary to study a model compound. N-phenylnadimide (for the chemical formula see Figure 5a) has been used very successfully (Hay, 1989) as a model compound for the norbornene endcap and the cross-linking reaction. As is evident from equation (1), the intensities of the bands in INS spectroscopy depend on the amplitude of vibration and the scattering cross-section of the atoms. Since the incoherent cross-section for hydrogen is $\sim$80 barns and that of C, N and O are all less than 5 barns,

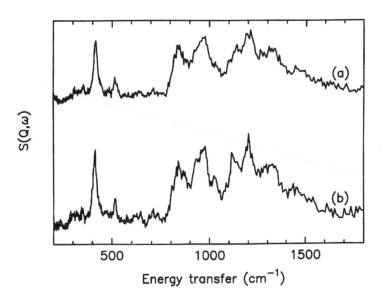

FIGURE 4. The INS spectrum of PMR-15 cured at (a) 330°C and (b) 270°C.

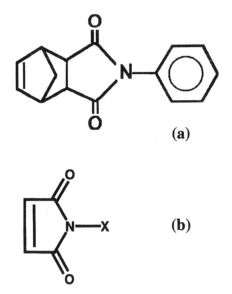

FIGURE 5. Structure of model compounds: (a) N-phenylnadimide, (b) N-substituted maleimide, for the parent molecules maleimide $X = H$.

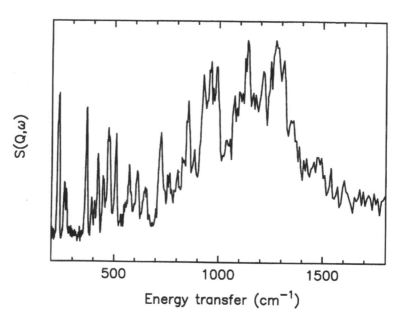

FIGURE 6. The INS spectrum of N-perdeuterophenylnadimide at 30 K.

the spectrum is dominated by the hydrogen atom motions. Moreover, the cross-section is isotope dependent and that of deuterium is also small. Thus by deuterating the phenyl ring it is possible to eliminate most of the spectral features of the phenyl ring hence highlighting the scattering from the norbornene endcap. This procedure provides an excellent model compound for the endcap and the resulting spectrum is shown in Figure 6. Comparison of Figures 4 and 6 suggests that the decrease in the 1114 cm$^{-1}$ and the changes in the 200–400 and 600–800 cm$^{-1}$ regions can reasonably be assigned to loss of the endcap. The 1031 cm$^{-1}$ band does not fit this pattern and may represent a different type of cross-link (Hay, 1989; Wells, 1987).

One possibility is that the new cross-link is an N-substituted maleimide (Figure 5b) linkage. In order to elucidate this problem further, the spectrum of the parent molecule maleimide and a spectrum calculated from a normal co-ordinate analysis using CLIMAX are compared in Figure 7. Clearly, to simulate spectra as complex as those presented in Figure 4 to test for the presence of particular types of end-groups and cross-links is a major undertaking. Nonetheless, it gives a good indication what it is, in principle, achievable using neutron spectroscopy.

The spectra in Figure 4 are complicated by the presence of the large number of aromatic and aliphatic C-H bonds in the backbone of the polymer. This means that the spectrum is dominated by the polyimide backbone and the features due to the end-cap and the cross-links produced by the reaction are difficult to detect. A way to simplify the spectra is to exploit the isotope dependence of the incoherent cross-section by synthesizing novel polymers where the "backbone hydrogen atoms" (i.e. those that originate from the BTDE and MDA

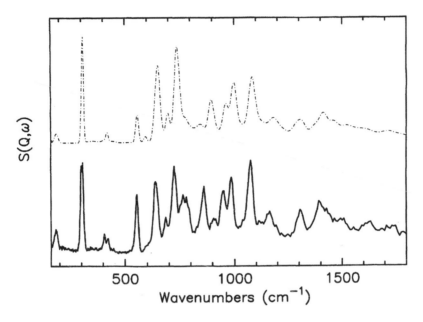

FIGURE 7. Comparison of experimental (lower) and calculated (upper) spectrum of maleimide.

components) are all replaced by deuterium. This method would virtually eliminate the backbone spectrum and allow the clear observation of the end-cap and its reaction products. The simplification of the spectrum by this approach is unique to INS, since deuterium substitution would only shift the bands in the infrared or Raman spectrum: it would not drastically alter their intensity.

## 9.4 POLYETHYLENE

Polyethylene was one of the first polymers to be discovered and is manufactured today on a multi-million tonne scale world-wide. The apparent simplicity of the molecular formula of polyethylene $(CH_2)_n$ belies the real complexity of the material. To understand the physical properties of the polymer it is necessary to consider at least a two-phase model consisting of crystalline blocks in an amorphous matrix (Young, 1991).

Vibrational spectroscopy has played a key part in the characterisation and understanding of this polymer since it is sensitive to both regions and also to the effects of finite chain length and side branches (Bower, 1985). The space group of polyethylene is centrosymmetric and some of the modes are infrared and Raman inactive. INS spectra (Lynch, 1968; Jobic, 1982) of polyethylene were recorded to detect the modes unseen by these techniques. By contemporary standards these were of low resolution and restricted in energy range (see Figure 2a).

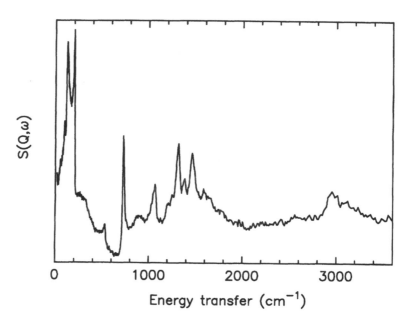

FIGURE 8.   The INS spectrum of polyethylene at 30 K.

The INS spectrum of a high density (950 kg/m$^3$) linear polyethylene (number average molecular weight $M_{\bar{\omega}} \sim 260{,}000$) at 30 K is shown in Figure 8 and was the first observation of the complete spectrum, 0–4000 cm$^{-1}$, by INS spectroscopy (Parker, 1996). For the purpose of discussion the spectrum will be divided into three regions: above 2000 cm$^{-1}$, 2000–600 cm$^{-1}$ and 600–0 cm$^{-1}$. The high energy region shows a broad feature centred at 3000 cm$^{-1}$ due to the symmetric and asymmetric C-H stretching modes. These are not resolved for two reasons: the instrumental resolution is insufficient ($\sim 100$ cm$^{-1}$ as compared to the splitting of ca. 50 cm$^{-1}$) and there is virtually complete transfer of intensity from the fundamental vibration into the phonon wings since at 3000 cm$^{-1}$ $\underline{Q}^2 \approx 190$. Thus the peak maxima are at 2965 and 3108 cm$^{-1}$ in the INS spectrum but are observed at 2851 and 2919 cm$^{-1}$ in the infrared spectrum and 2848 and 2883 cm$^{-1}$ in the Raman spectrum.

For a polymer such as polyethylene, equation (1) has to be modified to account for the dependence of the vibrational frequency on the wavevector. For polyethylene the partial differential cross section was derived (Myers, 1966) and this was subsequently reformulated (Jobic, 1982) in the one-phonon approximation, for neutron energy loss, as the molecular orientation average of:

$$\frac{d^2\sigma}{d\Omega dE} = \frac{k_f}{2k_0}\frac{\sigma_H}{4\pi}\sum_{d\alpha\beta}\exp[-2W_d(Q)]\frac{Q_\alpha Q_\beta G_d^{\alpha\beta}(\omega)}{m_H\omega[1-\exp(-\hbar\omega/kT)]} \qquad (6)$$

where $k_0$, $k_f$ and $\underline{Q}$ were defined earlier (equations (1) and (4)), $\exp[-2W_d(Q)]$ is the Debye-Waller factor and $G_d^{\alpha\beta}(\omega)$ is the amplitude (of vibration) weighted vibrational density of states for a hydrogen atom $d$:

$$G_d^\beta(\omega) = \frac{1}{N} \sum_{jq} \sigma_{dj}^{\alpha*}(k) \sigma_{dj}^\beta(k) \delta[\omega - \omega_j(q)] \tag{7}$$

where $\sigma_{dj}^\beta(k)$ is the polarisation vector of the $d$th proton in the $\beta$th direction for the phonon with wavevector $k$, branch $j$, and frequency $\omega_j$. For the internal modes at low temperature, the exponential term $\exp(-\hbar\omega/kT)$ will be very small so that the measured cross section will be simply proportional to $\exp(-2W)G(\omega)$. Thus INS spectroscopy gives information at all values of $k$, not simply those at $k = 0$ as is the case of optical spectroscopies.

It can be seen from equation (6) that the INS spectrum is directly proportional to the vibrational density of states weighted by the amplitude of vibration, the incoherent neutron scattering cross-section and the momentum transfer and is attenuated by the Debye-Waller factor. Unfortunately, there is no vibrational density of states of polyethylene derived from the best available dispersion curves (Barnes, 1978). However, to a first approximation, the vibrational density of states may be considered to be a one dimensional projection of the dispersion curves onto the energy axis (formally, it is proportional to $(d\omega/dk)^{-1}$). Thus maxima (Van Hove singularities) in the vibrational density of states will occur at energies corresponding to flat portions (critical points) of the dispersion curves. Hence, the dispersion curves themselves can be used to qualitatively predict the overall form of the INS spectrum. Dispersion curves are conventionally plotted over the first half of the Brillouin zone. In the case of a polymer the dispersion curve can be plotted in two ways: either as a function of the phase difference $\theta$ between adjacent oscillators (for polyethylene these are $CH_2$ groups) or as a function of $k$ the phase difference between adjacent translational units. Since the unit cell of polyethylene contains two chain segments each consisting of two $CH_2$ groups, the cell is twice as long for the adjacent translational units as that for the adjacent oscillator and consequently the Brillouin zone of the former is half the size of the latter. Thus dispersion curves as a function of $k$ can be obtained from those as a function of $\theta$ by folding back the right-hand half of the plot. For infrared and Raman spectroscopy, the allowed modes are those at $k = 0$, the Brillouin zone centre. In terms of $\theta$ the allowed modes are those at 0 and $\pi$ since the dispersion curves in this case span a complete Brillouin zone i.e. centre-to-centre.

Figure 9 shows a comparison of the dispersion curves for the optic branches and the INS spectrum of polyethylene at 30K for the 2000–600 cm$^{-1}$ region. The agreement is excellent; maxima occur at the energies corresponding to phase differences of 0 and $\pi$ (which are the infrared and Raman active energies). The strong band at ca. 1500 cm$^{-1}$ is assigned to the overlap of the $\nu_2$ $CH_2$ scissors mode and the first overtone of the 720 cm$^{-1}$ methylene rocking band. The broad feature at 900 cm$^{-1}$ is mainly due to the phonon wing of the 720 cm$^{-1}$ feature; there may also be a contribution from the minimum in the $\nu_4$ dispersion curve. The predicted splitting in the 720 cm$^{-1}$ methylene rocking band due to in-phase and out-of-phase coupling of motions on adjacent chains in the unit cell is not observed because of the instrumental resolution.

In the region below 600 cm$^{-1}$, two acoustic branches occur: $\nu_5$ is an in-plane skeletal mode and includes the longitudinal acoustic mode and $\nu_9$ is an out-of-plane skeletal mode. Portions of the dispersion curves have been mapped by studying the $n$-alkanes since in the solid state these have the same planar zig-zag conformation as the crystalline region

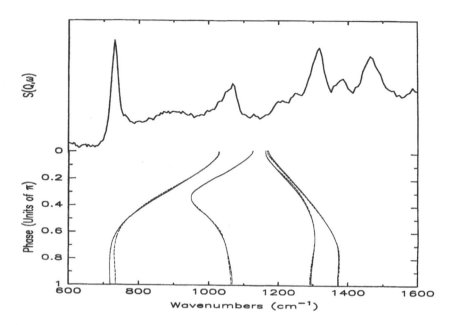

FIGURE 9. Comparison of the dispersion curves of Barnes and Franconi (Barnes, 1978) with the INS spectrum of polyethylene for the region 600–1600 cm$^{-1}$.

of polyethylene and hence are excellent model compounds (see (Bower, 1989; Painter, 1982) for reviews). The $\nu_5$ branch of perdeuteropolyethylene has been studied by coherent inelastic neutron scattering spectroscopy (Feldkamp, 1968; Twistleton, 1982; Pepy, 1978). The factor group splitting of polyethylene results in each branch giving two sub-branches. Each of the four sub-branches terminates at 0 and $\pi$ phase difference, thus there are eight modes to be considered. Three of these are the pure translations, hence have zero energy and are unobservable, one is inactive in both the infrared and Raman spectrum, two are Raman active and two are infrared active. However, the entire branches are directly observable with inelastic neutron scattering spectroscopy. Figure 10 shows a comparison of the calculated dispersion curves and the INS spectrum in the 600–0 cm$^{-1}$ region. The maximum in the $\nu_5$ dispersion curve occurs at 550 cm$^{-1}$ as compared to the experimental value of 525 cm$^{-1}$. The curves are calculated from data obtained at 90 K, however, the observed band positions are invariant with temperature in the range 5–100 K. Additional features are detected at 200 cm$^{-1}$ (maximum in ($\nu_9$), 130 cm$^{-1}$ ($\nu_{5a}$ at 0), 97 cm$^{-1}$ ($\nu_{9b}$ at 0) and 53 cm$^{-1}$ ($\nu_{5b}$ at 0). The last feature is particularly significant since this is both infrared and Raman inactive.

   As is clear from the preceeding discussion, it can be seen that the dispersion curves are in good agreement with the measured spectrum for the optic modes. For the acoustic modes the agreement is not so good, particularly for the features that reflect the central portions of the dispersion curves. This is not surprising: it is precisely in these regions that experimental data from the $n$-alkanes is not available.

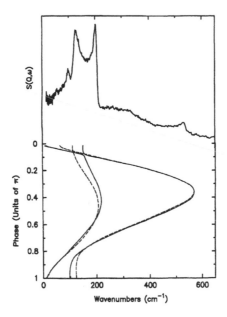

FIGURE 10. Comparison of the dispersion curves of Barnes and Franconi (Barnes, 1978) with the INS spectrum of polyethylene for the region 0–600 $cm^{-1}$.

This is an area that INS can also have an impact since it is not affected by the electronic effects and selection rules that make the low frequency region very difficult for infrared and Raman spectroscopies. For n-$C_{18}H_{38}$, see Figure 11, the spectrum below 600 $cm^{-1}$ shows several distinguishable features: the low frequency CCC in-plane bends, the higher members of which are seen in the range 300–400 $cm^{-1}$, (the longitudinal acoustic mode is at 124 $cm^{-1}$), the methyl torsions at 251 and 258 $cm^{-1}$ and the out-of-plane skeletal torsions below 200 $cm^{-1}$. To the author's knowledge, this is the first observation of *all* of these modes for an alkane chain longer than $\sim C_{10}$.

One of the surprising features of Figure 8 is that the spectrum can be accounted for simply by considering the crystalline region. However, $\sim 40\%$ of the polymer is amorphous and the absence of any features directly attributable to it requires comment. The usual picture of the amorphous region is of a random "spaghetti" of chains. A proportion of these will have segments in the *all-trans* conformation and these are likely to give INS spectra that closely resembles that of the crystalline region. Infrared and Raman spectra of polyethylene show features that can be assigned to the two regions, however, the vibrational frequencies for both regions are similar, but with the amorphous features usually much broader. It is possible that the amorphous features underlie the sharper crystalline features in the INS spectrum. This explanation is not wholly satisfactory since the total spectral intensity of the two regions should be comparable.

It is also possible to exploit the directionality of $Q$ by studying oriented samples to gain additional information. Uniaxially aligned polyethylene is obtained by stretching a sheet of

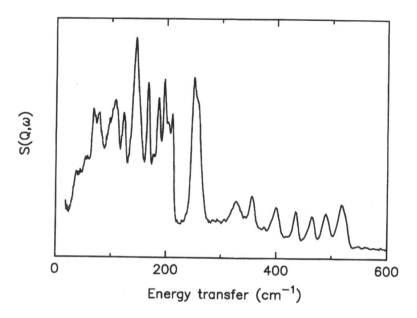

FIGURE 11. INS spectrum of octadecane, n-$C_{18}H_{38}$, at 30 K.

polycrystalline polyethylene to a high draw ratio. This results in almost complete alignment of the crystallites $c$-axis along the draw direction, with the $a$ and $b$ axes randomly oriented perpendicularly to the $c$-axis. For such samples equation (6) has to be modified. It has been shown (Lynch, 1968) that the longitudinal ($\underline{Q}//c$) and transverse ($\underline{Q}\perp c$) one-phonon cross sections are:

$$\left(\frac{d^2\sigma}{d\Omega dE}\right)_L = \frac{k_f}{2k_0}\frac{\sigma_H}{4\pi}\exp(-2W_L)\frac{Q^2}{m_H\omega[1-\exp(-\hbar\omega/kT)]}\sum_j G_L^I(\omega) \qquad (8)$$

and

$$\left(\frac{d^2\sigma}{d\Omega dE}\right)_T = \frac{k_f}{2k_0}\frac{\sigma_H}{4\pi}\exp(-2W_T)\frac{Q^2}{m_H\omega[1-\exp(-\hbar\omega/kT)]}\sum_j G_T^I(\omega) \qquad (9)$$

where

$$G_L^I(\omega) = \sum_j G_j^{CC}(\omega) \qquad (10)$$

and

$$G_L^I(\omega) = \frac{1}{2}\sum_j [G_j^{aa}(\omega) + G_j^{bb}(\omega)] \qquad (11)$$

where the subscripts $L$ and $T$ refer to the longitudinal and transverse materials, respectively. In the same way that equation (6) simplifies, (8) and (9) can be simplified so that the measured cross sections will be simply proportional to $\exp(-2W)G^I(\omega)$.

For the purpose of quantitative comparison, two calculations of the density of states of oriented polyethylene are available: by Lynch et al. (Lynch, 1968) based on the Schactschneider and Snyder forcefield (Schactschneider, 1963) and by Kitagawa and Miyazawa (Kitagawa, 1972) based on the Tasumi and Shimanouchi forcefield (Tasumi, 1965).

Figures 12 and 13 compare the calculated spectra of Lynch et al. (Lynch, 1968) with the experimental spectra with $\underline{Q}$ aligned perpendicular (transverse spectrum) and parallel (longitudinal spectrum) to the $c$-axis respectively. For both spectra in the 2000–600 cm$^{-1}$ region the qualitative agreement is good. An exact match would not be expected since the theoretical spectra are calculated at 16 cm$^{-1}$ (2 meV) intervals, whereas the resolution of the experimental spectra varies with energy transfer. In addition the theoretical spectra do not include the effect of phonon wings. For the transverse spectrum, the overall features are again reproduced, but the reversal of intensity between the 727 and 1452 cm$^{-1}$ features seems too large to be solely ascribed to the combined effects of resolution and phonon wings and suggests a deficiency in the model.

For the longitudinal spectra, Figure 13, the three features at 1296 ($\nu_7$ at $\pi$), 1369 ($\nu_3$ at $\pi$) and 1452 cm$^{-1}$ ($\nu_2$) are clearly resolved, Figure 13b; earlier work (Jobic, 1982) was unable to resolve these peaks. The largest difference between the calculated and experimental spectra is the intensity of the band at 720 cm$^{-1}$ ($\nu_8$ at $\pi$), the calculations indicate that this should have very low intensity in this orientation whereas the experimental spectrum shows a significant intensity in this region. A similar observation was made previously (Jobic, 1982) and assigned to incomplete orientation in the sample. From infrared spectroscopic measurements, it can be shown that the present sample is highly oriented (>90%) so this may be excluded. The momentum transfer vector $\underline{Q}$ is very well defined at this energy hence an effect directly analogous to the optical effect of polarisation leakage can also be excluded. In addition, this problem is minimised by the use of a thin sample to prevent the multiple scattering that causes this effect.

In the 600–0 cm$^{-1}$ region $\nu_5$ the out-of-plane C-C-C bending mode is correctly predicted to be transverse polarised. However, the experimental features below 200 cm$^{-1}$ are much stronger and have a different shape from those calculated. Thus the evidence suggests that there is significant deficiency in the model. This presumably lies in the force constants relating to the skeletal motions and the form of the intermolecular potential used to account for the interchain coupling in the unit cell.

Figure 14 compares the experimental spectra and the calculated spectra of Kitagawa and Miyazawa (Kitagawa, 1972). It can be seen that the trends are well reproduced, but the detail is incorrect. In particular, the intensity of the feature at 130 cm$^{-1}$ is seriously underestimated for the longitudinal orientation in the calculations. A degree of caution is required in this energy range since $\underline{Q}$ is not as perpendicular to the sample as in the higher energy regions. However, it is still 70–75° to the normal in the range 50–100 cm$^{-1}$ and hence still very orientation sensitive. This is demonstrated by the clear polarisation effect of the feature at 97 cm$^{-1}$.

At the time the calculations were made (Lynch, 1968) and (Kitagawa, 1972) the results compared very well with the best available experimental data. The quality of the spectra presented here emphasise the dramatic improvement in INS spectrometers acheived in the

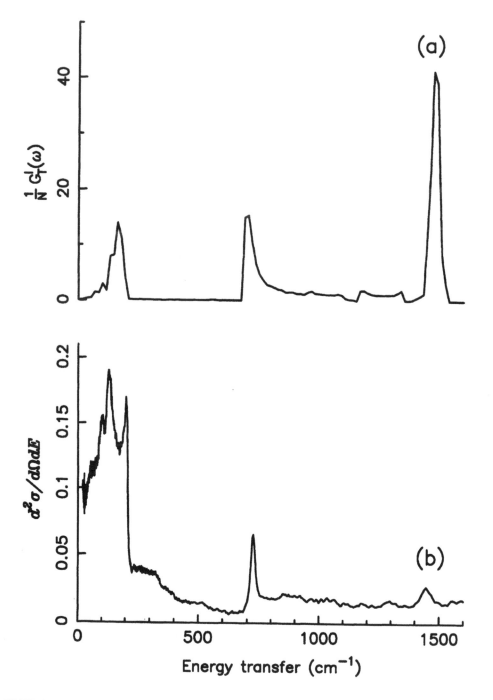

FIGURE 12. Comparison of calculated (Lynch, 1968) and experimental INS spectra of oriented polyethylene with the momentum transfer vector $\underline{Q}$ perpendicular to the $c$-axis.

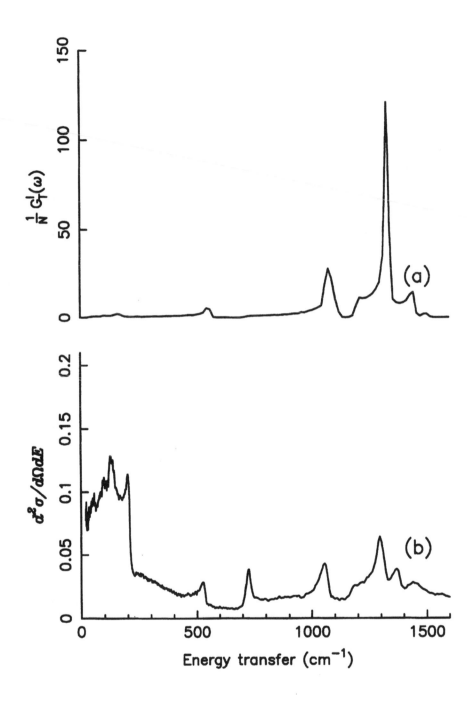

FIGURE 13. Comparison of calculated (Lynch, 1968) and experimental INS spectra of oriented polyethylene with the momentum transfer vector $\underline{Q}$ parallel to the $c$-axis.

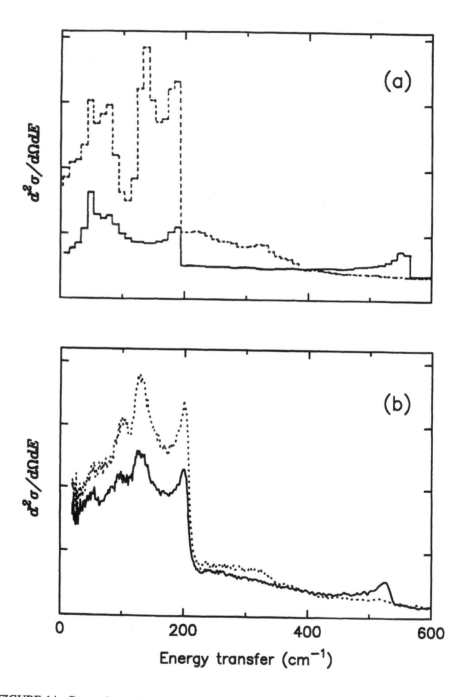

FIGURE 14. Comparison of calculated (a) (Kitigawa, 1972) and experimental (b) INS spectra of oriented polyethylene. Continuous line: $\underline{Q}$ parallel to the $c$-axis, dashed line $\underline{Q}$ perpendicular to the $c$-axis.

last few decades. The assignment of the polyethylene spectrum has also changed (Bower, 1968) as has the understanding of INS spectra, particularly the influence of phonon wings. It would therefore be very interesting to compare the experimental spectra with those that would be predicted today using current knowledge of the polyethylene forcefield and intermolecular potential combined with the improved understanding of INS spectra.

## 9.5 CONCLUSIONS

It is clear from the selected examples presented here that INS spectroscopy of polymers is a field that has been rejuvenated by the availability of spectrometers with good resolution, wide energy range and reasonable flux. Future prospects are encouraging: TFXA will be upgraded in the next few years to give a threefold improvement in resolution and at least an order of magnitude greater sensitivity. This will enable spectra to be obtained in an hour or two with good statistics, thus enabling systematic exploration of the effect of varying parameters such as cure time and temperature for the advanced composites or the effect of co-monomers on polyethylene properties.

### Acknowledgements

The author thanks the Rutherford Appleton Laboratory for access to neutron beam facilities. Encouraging and helpful discussions with Dr J. Tomkinson (RAL) are greatly appreciated.

### References

Barnes, J. and Franconi, B. (1978) Critical review of vibrational data and force field calculations for polyethylene, *J. Phys. Chem. Ref. Data*, **7**, 1309.

Bower, D.I. and. Maddams, W.F. (1989) The characterisation of polymers, chapter 5 in *The Vibrational Spectroscopy of Polymers*, Cambridge University Press, Cambridge.

Feldkamp, L.A., Vankataraman, G. and King, J.S. (1968) Dispersion relation for skeletal vibrations in deuterated polyethylene in *Neutron Inelastic Scattering*, Vol. 2, International Atomic Energy Authority, Vienna, p.159.

Hay, J.N., Boyle, J.D., Parker, S.F. and Wilson, D. (1989) Polymerisation of N-phenylnadimide: a model for the crosslinking of PMR-15 polyimide, *Polymer*, **30**, 1032.

Jobic, H. (1982) Neutron inelastic scattering from oriented and polycrystalline polyethylene: observation and polarization of the optical phonons, *J. Chem. Phys.*, **76**, 2693.

Kearley, G.J. (1986) A profile refinement approach for normal-coordinate analyses of inelastic neutron-scattering spectra, *J. Chem. Soc. Faraday Trans. II*, **82**, 41.

Kearley, G.J. (1995) A review of the analysis of molecular vibrations using INS, *Nucl. Inst. and Meth. in Phys. Res. A*, **354**, 53.

Kitagawa, T. and Miyazawa, T. (1972) Neutron scattering and normal vibrations of polymers in *Advances in Polymer Science*, Cantow, H.-J., Dall'asta, G., Ferry, J.D., Fujita, H., Kern, W., Natta, G., Okamura, S., Overberger. C.G., Prins, W., Schulz, G.V., Slichter, W.P., Staverman, A.J., Stille, J.K. and Stuart, H.A. (eds.) vol. 9, Springer-Verlag, Berlin, p.335.

Lovesey, S.W. (1987) *Theory of Neutron Scattering from Condensed Matter*, Oxford University Press.

Lynch, Jr., J.E., Summerfield, G.C., Feldkamp, L.A. and King, J.S. (1968) Neutron scattering in normal and deuterated polyethylene, *J. Chem. Phys.*, **48**, 912.

Myers, W., Summerfield, G.C. and King, J.S. (1966) Neutron scattering in stretch-oriented polyethylene, *J. Chem. Phys.*, **44**, 184.

Painter, P.C., Coleman, M.L. and Koenig, J.L. (1982) The normal vibrational analysis of polyethylene chapter 13 in *The Theory of Vibrational Spectroscopy and its Application to Polymeric Materials*, Wiley-Interscience, New York.

Parker, S.F. (1992) The application of vibrational spectroscopy to the study of polyimides and their composites, *Vib. Spec.*, **3**, 87.

Parker, S.F. (1996) Inelastic neutron scattering spectra of polyethylene, *J. Chem. Soc. Faraday Trans.*, **92**, 1941

Parker, S.F. (1995) Infrared spectroscopy: industrial applications in *Encyclopedia of Analytical Science*, A. Townsend (ed.), Academic Press, London, vol. 4, p.2232.

Penfold, J. and Tomkinson, J. (1986) *The ISIS Time Focused Crystal Spectrometer*, TFXA, RAL-86-019.

Pepy, G. and Grimm, H. (1978) Effect of interchain forces on acoustic phonon branches in deuterated uniaxial polyethylene in *Neutron Inelastic Scattering*, Vol. 1, International Atomic Energy Authority, Vienna, p.607.

Schactschneider, J.H. and Snyder, R.G. (1963) A valence force field for saturated hydrocarbons, *Spec. Acta*, **19**, 169.

Tasumi, M. and Shimanouchi, T. (1965) Crystal vibrations and intermolecular forces of polymethylene crystals, *J. Chem. Phys.*, **43**, 1245.

Tomkinson, J. and Kearley, G.J. (1989) Phonon wings in inelastic neutron scattering spectroscopy: the harmonic approximation, *J. Chem. Phys.*, **91**, 5164.

Twistleton, J.F., White, J.W. and Reynolds, P.A. (1982) Dynamical studies of fully oriented deuteropolyethylene by inelastic neutron scattering, *Polymer*, **23**, 578.

Wilson, D., Wells, J.K., Hay, J.N., Lind, D., Owens, G.A. and Johnson, F. (1987) Preliminary investigations into the microcracking of PMR-15-graphite composites 1. Effect of cure temperature, SAMPE J. May/June 35.

Wilson, D. (1988) PMR-15 processing, properties and problems — a review, *Brit. Polymer J.*, **20**, 405.

Young, R.J. and Lovell, P.A. (1991) *Introduction to Polymers*, Chapman, London.

# 10. Scattering From Dilute Solutions and Solid State Ionomers

ANNE M. YOUNG[1] and BARBARA J. GABRYS[2]

[1]*Biomaterials Department, St. Barts and Royal London Medical and Dental School, Queen Mary and Westfield College, London E1 4NS, UK*
[2]*Faculty of Mathematics and Computing, South Region, The Open University, Foxcombe Hall, Boars Hill, Oxford OX1 5HR, UK*

## 10.1 INTRODUCTION

Ionomers are macromolecules containing low levels of ionic groups chemically attached to a non-polar backbone. The addition of these charge groups can have pronounced effects on the properties of the polymer. In modified elastomers for example ionic interactions can greatly improve tensile properties, clarity or miscibility of an ionomer with other polymers and additives. Some of the first commercial products include ethylene-methacrylic acid (Surlyn®), and chlorosulfonated polyethylene (Hypalon®). In solution a few ionic groups attached to a non-polar polymer enhance its surface activity and its ability to control rheological properties. Ionomers used industrially in fluids include lightly sulfonated polystyrene in drilling muds and polyurethane ionomers in aqueous dispersions for coating and adhesive applications. The ability of ionomer membranes such as Nafion® to impose permselective control and carry large ionic fluxes has lead to their extensive employment in electrolytic cells. For more details on the applications there are several chapters in books devoted to the subject of ionomers (see for example Pineri and Eisenberg, 1987; Schlick, 1996; Tant *et al.*, 1997).

It was obvious already in the fifties and sixties that the addition of charges to a polymer could substantially alter its properties. The question that arose was: how could this influence be quantified and what was the topology of the charges? It was not until the late sixties that two phenomenological models were proposed to explain solid state properties of dry ionomers. One of these was based on a random distribution of charges (Otocka and Kwei, 1969). The other assumed the formation of clusters and multiplets (Eisenberg, 1970). The latter prevailed, and formed the basis of all subsequent models, as outlined by Mauritz (1996). The basic assumption was that the long-range electrostatic interactions are the driving force for ion pair aggregation. This is followed by 'multiplet condensation', in which several pairs of ions cluster together. Order of magnitude calculations of the average intercluster separation distance ranged from 44 to 95 Å (Eisenberg, 1970) for sodium sulfonated ethylene-methacrylate ionomers. According to the recently revised cluster / restricted mobility model (Eisenberg *et al.*, 1990) a typical multiplet would have dimensions of around 26 Å. In hydrated ionomer membranes, such as Nafion® the behaviour might be expected to be more complex.

The properties of ionomer solutions are usually split into two categories with respect to the dielectric constant of the solvent. In non-polar solvents such as aromatic and aliphatic hydrocarbons it has been proposed that the ionic groups are not dissociated and that intra- and inter-chain dipole interactions can occur leading to polymer aggregation and rheological features such as gelation and shear thickening (Witten, 1988). In polar solvents the rheological behaviour of ionomers is more typical of polyelectrolyte solutions (see for example Lantman *et al.*, 1988). In polyelectrolyte solutions ion pairs do ionize and the system is governed by electrostatic repulsions between charges on the polymer chains. In mixtures of polar and non-polar solvents unusual rheological effects such as large increases in viscosity with increasing temperature can be observed (Lundberg and Makowski, 1980). It has been suggested that these effects arise from changes in aggregation with temperature. In order to quantitatively explain the macroscopic properties of ionomers their microscopic structures both in solution and the solid state need to be determined. In this chapter recent work using scattering techniques to characterise the structuring of ionomers and determine their dimensions and aggregation in solution is reviewed.

Scattering of radiation on passing through a medium arises from fluctuations or inhomogeneities in refractive index (with light), electron density (in X-ray scattering) or the average neutron scattering lengths of atoms. These inhomogeneities must be of a similar magnitude to the radiation wavelength. The large differences in electron density that can be achieved between metal counterions and the polymer backbone with ionomers enables clustering of the ionic groups to be investigated using X-ray scattering. This technique has been applied to ionomers in both the solid state and semi-dilute solutions. The greater wavelength of light has meant that interpretation of the scattering of this radiation can provide information on the aggregation of ionomers in dilute solution. With small angle neutron scattering the large difference in scattering lengths of hydrogen ($b = -0.374 \times 10^{12}$ cm) and deuterium ($b = 0.667 \times 10^{12}$ cm) has made the technique ideal for investigating ionomers from the solid state down to very dilute solutions. In solution both the single chain structures and ionomer aggregation can be investigated. The following discussion will focus on the use of neutron scattering but will also briefly mention work using x-ray and light scattering in order to compare their respective merits.

The use of small angle neutron scattering (SANS) as well as small and wide angle X-ray scattering (SAXS and WAXS, respectively) in the study of perfluorinated ionomer membranes and solutions such as Nafion has recently been reviewed (Gebel and Loppinet,

1996). In general the membrane scattering is characterised by (i) a broad peak (the 'ionomer' peak), (ii) a small angle upturn, and (iii) scattering ascribed to a crystalline component. The first two effects are usually attributed to the ion pair associations mentioned above. There is still a great deal of dispute, however, over the interpretation of scattering effects observed with Nafion membranes. This is despite their commercial importance and therefore quite extensive study. The scattering from perfluorinated ionomers in polar solvents is slightly easier to interpret and indicates the formation of rod like structures. Again, however, the lack of information on the exact chemical structure of the polymers limits the analysis of the results.

Polystyrene ionomers (SPS) have proved to be better model systems for enabling a quantitative understanding of the scattering from ionomers. This arises because the level, type and counterion of the charged groups and molecular weight of the polymer backbone can all easily be varied making them a good model system to gain a basic understanding of the properties of ionomers in general. In addition, there is considerable data already available in the literature on the polystyrene and the corresponding ionomers can be readily obtained in a monodisperse and in a deuterated form making interpretation of scattering results easier. It has also been possible to prepare model telechelic polystyrene ionomers with charge groups only on the chain ends. All these materials are also soluble in a wide range of organic solvents. The preparation of such ionomers has been reviewed elsewhere (see for example Tant and Wilkes, (1988) or Fitzgerald and Weiss (1988)). Small angle X-ray scattering (SAXS) studies on such systems have also recently been reviewed (Chu (1996) and Vanhoorne and Jerome (1996)) and therefore are only briefly mentioned in this chapter. Here the focus is, to a large extent, on new neutron work on polystyrene ionomers carried out in the last decade.

## 10.2  IONOMERS IN SOLUTION

In what follows theories that can be used to interpret the scattering from aggregating polymer solutions will be described followed by the results obtained by fitting of these theories to the scattering from ionomer solutions. How the interpretation of scattering results can explain rheological features will also be discussed.

### 10.2.1  Scattering Theory for Polymer Solutions

10.2.1.1  Polydisperse, non-aggregating polymers

The Zimm expression (Zimm, 1948) gives the normalised coherent scattering cross section of neutrons versus wavevector, $q = (4\pi/\lambda)\sin\theta/2$, from a dilute polydisperse non aggregating polymer solution as

$$B\frac{c}{I(q)} = \left[\frac{1}{\sum_i\{w_i M_i S_i(q)\}}\right] + [2A_2c + 2A_3c^2 + \ldots] \tag{1}$$

where $\lambda$ is the wavelength of the radiation and $\theta$ the angle of scatter. The term in the first set of square brackets arises from intra-particle scattering and the terms in the second from inter-particle scattering effects. The quantities $w_i$ and $S_i$ are the weight fraction and particle scattering factor, respectively, for species of molecular weight $M_i$, and $c$ is the total polymer concentration. $A_2$ and $A_3$ are the second and third virial coefficients that can be related to the excess Gibbs free energy of the solvent. Inter-particle scattering effects can often be ignored either at high $q$, very low concentrations or for dilute solutions of polymers under $\theta$ (ideal) conditions. The contrast factor B is given by

$$B = \left(\frac{N_A}{m_0^2}\right)[a_p - a_s^*]^2 \qquad (2)$$

with

$$a_s^* = a_s \left(\frac{V_m}{V_s}\right). \qquad (3)$$

$V_m$ and $V_s$ are the molar volumes of the monomer and solvent, respectively. $N_A$ is Avogadro's number and $m_0$ the molecular weight of a monomer unit in the polymer. $a_p$ and $a_s$ are the average scattering lengths of the polymer monomer and solvent. They can be calculated from the average scattering lengths of all the atoms within the molecules provided the chemical structures of the monomers and solvent are known.

For $qR_{g_i} < 1$, where $R_{g_i}$ is the polymer root mean square radius of gyration, the particle scattering factor becomes

$$S_i(q) = \frac{1}{(1 + q^2 R_{g_i}^2 / 3)} \qquad (4)$$

and is independent of particle shape. Then from a Zimm plot of $I(q)^{-1}$ versus $q^2$ apparent inverse molecular weights at different concentrations defined as

$$M_{app}^{-1} = BcI(0)^{-1} = M_w^{-1} + 2A_2c + 3A_3c^2 \ldots \qquad (5)$$

and the apparent radii of gyration

$$R_{g_{app}}^2 = 3\frac{dI^{-1}}{dq^2}I(0)^{-1} = R_{g_z}^2 \frac{M_{app}}{M_w} \qquad (6)$$

can be obtained. The weight average molecular weight of the polymer

$$M_w = \sum_i w_i M_i \qquad (7)$$

can be calculated from the intensity extrapolated to zero angle and concentration. Once this is known the so called $z$-average radius of gyration

$$R_{g_z} = \left(\frac{\sum_i w_i M_i R_{g_i}^2}{\sum_i w_i M_i}\right)^{0.5} \approx \frac{\sum_i w_i M_i R_{g_i}}{\sum_i w_i M_i} \qquad (8)$$

is determined at a given concentration using equation (6).

If data at $qR_{g_i} > 1$ is used then a model for the shape of the scattering object is required. The Debye model (Debye, 1947) for Gaussian polymer coils for example gives the particle scattering factor (introduced in equation (1)) as

$$S_i(q) = \frac{2}{\mu^2}(e^{-\mu} - 1 + \mu) \tag{9}$$

where

$$\mu = q^2 R_{g_i}^2.$$

It has been suggested that a single polymer chain forms a Gaussian coil in a $\theta$-solvent and that this result can be extended to melts and solid polymers. It has also been shown theoretically to give a reasonable description of the scattering from high molecular weight polymers extended in good solvents (Witten and Schafer, 1981).

Equation (1) was initally developed for light scattering from dilute polymer solutions with only weak inter-particle interactions (although the contrast factor $B$ is defined differently). More recent theoretical work has shown, however, that similar expressions can be developed for neutron scattering from more concentrated solutions and even for bulk polymer scattering. In the latter theories, however, the virial terms are usually replaced by expressions containing an excluded volume parameter (see for example de Gennes, 1979, and Benoit and Benmouna, 1984). The Zimm expression is also the base for new equations used to interpret the scattering from polymers with long range inter-molecular interactions such as ionomers in polar solvents (Bodycomb and Hara, 1994).

### 10.2.2 Aggregating Polymers

A number of recent publications on model monodisperse ionomers in non-polar solvents, (to be described later) have shown that the intensity of intramolecular scattering can be taken as the sum of the intensities from all the different structures in the solution whether aggregates or single chains. Again the Zimm expression can be used to interpret the scattering but $w_i$, $M_i$ and $S_i$ refer to properties of both single chains and aggregates. For associating systems all these parameters can change with concentration due to varying extents of polymer aggregation. The subscript $i$ then characterises the number of chains in an aggregate. The quantities $M_w$ and $R_{g_z}$ refer to average values for all the single chains and aggregates in the solutions.

As an initial step in the analysis of scattering from aggregating polymers the Zimm plots are often used to determine apparent radii of gyration and molecular weights at various concentrations. If very dilute solutions are considered first then inter-particle scattering effects can be ignored. The apparent molecular weights and radii of gyration can then be assumed to be equal to the true weight and $z$-averages, respectively (see equations (7) and (8) above). Care must be taken, however, to ensure that the correct range of $q$ is used. The product $qR_{g_z}$ in all cases should be less than unity, but a sufficiently broad range of $q$ is required in order to accurately determine $Rg_z$. For this reason combining data from light scattering (where typically $q < 0.005$ Å$^{-1}$) and SANS ($0.001 < q(\text{Å}^{-1}) < 1$) can be beneficial in the investigation of dilute aggregating polymer solutions as scattering over a broader range of $q$ is then accessible.

Since the average aggregate dimensions tend to increase with concentration this can mean that lower $q$ ranges need to be used at higher concentrations. The variation in the molecular weights and radii of gyration with concentration can then be compared with what would be expected if there exists an equilibrium between single chains and aggregates. A summary of the main equations often used to interpret the variation of weight average molecular weight with concentration for aggregating systems at equilibrium is given below for two distinct model cases *closed* and *open association* models. (For further mathematical details see Elias, 1972.) Extension of these models to interprete the concentration dependance of the $z$-average radii of gyration in aggregating systems is also given. Moreover, the calculation of intra-particle scattering over a broad $q$ range for aggregating systems is discussed below.

### 10.2.2.1 Closed association model (CAM)

*Weight average molecular weights*    The closed association model assumes single chains are in equilibrium with aggregates consisting of one size only, i.e.

$$n P_1 \Leftrightarrow P_n$$

where $P_1$ represents a single chain, and $P_n$ an aggregate consisting of $n$ chains. The equilibrium constant for this aggregation is given by

$$K_c = \frac{[P_n]}{[P_1]^n} \tag{10}$$

where $[P_n]$ and $[P_1]$ are the molar concentrations of aggregates and uni-mers, respectively. If $M_1$ is the molecular weight of the single chains then the relationship between the ratio $x = M_w/M_1$ and total ionomer concentration is

$$gc = \left( \frac{[x-1]}{[n-x]^n} \right)^{1/(n-1)} \tag{11}$$

The constant $g$ is given by

$$g = \frac{(K_c n)^{1/(n-1)}}{M_1(n-1)} \tag{12}$$

The weight fractions of single chains at a given total polymer concentration, $c$, can be calculated using the following relationship

$$w_1 = \frac{n-x}{n-1} \tag{13}$$

Normalised plots of $gc$ against $gc$ are given in Figure 1 for various values of the integer $n$.

*Z-average radii of gyration*    If the CAM is valid then simple algebraic rearrangement of equations (7) and (8) with $w_1 = 1 - w_n$ gives (Pedley *et al.*, 1990b)

$$R_{g_z} x \approx \lfloor R_{g_1} + (1 - w_1)(n R_{g_n} - R_{g_1}) \rfloor \tag{14}$$

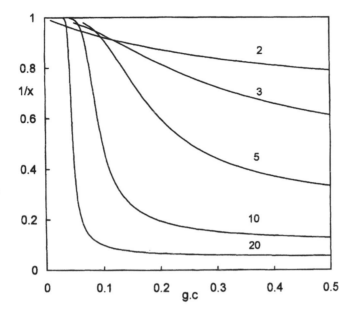

FIGURE 1. Weight average molecular weight versus concentration for polymers aggregating via the closed association model. The $x$- and $y$-axes have been normalised by the factor $g$ (see equation 12) and the single chain molecular weight, respectively. Integers above each curve indicate the number of chains in the aggregates.

Alternatively the expression (Pedley, 1990)

$$R_{g_z} \approx \left[\frac{1}{n-1}\right][(nR_{g_n} - R_{g_1}) - (nR_{g_n} - nR_{g_1})x^{-1}] \qquad (15)$$

also derived from equations (7) and (8) has been used to determine $R_{g_1}$ and $R_{g_n}$. These expressions show that if $n$ can be determined from the apparent molecular weights the radii of gyration of the uni-mers and n-mers ($R_{g_1}$ and $R_{g_n}$, respectively) can be estimated from a plot of either $R_{g_z}$ versus $1/x$ or $R_{g_z}x$ versus $(1 - w_1)$.

*Modelling of scattering*

If it is known that there is a mixture of single chains and aggregates of one size in a solution then the variation in intraparticle scattering versus $q$ can be calculated assuming a model for the shape of the scattering objects. Such curves can then be compared with the data collected. In Figure 2 the scattering function $S(q)$ for a gaussian polymer is shown as a function of $qR_{g_1}$ where $R_{g_1}$ is the single chain radius of gyration (see equation (9)). In addition examples of calculated intraparticle scattering ($\sum_i \{w_i M_i S_i(q)\}/M_1$) are given for two mixtures that contain equal weights of single chains and aggregates. In Figure 2 these curves are plotted against $qR_{g_1}$ and normalised by the single chain molecular weight.

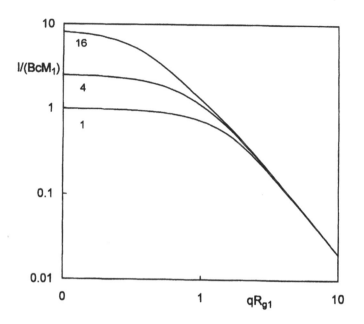

FIGURE 2. Intensity versus wavevector $q$ for mixtures containing equal weights of single chains and aggregates (assuming a gaussian coil and only one size of aggregate). The calculated curves shown are normalised by the single chain molecular weights and radii of gyration. The integers associated with each curve indicate the number of chains within an aggregate.

The aggregates in the examples consisted of either 4 or 16 chains. At high values of the scattering vector $q$ all the plots converge. This arises because the aggregates as well as the single chains were assumed to be Gaussian with a fractal dimension of 2. The fractal dimension, $D$, is the exponent in the relationship between the polymer molecular weight and its radius of gyration.

$$M \propto R_g^D \qquad (16)$$

For the aggregates in Figure 2 therefore the following relation holds

$$R_{g_n} = n^{0.5} R_{g_1}.$$

$D$ equals unity for rod-like structures and $D = 3$ for solid spheres. A generalization is often made that if scattering objects are fractal then $D$ can be obtained from the gradient of $\log I$ versus $\log q$ at sufficiently high $q$ since then

$$I(q) \propto q^{-D}. \qquad (17)$$

A Kratky plot of $I(q)q^D$ against $q$ or $q^2$ can often differentiate the range over which equation (17) is valid better than a double logarithmic plot of intensity versus $q$ (Glatter and Kratky, 1982). If a Kratky plot has a horizontal asymptote then the value of the exponent $D$ is confirmed.

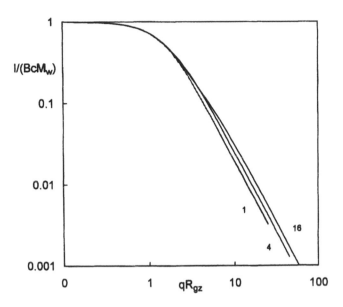

FIGURE 3. Intensity versus wavevector $q$ for mixtures containing equal weights of single chains and aggregates (assuming gaussian behaviour and only one size of aggregate). The calculated curves are shown normalised by the weight and $z$-average molecular weights and radii of gyration. The numbers associated with each curve indicate the number of chains within an aggregate.

In Figure 3 the calculated intensities given in Figure 2 are shown after normalisation by the weight average molecular weight (see equation (7)). These intensities are plotted against $q R_{g_z}$ where $R_{g_z}$ is the $z$-average radius of gyration of the aggregate and single chain as defined in equation (8). Below about $q R_{g_z} < 2$ the curves for the mixtures are identical with those for a monodisperse gaussian coil with a radius of gyration equal to the $z$-average of the mixture. If the Debye model (assuming monodispersity) were fitted to the calculated curves at higher $q$ then for the mixtures an ill-defined average radius of gyration that is smaller than the $z$-average would be obtained.

Provided $n$ is sufficiently large, however, the CAM model predicts a critical micelle concentration (see Figure 1) with primarily single chains in solution at low concentration ($w_1 \approx 1$) and only aggregates of one size at high concentration ($w_n \approx 1$). Then the fitting of a model for a monodisperse system over a broad range of $q$ at several concentrations should provide an estimate of the aggregate or single chain radius of gyration. This arises because significant mixtures of aggregates and single chains will only be present in solution over a narrow range of concentration. This is not true for systems where the open association model describes the extent of aggregation of the polymer chains which is described below.

#### 10.2.2.2  Open association model (OAM)

*Weight average molecular weights*   In the open association model it is assumed that single chains are in equilibrium with aggregates of all sizes, i.e.

$$P_1 + P_2 \Leftrightarrow P_2 \qquad K_2$$
$$P_1 + P_2 \Leftrightarrow P3 \qquad K_3$$
$$P_1 + P_n \Leftrightarrow P_{n+1} \qquad K_{n+1}$$

where $P_n$ is an aggregate consisting of $n$ chains. If the equilibrium constants for each step can be considered invariant to increases in the size of the aggregates, namely

$$K_0 = K_2 = \frac{[P_2]}{[P_1]^2}$$

$$= K_3 = \frac{[P_3]}{[P_2][P_1]}$$

then

$$x^2 = 1 + \frac{4K_0c}{M_1} \tag{18}$$

and

$$w_1 = \frac{M_1}{K_0c} \left[ \frac{x-1}{x+1} \right] \tag{19}$$

In this case a plot of $x^2$ (equal to $(M_w/M_1)^2$) versus concentration gives a straight line with an intercept of unity and gradient of $(4K_0/M_1)$. The weight fraction of aggregates is further given by

$$w_n = n \left( \frac{K_0c}{M_1} \right)^{n-1} w_1^n \tag{20}$$

Here $x$, $w_1$ and the various values of $w_n$ are all dependent upon the magnitude of $(K_0c/M_1)$ only. The distributions of aggregates for various values of this term are given in Figure 4. At sufficiently low concentrations it is possible to assume that with systems following this model only single chains and two chain aggregates will be in solution. As the polymer concentration increases, however, the average size of the aggregates will continually increase with aggregates of many more than two chains being formed.

*Z-average radii of gyration*

A recent method of interpreting the concentration dependence of $z$-average radii of gyration for ionomers associating via an open association process used equation (16) to define the aggregate dimensions (Young *et al.*, 1998b). Then it follows from equations (7) and (8) that

$$R_{g_z}x = w_1 R_{g_1} + \sum_i nw_n R_{g_n} \tag{21}$$

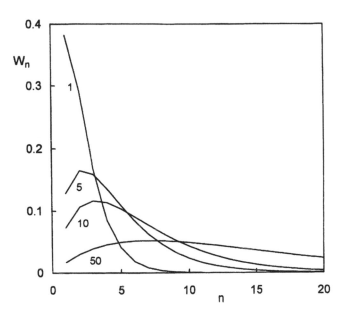

FIGURE 4. Weight distributions of aggregates and single chains at various concentrations predicted via the open association model. $K_0 c / M_1$ values for each curve are shown. (cf. equations (18–20)).

and from equation (16)

$$R_{g_n} \propto n^{1/D} \qquad (22)$$

## Modelling of scattering

As with the CAM, the scattering from polymers aggregating via an open association process can be modelled over a broad $q$ range assuming different shapes for the scattering objects. Calculated intensities for a system aggregating via an open association process, where both the single chains and aggregates behave like gaussian coils, are shown in Figures 5 and 6. In these figures the axes have been normalised either by the single chain (Figure 5) or weight and $z$-average (Figure 6) molecular weights and radii of gyration. The radii of gyration are all given by $R_{g_n} = n^{1/2} R_{g_1}$ as required for gaussian coils. From Figure 5 it is observed that at high $q$ all the curves are predicted to be concentration independent. If the aggregates and single chains have different fractal dimensions, however, this is not the case (cf. examples in Young et al., 1998b). At low $q$ the intensities are proportional to the total polymer concentration squared as required by equation (18).

## 10.2.2.3  Other association models

Other models such as combinations of the open and closed association models could also be feasible. In dilute polymer solutions, however, one of the above two models often explains

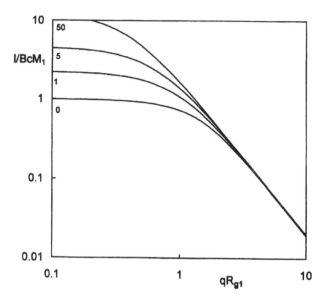

FIGURE 5. Intensity predicted for the OAM versus $q$ at various concentrations for gaussian single chains aggregating into gaussian aggregates. Values for $K_0 c / M_1$ are given on the plots. The intensities and wavevectors have been normalised by the single chain molecular weights and radii of gyration.

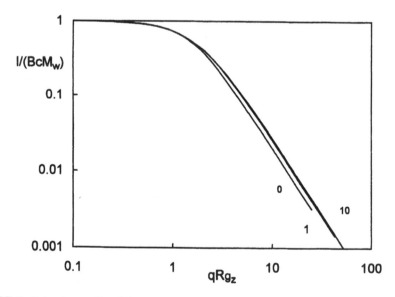

FIGURE 6. Intensity predicted for the OAM versus $q$ at various concentrations for gaussian single chains aggregating into gaussian aggregates. Values for $K_0 c / M_1$ are given on the plots. The intensities and wavevectors have been normalised by the weight and $z$ average molecular weights and radii of gyration.

well the variation of apparent molecular weights with concentration in aggregating systems (Hunter, 1987). The closed association model commonly fits results because only a slight maximum in the free energy change related to the addition of a single chain onto an aggregate as a function of the total number of chains in the aggregate is required in order for aggregates of one size to be dominantly observed in the solution. The open association model can be observed if the same process occurs each time a single chain joins onto an aggregate.

### 10.2.3  Contrast Match Neutron Scattering

The intensity of scattered neutrons from a mixture of otherwise idendical hydrogenous and deuterated monodisperse interacting polymer molecules in solution is given by

$$I(q) = Mc[AS(q) + BP(q)] \tag{23}$$

$P(q)$ is the pair scattering function and incorporates all the inter-single chain scattering effects (Ullman et al., 1986). The constant $B$ is as defined above (Eq. 2) but with the scattering length $a_p$ of the polymer equal to

$$a_p = a_D x_D + a_H (1 - x_D) \tag{24}$$

In the above equation $a_H$ and $a_D$ are the scattering lengths referring to the hydrogenous and deuterated monomers and $x_D$ is the mole fraction of deuterated monomers. The constant $A$, however, is given by

$$A = \frac{N_A}{m_0^2} [(a_D - a_s^*)^2 x_D + (a_H - a_s^*)^2 (1 - x_D)] \tag{25}$$

If the average scattering length of the solvent is changed by using different mixtures of hydrogenous and deuterated solvent then $A$ and $B$ vary independently. It is therefore possible to separate $S(q)$ and $P(q)$. In the contrast match method a ratio of deuterated to hydrogenous solvent can be chosen such that the average scattering from the polymer equals that of the solvent (i.e. $a_p = a_s^*$). Then, since $B = 0$, the contrast match intensity is given by

$$I(q) = McAS(q) \tag{26}$$

This method enables variation in single chain structures with concentration to be investigated.

If a monodisperse mixture of hydrogenous and deuterated polymer is aggregating then from the contrast matched intensity an average mean square radius of gyration of all the single chains $R_{g,sc}$ whether isolated or within an aggregate is obtained (Young et al., 1996b). In the range $qR_g < 1$ the contrast matched intensity is then

$$M_1 c A I(q)^{-1} = 1 + q^2 \langle R_{g,sc}^2 \rangle / 3 \tag{27}$$

From the contrast matched intensity at different concentrations the isolated and aggregated single chain dimensions can be estimated.

*10.2.3   Scattering Results for Ionomer Solutions*

10.2.3.1  Randomly sulfonated polystyrene (SPS) ionomers in non-polar solvents

Effects of Charge Levels

The aggregation behaviour in xylene of several monodisperse sodium sulfonated polystyrene (NaSPS) ionomers of molecular weight $10^5$ g/mol has recently been investigated by both light scattering and SANS. Determination of the $z$-average radii of gyration for these ionomers by light scattering is difficult due to the limited range of $q$ accessible. Apparent weight average molecular weights can be obtained by light scattering for these ionomers but care must be taken to ensure there is no time dependent "excess low angle scatter". The latter scattering can arise if the solutions are not given sufficient time to ensure complete polymer dissolution and hence are not at thermodynamic equilibrium. Both apparent molecular weights and $z$-average radii of gyration can be obtained from neutron scattering measurements as discussed above.

A combination of rheological measurements with dynamic and static light scattering has indicated that below 0.62 mol%, NaSPS ionomers of $10^5$ g/mole do not aggregate in xylene (Young *et al.*, 1998a). With a sulfonation level of 0.95 mol%, however, the inverse apparent molecular weight of the same molecular weight NaSPS ionomer in xylene obtained by SANS varies with concentration as shown in Figure 7 (Young *et al.*, 1995 and 1996a). These results clearly indicate that aggregation is occuring. In order to interpret the apparent molecular weights an estimate of the second virial coefficient (see equation (1)) is first required. This can be obtained from the gradient of the plot in Figure 7 at high concentrations. Since the polymer concentrations are low the third virial coefficient can be ignored. The results for this ionomer cannot be interpreted by the open association model even if virial terms are included in the analysis. An excellent fit of the closed association model with $n = 3$ to the results, however, is observed once the data at high concentrations are corrected for a small second virial coefficient term (see Figure 7).

The apparent radii of gyration for the above 0.95 mol% NaSPS ionomer obtained by SANS were corrected for the effects of inter-particle scattering using equation (6) (Young *et al.*, 1995). With this ionomer a plot of $R_{g_z}(M_w/M_1)$ versus the weight fraction of aggregates was found to be linear and used to obtain the aggregate and single chain radii of gyration. The reasonable linearity of the plot in Figure 8 confirms the applicability of the closed association model (CAM) for the system. The dimensions obtained for both radii of gyration were smaller than would be expected for the same molecular weight polystyrene in xylene but comparable with what would be observed for polystyrene in a $\theta$-state.

The apparent molecular weights obtained for NaSPS ionomers at sulfonation levels of between 1.25 and 1.65 mol% using light (Pedley *et al.*, 1990a) or SANS (Pedley *et al.*, 1990b; Young *et al.*, 1995) are in dilute xylene solutions ($<0.6$ g/dl) practically independent of sulfonation level. The apparent molecular weights squared are plotted against concentration for sulfonation levels of 1.39 mol% and 1.65 mol% in Figure 9. The linearity of the plot and intercept of close to unity indicate that the open association model can quantify the extent of aggregation at low concentrations well (see equation (18)). Above 1 g/dl the apparent molecular weights increase as the sulfonation level is raised.

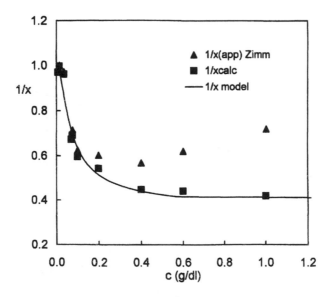

FIGURE 7. Apparent and true inverse weight average molecular weight versus concentration for 0.95 mol% NaSPS ionomers of $10^5$ g/mol in xylene. Results have been normalised by the single chain molecular weight. The curve is the best fit of the closed association model with $n = 3$ and $g = 4$.

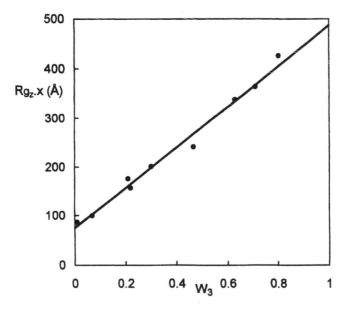

FIGURE 8. $Rg_z$ ($M_w/M_1$) versus weight fraction of aggregates for 0.95 mol% NaSPS ionomers of $10^5$ g/mol in xylene. The linearity of the plot confirms the applicability of the closed association model (see equation (14)).

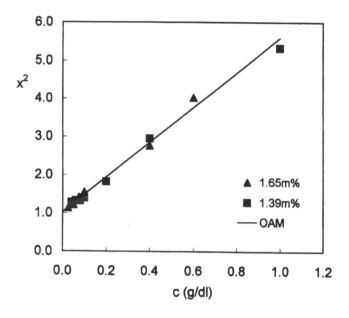

FIGURE 9. Weight average molecular weight squared versus concentration for 1.39 and 1.65 mol% NaSPS ionomers of $10^5$ g/mol in xylene. The $y$-axis has been normalised by the single chain molecular weight squared. The linearity of the plot and intercept of 1 confirms the applicability of the open association model (see equation (18)). From the gradient $K_0/M_1 = 1.1$ dl/g.

This has been interpreted as arising from inter-particle scattering effects that decrease as the sulfonation level is raised. A decrease in the virial terms with increasing sulfonation level (and excess free energy of the solvent) is not unexpected since xylene is a poor solvent for the ionic groups. At 1.65 mol% sulfonation the virial terms become negligible. At higher sulfonation levels the NaSPS ionomer becomes insoluble in xylene.

To interpret the concentration dependence of the $z$-average radii of gyration for ionomers associating via the OAM earlier studies tended to use low concentration data only. In this range it can be assumed that only single chains and two chain aggregates are present (Pedley $et$ $al.$, 1990b; Young $et$ $al.$, 1995). Then equations (14) and (15) valid for the CAM can be applied. A better method that accounts for aggregates larger than two chains using equations (21) and (22) was more recently applied to data for NaSPS with 1.39 mol% sulfonation (Young $et$ $al.$, 1998b). The values of $z$-average radii of gyration obtained for this polymer as a function of concentration are shown fitted to equations (21) and (22) in Figure 10. The fit gives a very compact single chain radius of gyration of only 55 Å which compares with a value of 93 Å for polystyrene of the same molecular weight in a $\theta$-state. The aggregates are however more expanded structures and have dimensions comparable with polystyrene of the aggregate molecular weight in a $\theta$-solvent. In this latter study intraparticle scattering curves were also calculated and compared with experimental data over a very broad $q$ range at various concentrations to confirm the aggregate and single chain structures.

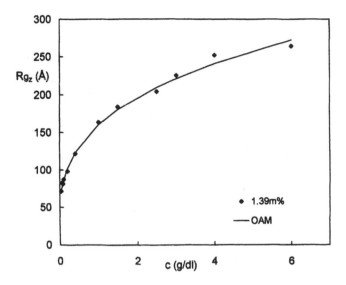

FIGURE 10. Z-average radii of gyration versus concentration for 1.39 and 1.65 mol% NaSPS ionomers of $10^5$ g/mol in xylene. The curve is the best fit of equations (21) and (22) with $Rg_1 = 55$ Å, $Rg_2 = 120$ Å and $D = 0.53$ assuming $K_0/M_1 = 1.1$ dl/g, as shown in Figure 9.

*Comparison of neutron scattering results with rheological measurements*

The reduced viscosity of a dilute polydisperse polymer solution can be described by

$$[\eta] = \frac{\eta_{\text{solutions}} - \eta_{\text{solvent}}}{\eta_{\text{solvent}}} c \approx \sum_i w_i [\eta]_i \tag{28}$$

provided interparticle interactions can be neglected. $\eta_{\text{solution}}$ and $\eta_{\text{solvent}}$ are the solution and solvent viscosities. $[\eta]_i$ is the reduced viscosity of a monodisperse polymer of molecular weight $M_i$. Most theories describing the rheology of dilute solutions assume $[\eta]_i$ is dependent upon the volumes occupied by the polymer chains. There have been many theoretical equations that relate polymer radii of gyration to dilute solution reduced viscosities (see for example Bohanecky and Kovar (1982) and references therein). Recent studies (Young *et al.*, 1998a) have shown that such theories can be extended to explain the rheological behaviour of ionomers in very dilute solution. In this case, however, $[\eta]_i$ is taken as a hypothetical reduced viscosity for an ionomer single chain or aggregate that each have weight fractions $w_i$. Since the extents of aggregation vary with concentration and $[\eta]_i$ can have differing values for single chains and aggregates the observed reduced viscosity of an ionomer solution can be concentration-dependent even in very dilute solutions. If, as with the case of the ionomers described above, there are large weight fractions of very compact single ionomer chains in the solutions then the viscosities with the ionomer can be below that of the polymer on which it is based.

*Single Chain Dimensions Within Ionomer Aggregates*

The average single chain dimensions as a function of concentration have been
for NaSPS ionomers with 0.98 (Young *et al.*, 1990a) and 1.25 mol% (Young *et al.*
degree of sulfonation in xylene using the contrast match method. With the (?
ionomer the average single chain dimension is at all concentrations measured (between
3 g/dl) greater than the isolated single chain dimension. If the aggregation behaviour of
ionomer is identical to that of the 0.95 mol% ionomer described above then primary
aggregates of constant size should be present in solution in the concentration range studied.
These results indicate that within the aggregates the single chain expand to dimensions
comparable with those observed for the base polystyrene in xylene.

The single chain dimensions for 1.25 mol% NaSPS ionomers in xylene have been
measured over a broad concentration range. The results at low concentrations for the
1.25 mol% ionomer confirm that the single isolated chains are compact but again also
suggest that within an aggregate chain expansion can occur. Using the weight fractions
of single chains and aggregates obtained above as a function of concentration, a model to
explain the variation of single chain size with concentration for the 1.25 mol% ionomer was
proposed (Young *et al.*, 1996b). In this model it was assumed that each time a single chain
joined onto an aggregate, since there was a constant increase in the free energy of aggregation
(an assumption implicit in the open association model), there was also a constant increase in
the average size of the single chains. The variation of average single chain dimensions with
concentration could be explained by this model and gave an extrapolated single chain radius
of gyration in agreement with the value obtained by extrapolation of aggregate dimensions
to zero concentration.

*Effects of Temperature on Aggregation*

The effect of temperature on the aggregation of the above NaSPS ionomers in xylene was
found to be negligible between 20 and 45°C (Pedley *et al.*, 1990; Young *et al.*, 1996a). From
the van't Hoff equation in thermodynamics this indicates that the enthalpy of aggregation
is small in comparison with the changes in entropy that occur. The rise in entropy could
occur through an increase in both polymer solvent mixing and configurational entropy
of a polymer chain when inter-molecular ion pair associations form at the expense of
intra-molecular ones.

Ordinarily, although the reduced viscosity may remain constant or perhaps increase or
decrease slightly with temperature, the actual viscosity of most polymer and ionomer
solutions decreases as the temperature is raised due to a reduction in solvent viscosity.
When low levels of alcohol co-solvent are added to ionomer solutions, however, anomolous
viscosity temperature dependencies can be observed: in some cases either distinct maxima
or minima in viscosity versus temperature plots can occur (Lundberg *et al.*, 1980). Light and
neutron scattering have shown that the addition of low levels of alcohol co-solvents breaks
up both intra and inter-molecular ion pair associations (Pedley *et al.*, 1990; Young 1997).
It has been suggested that the anomalous temperature dependence of some of these mixed
solvent ionomer solutions could arise through the co-solvent solvating the ionic groups
better at lower temperatures than at higher ones.

## Gelation and Shear Thickening

At polymer concentrations greater than 2 g/dl the above NaSPS ionomers with sulfonation levels greater than 1.25 mol% (Pedley *et al.*, 1989 and references therein) can cause xylene solutions to exhibit reversible shear thickening and a slow gelation with time. At these concentrations, however, the viscosities of the solutions are very sensitive to the exact method of preparation of the ionomer. It was initially proposed that shear caused extension of single chains, and that the shear thickening arose through inter-molecular ion pair associations forming at the expense of intra-molecular. The single chain dimensions of a shear thickening ionomer solution, were studied using SANS and found, however, to be invariant with shearing (Pedley *et al.*, 1989). It was therefore proposed that the reversible shear thickening arose through increased inter-aggregate interactions rather than changes in single chain dimensions.

## Effects of Ionomer Molecular Weight

The aggregation behaviour of several NaSPS ionomers of molecular weight lower than $10^5$ g/mol in xylene has also been investigated. For example, NaSPS ionomers with a molecular weight of $4.95 \times 10^4$ g/mol have been studied using SANS (Timbo, 1993). At a sulfonation level of 0.72 mol% no aggregation occurs below 1 g/dl. At 1.48 mol%, however, there is aggregation but no interpretation of the data by equilibrium models has as yet been attempted. Ionomers of $3.5 \times 10^4$ g/mol and sulfonation levels of 1.0 and 4.2 mol% were also investigated using static light scattering (Pedley, 1990). Although the latter results suggest aggregates are present in both cases neither of the simple equilibrium models described above can explain all the observed data. Complications in the data interpretation could possibly be arising through inter-particle scattering effects that cannot be modelled by the theories given earlier in this chapter. SANS measurements are required in order to obtain results both at lower concentrations and higher $q$ in order to assess any effects of inter-particle scattering. In order to interpret such results modelling of a broad range of $q$ as described in the earlier theory section will be required.

## Effects of Solvent Polarity

For relatively high molecular weight ionomers ($10^5$ g/mol) it has been found that the apparent intrinsic viscosity obtained by extrapolation of reduced viscosity versus concentration to zero concentration for a wide range of dilute SPS ionomer solutions can be scaled approximately using an expression of the form

$$[\eta] = [\eta]_{PS} F(\xi) \tag{29}$$

where

$$\xi = a_s a_a y.$$

Here $y$ is the mole fraction of ionic groups and $[\eta]_{PS}$ is the intrinsic viscosity of the parent polymer in the same solvent as the ionomer (Agarwal *et al.*, 1987). $a_s$ is a solvent factor given by

$$a_s \approx 1/\varepsilon$$

where $\varepsilon$ is the relative dielectric constant of the solvent. The parameter $a_a$ is dependent upon the counterion for the ionomer. Both $a_s$ and $a_a$ can be taken as a measure of the strengths of the ion-pair interactions. For small $\xi$ and only very weak interactions $F(x)$ is close to unity. As $\xi$ increases, however, $F(x)$ decreases until a critical level of charge is reached at which the ionomer becomes insoluble. With NaSPS ionomers in xylene this critical charge level is reached at 1.65 mol%, as mentioned above. In tetrahydrofuran (dielectric constant of 7.6), however, since the dielectric constant is three times that of xylene, NaSPS ionomers of $10^5$ g/mol with sulfonation levels up to 5.1 mol% are soluble. In non-polar solvents the viscosities of dilute ionomer solutions as shown above are determined primarily by the total volumes occupied by the polymer, whether in single collapsed chains or aggregates. This suggests that the average density of the aggregates and single chains present in a dilute ionomer solution should decrease if the dielectric constant of the solvent is raised. The rheological properties of ionomers in more concentrated solutions are more difficult to interpret fully due to inter-aggregate interactions. It appears that gelation is commonly observed, however, when the critical solubility sulfonation level is reached.

Light scattering results indicate that 1.15 mol% NaSPS ionomers of $10^5$ g/mol do not aggregate in THF (Lantman *et al.*, 1987). Contrast match measurements on a similar ionomer in THF also indicate that the single chain radius of gyration is comparable with the expanded dimensions of polystyrene in dilute THF or xylene solutions (Gabrys *et al.*, 1989). As indicated from SANS measurements described above 1.15 mol% NaSPS ionomers in xylene would be expected to form more compact structures than polystyrene. The results observed in THF, therefore, confirm rheological studies that suggest the average ionomer chain density decreases as solvent polarity is raised.

At sulfonation levels greater than 2 mol%, however, the apparent molecular weights and radii of gyration of NaSPS ionomers in THF obtained by either light scattering or SANS are much larger than the single chain values suggesting aggregation (Lantman *et al.*, 1988; Pedley, 1990). The results indicate that as the sulfonation level or polymer concentration is raised the average size of the aggregates increases. No simple association model can interpret, however, both the variations in molecular weight and radii of gyration with concentration for any of the aggregating ionomers in THF. It is possible that an inter-particle scattering effect is complicating interpretation of results. Further measurements over broader $q$ and concentration ranges and modelling as described in the earlier theory section of this chapter would be needed to confirm this hypothesis.

The single chain dimensions of 3.4 and 4.0 (Gabrys *et al.*, 1989) and 4.2 mol% (Lantman *et al.*, 1988) NaSPS have been determined over the concentration range of 0.5 to 4 g/dl using the contrast-match method. The results suggest that as in xylene within the aggregates the single chains expand to dimensions comparable with the expanded dimensions of polystyrene in a good solvent. There have also been a few studies on the effect of counterion on the single chain dimensions of ionomers detemined using the contrast-match method (Gabrys *et al.*, 1989) but further work on the effects of the counterion on the aggregation will be required in order to interpret the results.

A comparison of the results for ionomers in xylene and THF suggests that at low sulfonation levels or in more polar solvents the enthalpy change arising from ion pairs associating can be too small to overcome losses in entropy due to polymer configurational constraints that arise with the associations. When the level of ionic groups is sufficiently large, however, evidence for both intra and inter-molecular ion pair associations can be detected using SANS. Determination of the ionomer dimensions in solution using SANS has confirmed that the dilute solution rheology of ionomers is dependent only upon the total polymer volumes. Use of this technique has also indicated through determination of the second virial coefficients a measure of the extent to which the level of ionic groups reduces the excess free energy of mixing the ionomer and solvent. When this free energy approaches zero attractive interactions between the ionomer aggregates can result in rheological effects such as gelation and shear thickening being observed.

### 10.2.3.2 Telechelic ionomers in non-polar solvents

Telechelic ionomers have ionic groups present only at the chain ends. Their well-defined structure (when monodisperse) has lead authors to suggest that they can be used as model systems to explain the properties of random ionomers. In analogy to random ionomers in non-polar solvents semi-dilute solutions of telechelic ionomers can gel once a critical concentration is reached. The concentration at which this process occurs can either increase or decrease with increasing polymer molecular weight. This is likely to arise because increasing ionic concentration at a given total polymer concentration can only occur by decreasing the telechelic ionomer molecular weight. Both decreasing molecular weight and increasing ionic concentration are likely to have opposing effects on the gelation concentration. For reviews of the rheological and SAXS studies of telechelic ionomers the reader is refered to work by Chakrabarty et al. (1997) and Vanhoorne and Jerome (1996). SANS work on telechelic ionomers will be briefly reviewed below.

Much of the work on telechelic ionomers has been carried out on polymers with lower molecular weights than the SPS ionomers described above. The aggregation behaviour of sodium telechelic sulfonated polystyrene (NaTSPS) and carboxylate polystyrene (NaTCPS) ionomers in toluene for example have been investigated using SANS (Timbo et al., 1993). The NaTSPS ionomers had molecular weights of 9,000 and 18,000 g/mol. The molecular weights of the NaTCPS ionomers were 8000 and 20,000. With the lower molecular weight ionomers the authors suggested that an open association process might explain the observed continuous rise in apparent radii of gyration and molecular weights with concentration. The apparent molecular weights and radii of gyration for the larger ionomer, however, increased to a maximum value at concentrations exceeding 1.5 g/dl. This led the authors to suggest the closed association model might interpret the results. In none of the cases, however, was a fit of an association model attempted in the published work.

Similar NaTCPS ionomers have also been studied by other authors (Karayianni et al., 1995). These authors used several methods to interpret their scattering data including fitting the Zimm expression to the data at low $q$. The apparent molecular weights obtained via this method suggest that an open association process might interpret the scattering for both the 8,500 and 19,000 molecular weight ionomers studied. This is contrary to the conclusions of Timbo et al., (1993) obtained for almost identical ionomers described above. To obtain the apparent radii of gyration Timbo et al., fitted the Debye model for monodisperse chains over a broad $q$ range. As discussed in the earlier theory section in this chapter this procedure is likely to lead to radii of gyration that are smaller than the $z$-average ones and might explain the observation of misleading maxima in apparent molecular weights and radii of gyration with concentration. The data obtained by Karayianni et al., (1995), however, have not been obtained over a sufficiently broad concentration range to fit readily any association model. Although a full interpretation of scattering results for the above telechelic ionomers has not as yet been achieved all the results show clearly that at a given total polymer concentration smaller telechelic ionomers form aggregates consisting of many more chains than the larger molecular weight random ionomers described above. Further modelling is required in order to interpret the results for telechelic ionomers more fully.

The aggregation behaviour in toluene of a wide range of polystyrenes with a sulfonate group at just one end of the chain (Halato-Semi-Telechelic Sulfonated Polystyrene (STSPS)) has also been investigated by light scattering (Vanhoorne et al., 1995). In these cases the closed association model can describe all the results well. It was observed that for these semi-telechelics with 3,300 molecular weight that the number of chains in the aggregates varied only slightly as a function of counterion for the series Li, Na and K. This number decreased significantly, however, as the counterion size increased for the divalent counterions of Mg, Ca, and Ba. As the molecular weight of the polymer was raised from $10^3$ to $10^5$ g/mol the average aggregate size for lithium semi-telechelic sulfonated polystyrene (LiSTSPS) decreased from about seventeen down to three chains per aggregate. A lithium SPS telechelic ionomer of molecular weight 25,000 in toluene was also investigated using light scattering by these authors (Vanhoorne et al., 1995). In this case the results could not be interpreted in terms of a closed association model. Again, however, the results have not been obtained over a sufficiently broad concentration range to fit any other association model with confidence.

## 10.2.4   Ionomers in Polar Solvents

In polar solvents such as dimethyl formamide (DMF) and dimethyl sulfoxide (DMSO) a large increase in the reduced viscosity of ionomer solutions can be observed as the charge level is increased and/or the polymer concentration decreased (Agarwal et al., 1987). This behaviour is typical of polyelectrolytes and has been attributed to the ionization of ionic groups. It was proposed that this ionization results in intramolecular repulsion of the unshielded fixed ions on the polymer chain leading to chain expansion. Since rheological behaviour characteristic of polyelectrolytes has also been observed, however, for telechelic ionomers with only one ionic group it has been suggested that inter-molecular interactions must also be considered in any theoretical interpretation of scattering results.

In order to interpret the above rheological results light (Hara et al., 1984) and small angle neutron scattering (Lantman et al., 1988) have been used to determine the structure of the chains in dilute solution. In small angle neutron scattering a peak in intensity versus angle can be observed. This is also typical of polyelectrolyte solutions. Interpretation of this scattering or light scattering is complex, however, due to both a strong variation in single chain dimensions with concentration as well as strong inter-molecular interactions. Determining the scattering from solutions that are sufficiently dilute to be able to ignore inter-particle scattering would be impossible with light scattering and would require very long measurement times with SANS. The use of the contrast method to gain the single chain scattering, however, is relatively simple. The single chain dimensions of sulfonated polystyrene ionomers of molecular weight 50,000 g/mol in DMF have been determined using the subtraction method with SANS (Lantman et al., 1988). The very much larger radii of gyration of the ionomer single chains and the smaller fractal dimensions confirm that ionomer chains can expand significantly in polar solvents. Further measurements of this type should enable a much better understanding of the scattering from polyelectrolyte solutions.

### 10.2.5 Dynamical Experiments on SPS Ionomers

To date, there are only a few reports on the influence of sulfonation on the segmental and local dynamics. Neutron spin echo technique (Mezei, 1972) is particularly well suited to examine dynamical processes occuring in polymers simultaneously in space and time on a semi-local scale. In analogy with the photon correlation spectroscopy, neutron spin-echo measures the same physical quantity, but in a shorter time (of the order of nanoseconds) and at larger momentum transfer. That is, the experiment yields the normalized coherent dynamic structure factor $S_N(q, t) \equiv S(q, t)/S(q, 0)$, where $S(q, 0)$ is the coherent static structure factor. For more information, the reader is directed to the classic collection of papers (Mezei, 1980), the most lucid exposition of the technique given by Nicholson (1981), and recent review of this subject by Ewen and Richter (1997).

Nyström et al. (1986) studied solutions of sodium sulfonated polystyrenes (SPS) as a function of degree of sulfonation in three different deuterated solvents, namely dimethylsulfoxide (DMSO), dimethylformamide (DMF), and tetrahydrofuran (THF). While the former solvents have high polarity, the latter is of low polarity, hence different interactions between ionomeric chains are expected, as discussed above. To recapitulate, in polar media a sufficient number of the metal counterions of the sulfonate groups are dissociated: the chain becomes extended due to the Coulomb repulsion of similar charges and exhibits classical polyelectrolyte behavior even at low (ca 2 mol%) sulfonate levels. THF does not have sufficient polarity to cause ion pair dissociation, and the interactions are controlled by aggregation of ion pairs. The spin-echo results were combined with viscosity measurements, and interpreted in terms of polyelectrolyte models. The following quantities were obtained: (i) a characteristic relaxation frequency $\Omega$ taken from the initial slope of the logarithm of the normalized coherent intermediate scattering function $S_N(q, t)$, i.e.

$$\Omega \equiv -\lim_{t \to 0} \frac{\partial \ln S_N(q, t)}{\partial t} \tag{30}$$

The function $S_N(q, t)$ can be related to an effective, $q$-dependent, diffusion coefficient $D(q)$

$$S_N(q, t) = \exp(-\Omega t) \tag{31}$$

where

$$\Omega = D(q)q^2 \tag{32}$$

and $D(q)$ has the following form:

$$D(q) \equiv \frac{\Omega}{q^2} = k_B T \frac{\mu(q)}{S(q)} \tag{33}$$

In the above equations, $k_B$ is the Boltzmann constant; $\mu(q)$ is a mobility which depends on the wave vector, introduced to account for the internal structure and for the small-scale rigidity of a polyelectrolyte or an ionomer chain.

Finally, the quantity $\eta\Omega/k_B \cdot Tq^2$ which is proportional to the effective diffusion coefficient was plotted as a function of the scattering vector $q$. The differences in the dynamical behavior of SPS ionomers dissolved in solvents of various polarity were observed, and can be summarized as follows: (a) The characteristic frequency $\Omega$ displayed a marked upsweep at low sodium sulfonate contents of SPS in a polar solvent; this effect is attributed to changes in inter-molecular interactions at low levels of sulfonation. (b) The diffusion related parameter exhibited a strong increase at low $q$ for the highly polar solvent DMSO and became practically constant at higher $q$ values. (c) The viscosity reduced time correlation functions, at low $q$, decay faster for SPS in polar solvents than for SPS in solvents of low polarity; the difference is more marked at low $q$ values. At higher $q$ no dependence on solvent polarity was observed.

The features observed in (b) and (c) were explained in terms of the 'correlation hole' model commonly used for analyzing dynamics of aqueous polyelectrolyte solutions (Hayter 1980). Such a hole possesses a diameter $\xi \sim a \cdot c^{-1/2}$, where $\xi$ represents an average distance between neighboring chains and $a$ is the monomer or segment size. The correlation hole has a size comparable with the overall size of one chain and is defined from a repulsive potential produced by the polyions in the region which surrounds it. Each polyelectrolyte chain is surrounded by a correlation tube, of an approximate radius $\kappa^{-1}$ (the Debye-Hückel screening length) from which other chains are strongly expelled.

## 10.3  SCATTERING FROM BULK IONOMERS

It is clear from Part I that the properties of very dilute ionomer solutions in relation to the large scale structures such as aggregates are now well understood. There are some reservations, though: the equilibrium conditions must be met, and the complications brought about by the interactions between the chains are hardly accounted for. The latter appear already in semi-dilute solutions, and will dominate the concentrated solutions which form an intermediate state between solutions and solid ionomers. The question arises what is the dimension of a single ionomer chain in a solid, and how does it relate to the dimension of an equivalent chain in solution?

While the neutron scattering is the direct probe of density distributions in the matter, the interpretation of the scattering pattern will be different for different length scales. In what follows we make a clear distinction between the results obtained using the small angle scattering, and the wide angle neutron scattering (WANS) techniques. What kind of scattering pattern is to be expected from ionomers in the bulk?

### 10.3.1  Small angle scattering: SANS and SAXS

As mentioned in the Introduction, ionomer structures were thought to consist of hard regions interspersed in a soft matrix. Such hard zones are in general referred to as 'phases'. If such phases form regions with different local order, *and* different scattering length density, then the diffraction techniques can be employed meaningfully. Until recently, mainly large scale structures formed by both neutral polymers and ionomers were investigated extensively using small angle neutron and x-ray scattering (SANS and SAXS, respectively) (cf. Lindner, 1991; Baruchel *et al.*, 1994; Schlick, 1996).

### 10.3.1.1 Early SANS experiments

The interpretation of the scattering pattern dates back to Flory (1949) who proposed that a single chain forms a Gaussian coil in a $\theta$-solvent, and that this result can be extended to melts and solids (Flory, 1969). The first SANS experiments on amorphous polymers in bulk (Kirste *et al.*, 1972; Benoit *et al.*, 1973; Wignall *et al.*, 1973) were accepted as the needed support for this hypothesis. Because of a good agreement between the calculated and measured radii of gyration (large scale), it was *assumed* that the monomers were randomly arranged on the *local scale*. However, two problems were overlooked: (i) there is evidence of some local order coming from techniques such as electron microscopy, and (ii) the question is not how a single polymer behaves in a solid, but how a collection of chains is packed. In addition, for ionomers, the possible effects of crystallinity on the diffraction pattern were purposely excluded from the beginning (Eisenberg, 1970).

In the scattering patterns obtained by Pineri *et al.* (1970) using SAXS a peak at low angles of scatter was observed in the telechelic carboxylic polymers neutralised with manganese acetate. Since the peak appeared above 10% manganese salt content, and its intensity increased with the increasing salt content, it originated the term 'ionomer peak'. This peak was interpreted as the evidence of ion clustering, and a value of a mean radius of 5.6 Å for the clusters and a value of 70 Å for the distance between them was obtained. Subsequently a combination of SANS and SAXS was used to study structure of a butadiene, styrene, 4-vinylpiridine polymer crosslinked by coordination with iron(III) chloride by Meyer *et al.*, (1978). The question posed was whether the metallic complexes dispersed uniformly in the polymer matrix or did they cluster together? Although no clear ionomer peak was observed, it was concluded from the scattering curves that there is a very broad distribution of cluster sizes, with the majority of clusters of radius smaller than 30 Å.

Not surprisingly, the choice of samples is often dictated by commercial interest. For example, Nafion® membranes were studied by the group of Pineri (1981) since the eighties: an in-depth review is given by Gebel and Loppinet (1996). There are some problems with interpretation of data collected from such systems, since the SANS technique is much more sensitive to the presence of water than to the aggregation of ionic groups due to the high scattering cross-section of $D_2O$. The scattering pattern showed two peaks, one of them attributed to structures in the crystalline region (the Bragg spacing of 180 Å), and a second peak attributed to the ion-containing regions which may be swollen with water (Pineri, 1981). Further study by Dreyfus *et al.* (1990) concluded that there is a distribution of hydrated 'micelles' in the perfluorinated matrix, and that there exists a locally ordered structure for these micelles. Such a structure has four first neighbours, located at a well-defined distance, embedded in a completely disordered gas of micelles. The Hsu-Gierke model (1982) assumed an existence of a dry cluster which was allowed to swell in water if the number of ion-exchange groups within the cluster remains constant. These results and their interpretation were disputed in the nineties by Lee *et al.* (1992) who measured nickel-substituted amorphous Nafion membrane using a combination of small- and wide-angle neutron scattering. Their data was found to be consistent with a continuous network of, for example, rodlike aqueous regions intersecting at nodes to form a continuous structure.

Roche *et al.* (1980) studied cesium salts, dry and water-saturated, of an ethylene-methacrylic acid (E-MAA) copolymer. The results were declared consistent with the presence of the separate phase containing water molecules and ions in a matrix of the non-ionic units.

The origin of the ionomer peak was investigated by Earnest *et al.*, using model sulfonated polystyrene (PS) (Earnest *et al.*, 1981) and polypentenamer (Earnest *et al.*, 1982) ionomers. Polypentenamer sulfonate ionomers (Earnest *et al.*, 1982) when wetted showed a peak around 0.15 Å$^{-1}$. The peak position shifts markedly to lower angles above a water-to-salt ratio of about six. This behaviour is similar to that exhibited by perfluoro-sulfonate ionomers; it is consistent with a phase-separated model where absorbed water is incorporated into the ionic phase (Roche, 1981). The sodium sulfonate PS data showed a beginning of a peak at higher $q$ values (0.2 Å$^{-1}$) but nothing at low $q$ values. Yet it was concluded that ionic clusters occur in sulfonated PS ionomers, and that clustering in these ionomers is accompanied by considerable chain expansion with respect to the chain dimensions observed in neutral polymers (Forsman, 1982; Forsman *et al.*, 1984). The conclusions drawn in the latter paper were contested by Squires *et al.* (1987) who argued that for ionomers with a truly random distribution of clusters, linked by subchains that obey Gaussian statistics, there is no chain extension term. Few years later Visser *et al.* (1991) concluded from their SANS and SAXS studies of sodium sulfonated model polyurethane ionomers that no chain expansion occurs in these ionomers upon aggregation.

### 10.3.1.2 Small angle x-ray scattering (SAXS)

The pioneering papers of Longworth and Vaughn (1968) and by Delf and MacKnight (1969) showed the usefulness of this technique for the investigation of the structure of ionomers. In the first paper, a peak appeared in the SAXS pattern of an ethylene/methacrylic acid copolymer upon neutralization with sodium. Delf and MacKnight (1969) used low angle x-ray to measure the structural differences between un-ionized copolymer of ethylene/4.1% methacrylic acid groups and partially ionized sodium, cesium, and lithium salts. They interpreted the data obtained in terms of the 'three-phases' model, i.e. polyethylene crystalline phase, polyethylene amorphous phase and an ionic phase consisting of clusters containing the ionized carboxylate groups. (This model later got transformed into the 'cluster model' by Eisenberg (1970).)

With the event of synchrotron radiation, this type of study became very popular, and soon anomalous SAXS (ASAXS) was born: an excellent overview is given by Chu (1996). Chu has also combined USAXS (ultra-small angle x-ray scattering) with light scattering to access the spatial region below $q = 0.1$ nm$^{-1}$ (Li *et al.*, 1993). The intention was to identify the origin of a small angle upturn which is much stronger for metal sulfonated polystyrenes than for their parent pure polystyrene. The upturn behavior was described by a power law suggesting an inhomogenous structure with the length scale variation from nanometers to microns, hence no single characteristic length could be identified.

There are more SAXS or combined SAXS and SANS studies on ionomers in the bulk than SANS studies alone. From our point of view, a judicious combination of two or more techniques is a better choice than one technique on its own. SAXS and SANS complement each other naturally: while SAXS is sensitive to electron differences in materials SANS responds to density fluctuations of spin-bearing nuclei. The most obvious example is determination of the position of hydrogen or deuterium by neutrons 'missed' by x-rays. In contrast x-rays are good at detection of metal ions due to significant number of their electrons, hence should be ideal for detecting clusters and aggregates in ionomers.

The drawback of x-rays is that there is not enough contrast in parent (neutral) ionomer for a meaningful comparison of its structure with that of sulfonated 'child'. This has led to a practice of fitting the data to theoretical models using adjustable parameters whereupon the calculated and measured SAXS curves can be compared. However, the SAXS profiles of ionomers are not well structured: as a result it is very tricky to distinguish between different models since all of them give the 'ionomer peak' (Chu, 1996). In addition, the issue of clustering in bulk ionomers is still controversial: the existence of ionic aggregates is generally accepted, however there is no agreement on their size, internal structure, and distribution in the material. This is especially visible in the papers published in the eighties. An extension to anomalous SAXS (ASAXS) technique makes it possible to use the variation of the x-ray wavelength in synchrotron radiation. ASAXS profiles are collected very near and below the absorption edge of the metal cation of interest. It is the *difference* SAXS profile determined from two different x-ray wavelengths that provides information on the net cationic structure in an ionomer. This approach has been used successfully by the group of Cooper to measure $L_3$ absorption edge of lead (Ding *et al.*, 1988), and the K-edge of $Ni^{2+}$ (Register and Cooper, 1990a; Register and Cooper, 1990b). Chu *et al.* (1993a) looked at K-edge of $Zn^{2+}$. The problem of ASAXS is low signal-to-noise ratio; on the other hand it is from ASAXS that the contribution of metal ions to the long-range inhomogeneities in ionomers was demonstrated (Chu, 1996).

The ionomers investigated by SAXS can be divided into several classes. Most relevant to this chapter are studies of model sulfonated polystyrenes which are random ionomers. Measurements are typically carried out over a very broad $q$ range, $0.03 \text{ nm}^{-1} < q < 8 \text{ nm}^{-1}$. Data analysis is based on the assumption that the presence of ionic aggregates should not change the structure of amorphous backbone in the SAXS range. Studies of narrow molecular weight sodium- and zinc-sulfonated samples, both about 4.5 mol% sulfonation (Chu *et al.*, 1988) have recorded several interesting features in the SAXS profiles. Firstly, the 'ionic peak' height is a factor of hundred smaller than the small angle upturn (the intensity extrapolated to zero scattering angle), making it an important contribution to the long-range inhomogeneities. Secondly, the curves of both sodium and zinc salts of SPS show similar $q_{max} \sim 1.7 \text{ nm}^{-1}$ while the zinc peak is broader. Finally, no fine structure and no secondary peak were observed.

Another very important class of ionomers, the telechelic ionomers, were also studied by Williams *et al.* (1986), Horrion *et al.* (1988), and Register *et al.* (1990d). In the first case, a series of halato-telechelic butadiene and isoprene polymers were studied as a function of the metal cation, degree of neutralization, swelling by both polar and non-polar solvents, and temperature. The main finding from these investigation was that the SAXS profile is principally determined by the configuration of the polymer molecule between the ionic groups located at the ends of each chain whereas the nature of the cation plays only a secondary role. They concluded that for a narrow molecular weight distribution an almost complete microphase separation occurs between the ionic groups and the hydrocarbon chain. The results were further interpreted in terms of multiplets (which are supposed to be small clusters containing up to eight ions). These mini-clusters were surrounded by a volume from which other ionic domains are excluded. However, no evidence was found for the existence of larger ionic clusters. Horrion *et al.* (1988) studied halato-telechelic low molecular weight polybutadiene bearing alkaline carboxylate end groups, and combined the results with these obtained using dynamical mechanical measurements. The agreement was reached that sodium carboxylates form the biggest ionic domains with an approximate radius of 12 Å, and that a small extension of the chains occured in the sample. Telechelic polystyrenes were found to behave according to the clustering model (Eisenberg, 1970) as shown by Register *et al.* (1990d) for carboxy-telechelic variety. The SAXS profiles were measured for sodium-substituted ionomer, and for methyl ester forms. The chain dimensions were found to be the same for the ionomer and ester which did not confirm Forsman (1982) model nor predictions of Dreyfus (1985). The data were satisfactorily described by a polydisperse assembly of wormlike chains, with a Kuhn statistical length comparable to that of neutral polystyrene.

Differences between model-fitted curves can be very subtle, and there is no real reason why the behavior of the long-range inhomogeneities should follow the Debye-Bueche model. SAXS data analysis for ionomers is very complex, and the reader is referred to Chu's review (1996) and original papers for a more complete picture.

### 10.3.2  Wide Angle Neutron Scattering (WANS) Experiments with Spin Polarized Neutrons

The strength of any diffraction technique lies in the access to space on the scale comparable to inverse wavelength of radiation used. It is recognised that wide angle x-ray diffraction is suitable for semi- and crystalline polymers only, and it does not give the position of hydrogens. WANS can be used for non-crystalline polymers in conjunction with selective deuteration, but even then the amount of information extracted from the scattering pattern of an amorphous polymer is relatively small. The reason is that in a standard neutron scattering experiment one measures coherent and incoherent scattering at the same time. Subsequently, an arbitrary subtraction of an incoherent 'background' is carried out which may give problems in subsequent model fitting of the coherent spectrum.

However, this situation changes dramatically if the experiment is performed using spin polarized neutrons with spin polarization option. The coherent and incoherent spectra are *experimentally* separated: this way, the clear-cut evidence of short range order in amorphous poly(methyl methacrylate) (Gabrys *et al.*, 1986) and polycarbonates (Lamers *et al.*, 1992) was obtained. Further studies on series of polystyrene and sodium sulfonated polystyrene samples confirmed the validity of this technique for study of ordering in amorphous polymers (Schärpf *et al.*, 1990).

A short note on the importance of spin polarization analysis is in place here. By yielding a purely coherent spectrum suitable for model fitting, polarization analysis delivers as direct proof as is possible of any ordering present. There are two controversial issues to be considered: (i) is there a short range order in amorphous polymers? (ii) if it exists, how would the models describing clustering in ionomers be affected? The critical review of the first dispute was given by Gabrys (1994) where the essential sources are referenced. More developments in this area, especially using computer modelling, are presented in the special issue on prediction of polymer form and properties from molecules to microstructure (Faraday Transactions, 1995). In any case, the existence of short range order in non-crystalline polymers is now widely accepted.

This, of course, has bearing on the interpretation of the scattering pattern obtained for ionomers in bulk. One has to bear in mind that the clustering models of Eisenberg (1970, 1990) ignore the existence of crystallinity or short range order, therefore can be applied *only* once these effects have been excluded. The way to do it was to investigate a series of parent polymers first, and then their sulfonated analogs (Schärpf *et al.*, 1990; Gabrys *et al.*, 1993; Gabrys *et al.*, 1996). The essence of these studies is given below, and the reader is referred to these publications for description of the technique and detailed study of the samples.

The scattering patterns from polystyrenes, especially from isotactic samples, show peaks with increasing width in $q$ (Figure 11). This indicates that they result from disordered crystallites, and therefore the best description is given in terms of paracrystallinity. In this description short-range order with a distribution of distances between the scattering units is quantitatively given by: the average distance of the scattering units $\langle a \rangle$, the width of the distribution of the distances $\Delta$, the measure of disorder $\Delta / \langle a \rangle \cdot 100$ [%], and the interaction radius $x_m$ (Schärpf *et al.*, 1990; Vainshtein, 1966). The mean distance $\langle a \rangle$ between nearest neighbors enters a bell-shaped probability distribution function $H_1(x)$ giving the probability of finding any particular distance between neighbors. This function is normalised: $\int H_1(x)dx = 1$, and has the condition $\langle a \rangle = \int x H_1(x)dx$. For a Gaussian distribution function, one obtains

$$H_1(x + \langle a \rangle) = (\Delta \sqrt{2\pi})^{-1} \exp \left( -\frac{1}{2} \frac{x^2}{\Delta^2} \right) \tag{34}$$

The width of the distribution of the distances $\Delta$ is governed by the degree of order in the object as a whole. This, of course, is valid for each neighbor, hence one obtains the probability distribution function for any neighbor by the self-convolution of the $H_1$ function $m$ times:

$$H_m = H_1 \oplus H_1 \oplus H_1 \ldots \oplus H_1$$

For example, for the second neighbor $H_2(x) = H_1 \oplus H_1$. The average distance of the $m$-th scattering unit from the origin is given by $m.\langle a \rangle = \int x H_1(x)dx$.

Effectively, these functions replace the usual $\delta$-distribution functions used to describe the long-range order in crystals, hence the distribution function for the whole paracrystal can be written. It is a probability distribution with increasing width at each next point with an average distance $\langle a \rangle$ between each two neighboring points; this represents a periodic distribution with separations of $\langle a \rangle$. An interesting case is when this procedure is extended to two dimensions by assuming the "ordering" of chains parallel to each other normal to a plane: then there is an average $\langle b \rangle$ distance in another direction. In the paracrystalline model there are directions where the distances are well defined, for example chain periodicities along the chain normal to the above-assumed plane. This type of ordering in an 'amorphous' polymer bears a strong resemblance to that observed in liquid crystals.

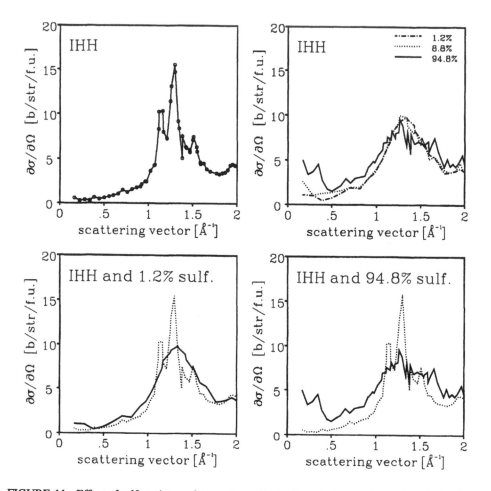

FIGURE 11. Effect of sulfonation on the structure of isotactic polystyrene, denoted IHH. The degree of sulfonation for each sample is marked in the figure. Error bars (3%) are not visible on this scale.

How does it relate to the measured spectrum? The Fourier transform of such a distribution is also a periodic function with increasing width of the peaks corresponding to the Fourier transform of $H_m(x)$. For a Gaussian function we obtain

$$|F(q) = \exp(-2\pi^2 q^2 \Delta^2)| \tag{35}$$

and

$$|F_m(q)| = |F|^m = \exp(-2\pi^2 q^2 \Delta^2 m) \tag{36}$$

Hence the width of the peak increases in the $m$-th order with $\sqrt{m}$:

$$\Delta_m = \Delta \sqrt{m} \tag{37}$$

The order peaks disappear once $H_m$ becomes so broad that it attains half its peak value at half the distance between two peaks, i.e. at $x = \langle a \rangle / 2$. The limit of the number of visible peaks $M$ is reached when $\exp(-y) = \frac{1}{2}$, as seen below:

$$y = \frac{1}{2} \frac{\langle a \rangle^2}{2} (M \Delta^2)^{-1} \approx 0.7 \tag{38}$$

Then $x_m$, the interaction radius, is given by $x_m = M.\langle a \rangle$, where $M = (2.5\Delta/\langle a \rangle)^{-2}$.

One more relation is needed for completeness, that between the peak width $\Delta q$ and the width of the distribution of the distances in object space:

$$\Delta q = \frac{2\pi}{\langle a \rangle} \pi^2 h^2 \left( \frac{\Delta}{\langle a \rangle} \right)^2 \tag{39}$$

where $h$ is the order of the peak.

*Application to isotactic polystyrene*   The above approach was applied to analyze diffraction patterns obtained from isotactic polystyrene, sulfonated and unsulfonated, shown in Figure 11a–d. The pattern of isotactic polystyrene shows distinct order peaks (Figure 11a); a good agreement was achieved between data and model calculations in terms of paracrystallinity (Gabrys *et al.*, 1996 and references therein). In this model, the calculated single peaks contribute to the total pattern in the manner described above: each peak consists of a broad and a narrow component. The peak width increases with increasing values of the scattering vector $q$ which is an effect caused by disorder. Such data analysis is very powerful: the purely coherent scattering is experimentally separated and there is an absolute intensity available due to internal calibration. On the side of the model, the assumption that the structure is determined by a distribution of distances between the neighboring chains gives a better agreement than other models (Schärpf *et al.*, 1990). Hence a description of this broadening, and the relative weight of the narrow and wide distributions of the distances, can be *quantitatively* given in terms of the quantities $\langle a \rangle$, $\Delta$ and $x_m$ defined above. From the position of the first peak at $q_1 = 0.6 \ \text{Å}^{-1}$, one average distance between neighboring atoms or groups of atoms or chains is found to be $\langle a_1 \rangle = 5.5 \ \text{Å}$ and $\Delta q_1 = 0.06 \ \text{Å}^{-1}$. It then follows from equations (33–35) the width of this distribution is $\Delta_1 = 0.4 \ \text{Å}$ and the interaction radius $x_{m1} = 166 \ \text{Å}$. The corresponding disorder is $\Delta_1/\langle a_1 \rangle = 0.073$, or 7% for the part of the sample yielding narrow peaks. There is a second average distance $\langle a_2 \rangle = 4.83 \ \text{Å}$ (the position of the peak is $q_2 = 1.3 \ \text{Å}^{-1}$, a width $\Delta_2 = 0.91 \ \text{Å}$, and an interaction radius $x_{m2} = 21.6 \ \text{Å}$). The measure of disorder in this region is given by the ratio $\Delta_2/\langle a_2 \rangle = 0.188$, or 19%. A graphical representation of these results is shown in Figure 12. The tail of the scattering at higher $q$ looks exactly like the smeared peaks of a paracrystal.

The effect of sulfonation on the diffraction spectrum is striking. Even with a small degree of sulfonation, 1.2 mol%, the sharp order peaks disappear (Figure 11c). Increasing the degree of sulfonation does not change the spectra significantly, as plotted in Figure 11b. The short range order peak only becomes broader and lower, indicating a further increase in disorder. For 1.2 mol% sulfonation the peak width is $\Delta q = 0.45 \ \text{Å}^{-1}$, yielding $\Delta = 0.9 \ \text{Å}$.

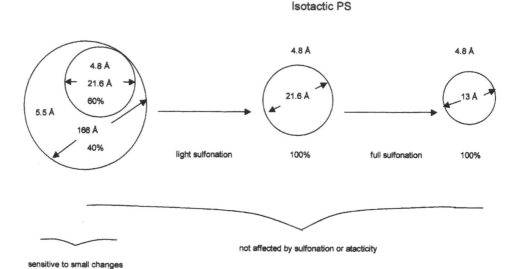

FIGURE 12.  A graphical representation of disorder setting in isotactic PS upon sulfonation.

This corresponds to 18.9% disorder and an interaction radius of 23.2 Å. On the other end of the sulfonation spectrum there is a broad peak present at $q = 1.28$ Å in the 94.8 mol% sulfonated sample which implies an average periodicity distance equal to 4.9 Å; it has a width $\Delta q = 0.7$ Å$^{-1}$, hence $\Delta = 1.2$ Å, a disorder of 25% and an interaction region of $x_m$ equal to 13 Å. Something interesting happens at lower $q$-values, near $q = 0.5$ Å$^{-1}$: an increasingly broader peak is observed, centered around $q = 0$, which is well noticeable for 94.8 mol% sulfonated sample. This means that the ordered regions become so small (a diameter about 12 Å or smaller) that the small angle scattering peak, centered around $q = 0$, shows through the shape factor at $q = 0.5$ Å$^{-1}$.

These measurements have shown conclusively that the change of disorder observed in the sample with 1.2 mol% sulfonation compared to that with 94.8 mol% sulfonation is not a large one; the respective parameters for the other samples fall between these values. A systematic comparison of the relevant numbers shows, that in the neutral sample 40% of the material consists of regions with an average diameter of 166 Å, and a preferential average periodicity of $5.5 \pm 0.4$ Å. The remaining 60% contains regions of 21.6 Å average diameter with an average periodicity of $4.83 \pm 0.91$ Å. The respective disorders are 7 and 19%. The process of sulfonation causes the large regions with the periodicity of 5.5 Å to disappear, and starting from the 1.2 mol% sulfonated sample the total scattering is now due to smaller regions, with periodicity of 4.8 Å. In the 94.8 mol% sulfonated material, the low-angle scattering and the observed peak results from the regions of 12.5 Å with the nearly the same periodicity of $4.9 \pm 1.2$ Å, disorder 25% (cf Figure 12). Regions with 4.8 Å distances shrink, yet the average distance remains the same; their distribution becomes broader, indicating an increasing disorder. The persistent distance of $4.9 \pm 1.2$ Å is resistant to even dramatic changes, such as an addition of nearly 100% of the large sulfonated groups. Therefore it must be a distance within the molecular chain itself — it is equal to two monomer length. Perhaps 60% of the isotactic material is really atactic? The other distance, 5.5 Å, is very sensitive to small changes of the sulfonate groups' content; hence it must result from conformation.

This type of analysis has been also applied to selectively deuterated atactic samples. For full details the reader is referred to Gabrys *et al.* (1996).

### 10.3.3  Dynamical Neutron Scattering Experiments

The ultimate aim of fundamental studies is to find a link between macroscopically observable properties such as elasticity and viscosity and chain microstructure. A technique worth exploring in this context is that of inelastic incoherent neutron scattering which yields information about the local segmental and side chain motion. To our knowledge there is only one such attempt at capturing the dynamics of ionomers on this level (Gabrys *et al.*, 1993). Atactic and isotactic and sodium sulfonated polysterene (PS) samples were investigated using the time-focused crystal analyzer TFXA (Penfold and Tomkinson, 1986) at the spallation source ISIS. Vibrational modes due to the ring and chain moieties were identified by the process of selective deuteration: it was found that the process of sulfonation had an analogous effect on the spectrum of isotactic PS as that observed in atactic PS. The addition of the sulfonate group clearly enhances the local chain mobility — a new mode appeared at 500 cm$^{-1}$ upon sulfonation which was interpreted as a mixed mode of a benzene ring vibration and a phonon wing. Hence the process of sulfonation results in releasing a low frequency "breathing mode" between the chains, a finding which corroborates the structure changes observed in these ionomers by spin polarized neutrons with spin polarization analysis (Schärpf *et al.*, 1990; Gabrys *et al.*, 1996). At the time of writing there is no quantitative microscopic theory about the modification of the normal vibrations in styrene upon sulfonation, therefore it is difficult to further quantify these data. The qualitative approach by Eisenberg *et al.* (1990) to the influence of ionic clusters on the overall chain dynamics predicts restricted 'mobility' of the chains surrounding the ionic microphase-separated regions (long range, slow motion). Therefore the extent of this disagreement is not clear. What is clear that a vast area of careful studies has been opened that way.

## 10.4  CONCLUSIONS ON THE USE OF SANS TO IONOMER STUDIES

The study of dilute ionomer solutions by SANS can provide a great deal of information on the structure of the polymers in different solvents. This information can be used to explain the rheological behavior, for example. There are reasonable expectations that this work would form a basis of the microscopic interpretation of macroscopic properties of more concentrated solutions and solid state samples of ionomers. There are many difficulties, however, in the interpretation of scattering results when solutions become more concentrated. For this reason data should be obtained over a broad $q$ and concentration range. It is believed that more use of modelling over a broad $q$ range as described in the theory section shall lead in the future to less ambiguous interpretation of the scattering from aggregating systems.

The situation is at present even less clear for the bulk ionomers. Here it is the combination of two or more methods in similar spatial region (SANS and SAXS, for example), extending the spatial region of interest through methodical use of different methods — ASAXS, SAXS, SANS and WANS would be ideal — that can give insight into their complexity. These methods, backed by the systematic use of absolute intensity, and combined with computer modelling, are the way forward. The results obtained must, however, be combined with available macroscopic property data in order to improve our understanding how to optimise physical and chemical properties of polymers in the future.

## References

Agarwal, P.K., Garner, R.T. and Graessley, W.W. (1987) Counterion and solvent effects on the dilute solution viscosity of polystyrene ionomers. *J. Polymer Sci.: Polymer Phys.*, **25**, 2095.

Arai, T., Abe, F., Yoshizaki, T., Einaga, Y. and Yamakawa, H. (1995) Hydrodynamic radius expansion factor of polystyrene in cyclohexane near the theta temperature, *Macromolecules*, **28**, 5458.

Ballard, D.G.H., Wignall, G.D. and Schelten, J. (1973) Measurement of molecular dimensions of polystyrene chains in the bulk polymer by low angle diffraction. *Eur. Polym. J.*, **9**, 965.

Baruchel, J., Hodeau, J.L., Lehmann, M.S., Regnard, J.R. and Schlenker, C., Eds. (1994) *Neutron and Synchrotron Radiation for Condensed Matter Studies*, Springer-Verlag, Heidelberg.

Benoit, H. and Benmouna, M. (1984) Scattering from a polymer solution at an arbitrary concentration. *Polymer*, **25**, 1059.

Benoit, H., Decker, J.S., Picot, C., Cotton, J.P., Farnoux, B., Jannik, G. and Ober, R. (1973) Dimensions of a flexible polymer chain in the bulk and in solution. *Nature*, **245**, 13.

Bohdanecky, M. and Kovar, J. (1982) *Viscosity of Polymer Solutions*, Elsevier, Oxford.

Bodycomb, J. and Hara, M. (1994) Light scattering on ionomers in solution 4, *Macromolecules*, **27**, 7369.

Chu, B. (1996) Small-Angle X-ray Scattering (SAXS) Studies of Bulk Ionomers Chapter 3 in *Ionomers Characterization, Theory and Applications*, Schlick, S. (Ed) CRC Press, New York.

Chu, B., Wu, D.Q., Lundberg, R.D. and MacKnight, W.J. (1993a) Small-angle x-ray scattering (SAXS) studies of sulfonated polystyrene ionomers. 1. Anomalous SAXS, *Macromolecules*, **26**, 994.

Chu, B., Wu, D.Q., Lundberg, R.D. and MacKnight, W.J. (1993b) Small-angle x-ray scattering (SAXS) studies of sulfonated polystyrene ionomers. 2. Correlation function analysis, *Macromolecules*, **26**, 1000.

Debye, P.J. (1947) Molecular-weight determination by light scattering. *J. Phys. Colloid Chem.*, **51**, 18.

Delf, B.W. and MacKnight, W.J. Low angle x-ray scattering from ethylene-methacrylic acid copolymers and their salts, *Macromolecules*, **2**, 309.

De Gennes, P.G. (1979) *Scaling Concepts in Polymer Physics*, Cornel University, Ithaca, NY.

Ding, Y.S., Hubbard, S.R., Hodgson, K.O., Register, R.A. and Cooper, S.L. (1988) Anomalous small-angle x-ray scattering from a sulfonated polystyrene ionomer, *Macromolecules*, **21**, 1698.

Ding, Y.S., Yarusso, D.J., Pan, H.K.D. and Cooper, S.L. (1984) Extended x-ray absorption fine structure: studies of zinc-neutralized sulfonated polystyrene ionomers. *J. Appl. Phys.*, **56**, 2396.

Dreyfus, B. (1985) Model for the clustering of multiplets in ionomers, *Macromolecules*, **18**, 284.

Dreyfus, B., Gebel, G., Aldebert, P., Pineri, M., Escoubes, M. and Thomas, M. (1990) Distribution of the "micelles" in hydrated perfluorinated ionomer membranes from SANS experiments. *J. Phys. France*, **51**, 1341.

Earnest, T.R., Jr., Higgins, J.S., Handlin, D.L. and MacKnight, W.J. (1981) Small-angle neutron scattering from sulfonate ionomers, *Macromolecules*, **14**, 192.

Earnest, T.R., Jr., Higgins, J.S. and MacKnight, W.J. (1982) Small-angle neutron scattering from polypentenamer sulfonate ionomers, *Macromolecules*, **15**, 1390.

Eisenberg, A. (1970) Clustering of ions in organic polymers. A theoretical approach, *Macromolecules*, **3**, 147.

Eisenberg, A., Hird, B. and Moore, R.B. (1990) A new multiplet-cluster model for the morphology of random ionomers, *Macromolecules*, **23**, 4098.

Elias, H.G. (1972) Chapter 9 in *Light Scattering from Polymer Solutions*, Huglin, M.B. (Ed) Academic Press, London and New York.

Ewen, B. and Richter, D. (1997) Advances in Polymer Science, Vol. 134, 1, Springer, Berlin.

Faraday Transactions (1995) Special Issue on Prediction of Polymer Form and Properties: From Molecules to Microstructure, *J. Chem. Soc. Faraday Trans.*, **91**, 2355–2680.

Fitzgerald J.J. and Weiss, R.A. (1988) Synthesis, properties and structure of sulfonate ionomers. *JMS-Rev. Macromol. Chem. Phys.*, **C28**, 99.

Flory, P.J. (1949) The configuration of real polymer chains. *J. Chem. Phys.*, **17**, 303.

Flory, P.J. (1969) *Statistical Mechanics of Chain Molecules*, Interscience Publishers.

Forsman, W.C. (1982) Effect of segment-segment association on chain dimensions, *Macromolecules*, **15**, 1032.

Forsman, W.C., MacKnight, W.J. and Higgins, J.S. (1984) Aggregation of ion pairs in sodium poly(styrenesulfonate) ionomers: Theory and experiment, *Macromolecules*, **17**, 490.

Gabrys, B., Higgins, J.S. and Schärpf, O. (1986) Short range order in amorphous poly(methyl methacrylate), *J. Chem. Soc. Far. Trans. I*, **82**, 1929.

Gabrys, B., Higgins, J.S. and Peiffer, D.G. (1987) Effects of inter- or intra-molecular interactions on dynamics of ionomers in dilute solutions, in *Structure and Properties of Ionomers*, NATO ASI Series C198, Reidel, Dordrecht.

Gabrys, B., Higgins, J.S., Lantman, C.W., MacKnight, W.J., Pedley, A.M., Peiffer, D.G. and A.R. Rennie, A.R. (1989) Single chain dimensions in semi-dilute ionomer solutions: small angle neutron scattering study, *Macromolecules*, **22**, 3746.

Gabrys, B., Huang, D., Nardi, F., Peiffer, D.G. and Tomkinson, J. (1993) Dynamics of ionomers studied by inelastic neutron scattering, *Macromolecules*, **26**, 2007.

Gabrys, B., Schärpf, O. and Peiffer, D.G. (1993) Short range order in noncrystalline polymers examined with spin polarised neutrons, *J. Polym. Sci.: Part B: Polymer Physics*, **31**, 1891.

Gabrys, B. (1994) Long history of short range order in polymers Trends in *Polymer Science*, **2**, 2.

Gabrys, B., Schärpf, O. and Peiffer, D.G. (1996) Model ionomers studied with polarised neutrons with spin polarisation analysis, Chapter 4 in *Ionomers Characterization, Theory and Applications*, Schlick, S. (Ed) CRC Press, New York.

Galambos, A.F., Stockton, W.B., Koberstein, J.T., Sen, A., Weiss, R.A. and Russell, T.P. (1987) Observation of cluster formation in an ionomer, *Macromolecules*, **20**, 3091.

Gebel., G., Aldebert, P. and Pineri, A. (1987) Structure and related properties of solution-cast perfluorosulfonated ionomer films, *Macromolecules*, **20**, 1425.

Gebel, G. and Loppinet, B. (1996) Small-angle scattering study of perfluorinated ionomer membranes and solutions, Chapter 5 in *Ionomers Characterization, Theory and Applications*, Schlick, S. (Ed) CRC Press, New York.

Glatter, O. and Kratky, O., (1982) *Small Angle X-ray Scattering*, Academic Press, London and New York.

Gouin, J.P., Williams, C.E. and Eisenberg, A. (1989) Microphase structure of block ionomers. 1. Study of molded styrene-4-vinylpiridinium ABA blocks by SAXS and SANS, *Macromolecules*, **22**, 4573.

Gouin, J.P., Eisenberg, A. and Williams, C.E. (1992) Microphase structure of block ionomers. 2. Nonionic segment conformation in molded styrene-4-vinylpyridinium ABA copolymers, *Macromolecules*, **25**, 1368.

Gouin, J.P., Bosse, F., Nguyen, D., Williams, C.E. and Eisenberg, A. (1993) Microphase structure of block ionomers. 3. A SAXS study of the effects of architecture and chemical structure, *Macromolecules*, **26**, 7250.

Grady, B.P., Matsuoka, H., Nakatani, Y., Cooper, S.L. and Ise, N. (1993) Influence of the sample preparation method on the ultra-small-angle x-ray scattering of lightly sulfonated polystyrenes, *Macromolecules*, **26**, 4064.

Hayter, J., Jannik, G., Brochard-Wyart, F. and de Gennes, P.G. (1980) *J. Phys. Lett (Paris)*, **41**, L-541.

Higgins, J.S. and Benoit, H.C. (1994) *Polymers and Neutron Scattering*, Oxford University Press, Oxford.

Horrion, J., Jerome, R., Teyssie, Ph., Marco, C. and Williams, C.E. (1988) Halato-telechelic polymers. XIII. Viscoelastic properties and morphology of low molecular weight polybutadiene bearing alkaline carboxylate end-groups, *Polymer*, **29**, 1203.

Hsu, W.Y. and Gierke, T.D. (1982) Elastic theory for ionic clustering in perfluorinated ionomers, *Macromolecules*, **15**, 101.

Huglin, M.B. (Ed) *Light Scattering from Polymer Solutions*, Academic Press, London and New York.

Hunter, R.J. (1987) Chapter 10 in *Foundations of Colloid Science*, **1**, Oxford University Press, New York.

Ishioka, T. and Kobayashi, M. (1990) Small- angle x-ray scattering study for structural changes of the ion cluster in a zinc salt of an ethylene-methacrylic acid ionomer on water absorption, *Macromolecules*, **23**, 3183.

Karayianni, E., Jerome, R. and Cooper, S.L. (1995) Small-angle neutron scattering studies of low-polarity telechelic ionomer solutions. 1. Total scattering, *Macromolecules*, **28**, 6494.

Kirste, R.G., Kruse, W.A. and Schelten (1972) *J. Makromol. Chem.*, **162**, 299.

Lamers, C., Schärpf, O., Schweika, W., Batoulis, J., Sommer, K. and Richter, D. (1992) Short-range order in amorphous polycarbonates, *Physica B*, **180&181**, 515.

Lantman, C.W., MacKnight, W.J., Peiffer, D.G., Sinha, S.K. and Lundberg, R.D. (1987) *Macromolecules*, **20**, 1096.

Lantman, C.W., MacKnight, Higgins, J.S., Peiffer, D.G., Sinha, S.K. and Lundberg, R.D. (1988) SANS from sulfonate ionomer solutions 2. Polyelectrolyte effects in polar solvents, *Macromolecules*, **21**, 1344.

Lantman, C.W., MacKnight, Higgins, J.S., Peiffer, D.G., Sinha, S.K. and Lundberg, R.D. (1988) SANS from sulfonate ionomer solutions 1. Associating polymer behaviour, *Macromolecules*, **21**, 1339.

Lee, E.M., Thomas, R.K., Burgess, A.N., Barnes, D.J., Soper, A.K. and Rennie, A.R. (1992) Local and long-range structure of water in a perfluorinated ionomer membrane, *Macromolecules*, **25**, 3106.

Li, Y., Peiffer, D.G. and Chu, B. (1993) Long-range inhomogeneities in sulfonated polystyrene ionomers, *Macromolecules*, **26**, 4006.

Lindner, P. and Zemb, Th., Eds. (1991) *Neutron, X-ray and Light Scattering: Introduction to an Investigative Tool for Colloidal and Polymeric Systems*, North-Holland, Amsterdam.

Longworth, R. and Vaughn, D.J. (1968) *Nature*, **218**, 85.

Lundberg, R.D. and Makowski, J.S. (1980) Solution behavior of ionomers. 1. Metal sulfonate ionomers in mixed solvents J. *Polymer, Sci. Polymer, Phys. Ed.*, **18**, 1821.

Mauritz, K.A. (1988) Review and critical analyses of theories of aggregation in ionomers, *JMS-Rev. Macromol. Chem. Phys.*, **C28**, 65.

Mezei, F. (1972) Neutron spin echo: a new concept in polarized thermal neutron techniques, *Z. Physik*, **255**, 146.

Mezei, F. Ed., (1980) Proceedings of an International Workshop on Neutron Spin-echo, *Lecture Notes in Physics*, **128**, Springer-Verlag, Berlin.

Meyer, C.T. and Pineri, M. (1978) Ion clustering in a butadiene, styrene, 4-vinylpyridine terpolymer crosslinked by coordination with iron(III) chloride, *J. Polymer Sci. Polymer Phys. Ed.*, **16**, 569.

Nicholson, L.K. (1981) The neutron spin-echo spectrometer: a new high resolution technique in neutron scattering, *Contemp. Phys.*, **22**, 451.

Nyström, B., Roots, J., Higgins, J.S., Gabrys, B., Peiffer, D.G., Mezei, F. and Sarkissian, B. (1986) Dynamics of polysterene sulfonate ionomers in solution. A neutron spin-echo study, *J. Polym. Sci.: Part C: Polym. Lett.*, **24**, 273.

Otocka, E.P. and Kwei, T.K. (1968) Properties of ethylene-metal acrylate copolymers, *Macromolecules*, **1**, 401.

Pedley, A.M. (1990) PhD Thesis, Imperial College, London.

Pedley, A.M., Higgins, J.S., Peiffer, D.G., Rennie, A.R. and Staples, E. (1989) Single chain dimensions in ionomer solutions under quiescent and shear conditions as determined by small angle neutron scattering, *Polymer Communications*, **30**, 162.

Pedley, A.M., Higgins, J.S., Peiffer, D.G. and Rennie, A.R. (1989) Single chain dimensions in semidilute ionomer solutions, *Macromolecules*, **22**, 3746.

Pedley, A.M., Higgins, J.S., Peiffer, D.G. and Burchard, W. (1990a) Light scattering from sulfonate ionomers in xylene, *Macromolecules*, **23**, 1434.

Pedley, A.M., Higgins, J.S., Peiffer, D.G. and Rennie, A.R. (1990b) Thermodynamics of the aggregation phenomenon in associating polymer solutions, *Macromolecules*, **23**, 2494.

Penfold, J. and Tomkinson, J. (1986) *The ISIS Time Focused Crystal Analyzer Spectrometer*, TFXA (RAL report); Rutherford Appleton Laboratory: Chilton, Didcot, OX11 0QX, U.K..

Pineri, M. and Eisenberg, A. Eds. (1987) *Structure and Properties of Ionomers*, NATO ASI Series C198, Reidel, Dordrecht.

Pineri, M., Meyer, C.T., Levelut, A.M. and Lambert, M. (1974) Evidence for ionic clusters in butadiene-methacrylic acid copolymers neutralized with various salts, *J. Polymer Sci. Polymer. Phys. Ed.*, **12**, 115.

Register, R.A., Cooper, S.L., Thiyagarajan, P., Chakrapani, S. and Jerome, R. (1990a) Effect of ionic aggregation on ionomer chain dimensions. 1. Telechelic polystyrenes, *Macromolecules*, **23**, 2978.

Register, R.A., Pruckmayr, G. and Cooper, S.L. (1990b) Effect of ionic aggregation on ionomer chain dimensions. 2. Sulfonated polyurethanes, *Macromolecules*, **23**, 3023.

Register, R.A. and Cooper, S.L. (1990c) Anomalous small-angle x-ray scattering from nickel-neutralized ionomers. 1. Amorphous polymer matrices, *Macromolecules*, **23**, 310.

Register, R.A. and Cooper, S.L. (1990d) Anomalous small-angle x-ray scattering from nickel-neutralized ionomers. 2. Semicrystalline polymer matrices, *Macromolecules*, **23**, 318.

Register, R.A., Foucart, M., Jerome, R. Ding, Y.S. and Cooper, S.L., (1988) Structure-property relationships in elastomeric carboxy-telechelic polyisoprene ionomers neutralized with divalent cations, *Macromolecules*, **21**, 1009.

Register, R.A., Sen, A., Weiss, R.A. and Cooper, S.L. (1989) Effect of thermal treatment on cation local structure in manganese-neutralized sulfonated polystyrene ionomers, *Macromolecules*, **22**, 2224.

Roche, E.J., Stein, R.S. and MacKnight, W.J., (1980) Small-angle x-ray and neutron scattering studies of the morphology of ionomers, *J. Polymer Sci. Polymer. Phys. Ed.*, **18**, 1035.

Roche, E.J., Pineri, M., Duplessix, R. and Levelut, A.M. (1981) Small-angle scattering studies of nafion membranes, *J. Polymer Sci. Polymer. Phys. Ed.*, **19**, 1.

Schärpf, O., Gabrys, B. and Peiffer, D. G. (1990) *Short range order in isotactic, atactic and sulfonated polystyrene measured by polarised neutrons*, (Parts A and B) ILL report 90SC26T.

Schlick, S., Ed (1996) *Ionomers Characterization, Theory and Applications*, CRC Press, New York.

Schulz, G.V. and Lechner, M. (1972) *Chapter 12 in Light Scattering from Polymer Solutions*, Huglin, M.B. (Ed) Academic Press, London and New York.

Squires, E., Painter, P. and Howe, S. (1987) Cluster formation and chain extension in ionomers, *Macromolecules*, **20**, 1740.

Tant, M.R., Mauritz, K.A. and Wilkes, G.L., Eds. (1997) *Ionomers: Synthesis, Structure, Properties and Applications*, Chapman and Hall, New York.

Tant, M.R. and Wilkes, G.L. (1988) An overview of the viscous and viscoelastic behavior of ionomers in bulk and solution, *JMS-Rev. Macromol. Chem. Phys.*, **C28**, 1.

Timbo, A.M. (1993) PhD thesis, Imperial College, London.

Timbo, A.M., Higgins, J.S., Peiffer, D.G., Maus, C., Vanhoorne, P. and Jerome, R. (1993) Aggregation behaviour of halato-telechelic polymers in non polar solvents, *Journal de Physique IV*, **3**, 71.

Tomita, H. and Register, R.A. (1993) Morphology of lightly carboxylated polystyrene ionomers, *Macromolecules*, **26**, 2791.

Ullman, R., Benoit, H. and King, J.S. (1986) Concentration effects in polymer solutions as illuminated by neutron scattering, *Macromolecules*, **19**, 183.

Vanhoorne, P. and Jerome, R. (1996) Contribution of Halato-Telechelic Polymers to the Modelization of Morphology and Solution Properties of Ionomers, Chapter 9 in *Ionomers Characterization, Theory and Applications*, Schlick, S. (Ed) CRC Press, New York.

Vanhoorne, P. and Jerome, R. (1995) Aggregation behaviour of omega metal and alpha omega metal sulfonato polystyrene in toluene, *Macromolecules*, **28**, 5664.

Visser, S.A., Pruckmayr, G. and Cooper, S.L. (1991) Small-angle neutron scattering analysis of model polyurethane ionomers, *Macromolecules*, **24**, 6769.

Wang, J., Li., Y., Peiffer, D.G. and Chu, B. (1993) Small-angle x-ray scattering investigation of temperature influences on microstructures of an ionomer, *Macromolecules*, **26**, 2633.

Williams, C.E., Russell, T.P., Jerome, R. and Horrion, J. (1986) Ionic Aggregation in Model Ionomers, *Macromolecules*, **19**, 2877.

Witten, T.A. (1988) Associating polymers and shear thickening. *J. Phys. (France)*, **49**, 1055.

Witten, T.A. and Schafer, L. (1981) Spatial monomer distribution for a flexible polymer in a good solvent. *J. Chem. Phys.*, **74**, 2582.

Yarusso, D.J. and Cooper, S.L. (1983) Microstructure of ionomers: interpretation of small-angle x-ray scattering data, *Macromolecules*, **16**, 1871.

Young, A. M., Higgins, J.S., Peiffer, D.G. and Rennie, A.R. (1995) Effect of sulfonation level on the single chain dimensions and aggregation of sulfonated polystyrene ionomers in xylene. *Polymer*, **36**, 691.

Young, A. M., Timbo, A.M., Higgins, J.S., Peiffer, D.G. and Lin, M.Y. (1996a) Thermodynamics of aggregation in associating polymer solutions. *Polymer*, **37**, 2701.

Young, A. M., Higgins, J.S., Peiffer, D.G. and Rennie, A.R. (1996b) Effect of aggregation on the single-chain dimensions of sulfonated polystyrene ionomers in xylene. *Polymer*, **37**, 2125.

Young, A. M., Garcia, R., Higgins, J.S., Timbo, A.M.,Peiffer, D.G. (1998a) Dynamic light scattering and rheology of associating SPS ionomers in non polar solvents. *Polymer*, **39**, 1525.

Young, A. M., Brigault, C., Higgins, J.S., Heenan, R. and Peiffer, D.G. (1998b) SANS from ionomer gels. *Polymer*, **39**, 6685.

Zimm, B.H. (1948) The scattering of light and the radial distribution function of high polymer solutions. *J. Chem. Phys.*, **16**, 1093.

# 11.  Neutron Scattering in Pharmaceutical Sciences

C. WASHINGTON[1], M.J. LAWRENCE[2] and D. BARLOW[2]

[1]*School of Pharmacy, University of Nottingham, Nottingham and*
[2]*Department of Pharmacy, King's College London, London, UK*

## 11.1  INTRODUCTION

Neutron scattering is not a subject that one would normally expect pharmaceutical scientists to pursue. This is unfortunate since neutron techniques have the power to illuminate many problems of pharmaceutical interest. In this review we will begin by briefly examining some of the pharmaceutically relevant areas in which neutrons have made a significant contribution, and then look more closely at a number of areas of particular pharmaceutical interest.

Pharmaceutics is an extremely broadly based discipline which takes input from areas as diverse as molecular biology, thermodynamics, and materials engineering, in order to convert pharmacologically active chemicals to medicines. Increasingly these medicines are not simple tablets or syrups, but consist of highly sophisticated systems designed to deliver drugs to specific sites and to regulate the dose of drug to the patient. Examples include colloidal particles which specifically target drugs to tumour sites, polymer implants which release sustained doses of hormones over many months, and transdermal patches which deliver controlled doses of drug through the skin. As a result of this diversity pharmaceutics has benefited indirectly from studies using neutrons in many scientific areas. Many of these areas have been described in greater detail elsewhere, (or even in this volume), so we will only briefly survey them here prior to looking in more detail at two areas of specific importance.

In order to understand the use of neutrons it is helpful to recall a few simple facts. Neutrons produced by either reactors or spallation sources have a wavelength in the region of 1–10 Å, and so scattering experiments provide structural information on length scales from a fraction of an angstrom to tens or hundreds of angstroms, i.e. covering the atomic to macromolecular scale, and the smaller end of the colloidal size range. Neutrons interact weakly with matter, and hence do not disturb the experimental system; as a consequence they are also highly penetrating.

Unlike X-rays, which interact with the electrons of an atom, neutrons interact with the nucleus through the strong nuclear force. This makes X-rays and neutrons complementary, each technique revealing details that the other cannot. In particular, neutrons are strongly scattered by hydrogen, and therefore are a useful probe in hydrogenous systems. The ability of a system to scatter neutrons is described in terms of its *scattering length density*; typical values have been given in Chapter 1.

Because neutrons interact with the nucleus, they are scattered differently by different isotopes of the same element. As a result it is often possible to vary the visibility, or *contrast*, of specific parts of the experimental system by isotopic substitution. This is the basis of the so-called contrast variation method, and the commonest example of this technique is the variation of the hydrogen-deuterium ratio in hydrogenous materials. For example, a complex system can be prepared as a solution or suspension in an $H_2O/D_2O$ mixture which has the same scattering length density as a part of the system; that part would then not contribute to the observed scattering. A simple analogy is the manner in which a glass bead will vanish from view when immersed in a liquid having a similar refractive index.

## 11.2  PHARMACEUTICAL EXAMPLES

### 11.2.1  *Molecular and Crystal Structure Determination*

In the course of the last decade or so it has repeatedly been demonstrated that a detailed knowledge of macromolecular and drug structures can greatly facilitate and rationalize the process of drug design. For the most part the required structural data has been sought and accumulated through the use of X-ray diffraction techniques, although a growing number of studies are now also made using high resolution multidimensional NMR spectroscopy. It is also true, however, that there have been some very worthwhile contributions afforded from neutron diffraction experiments.

In the case of single crystal neutron diffraction studies, there have been data obtained not only on drug structures, but also their macromolecular target 'receptors'. With the small molecule studies, the neutron diffraction data are not generally employed in *ab initio* determinations of structure, but rather for the refinement of known X-ray crystal structures. Used in conjunction with X-ray data the neutron data provide for far greater accuracy in the determination of atomic positions (and this is particularly significant for lighter atoms because their scattering power is much the same as for heavy atoms); at the same time they permit more accurate and precise measurement of anisotropic temperature factors and site occupancies (because of the Q-independent scattering of neutrons which means that the refinements benefit from more high-Q data). The neutron diffraction data are thus especially useful for obtaining detailed information on hydrogen bonding interactions, and also for highlighting the solvent structure in crystals. On the down side, the neutron data analysis is made problematic by incoherent scattering arising from any hydrogen atoms in the system, and this unfortunately dictates that the (high resolution) neutron diffraction studies must be performed using deuterated compounds. The experiments also demand very large crystal sizes, with the required volumes of the order of $1 \text{ cm}^3$.

Despite these limitations, there have been a number of systems studied that have relevance to pharmaceutical sciences. These include creatine monohydrate (Frampton *et al.*, 1997), racemic ibuprofen (Shankland *et al.*, 1997b), $S(+)$ ibuprofen (Freer *et al.*, 1993), paracetamol (Wilson, 1997) and acetylcholine bromide (Shankland *et al.*, 1997a). The latter has provided the most detailed picture to date of the interatomic contacts around the acetylcholine cation, and on the basis of their observations the authors discuss the binding of acetylcholine to its receptor. With the ibuprofen structure analysis, Shankland *et al.* were able to use their structure data to predict the crystal morphology (Shankland *et al.*, 1996a).

The single crystal neutron diffraction studies of macromolecular systems are, as with the small molecule studies, usually employed in combination with X-ray structure determinations. Here again the crystal volumes required are of the order of $1 \text{ cm}^3$, and the acquisition of the most useful information from the experiments depends upon the incorporation of deuterated species in the system studied. In all cases so far, however, it has not been the macromolecule that has been deuterated but rather some other component(s) of the crystal, such as solvent, surfactant or a bound ligand molecule. For example, Roth *et al.* (1991) determined the structure of the surfactant (detergent) phase in crystals of the *Rhodobacter sphaeroides* photosynthetic reaction centre (RC), which was grown in the presence of *n*-octyl $\beta$ D-glucoside ($\beta$-OG). In order to obtain this information they carried out a series of low resolution neutron diffraction experiments using different contrasts obtained by varying the ratio of $H_2O$ and $D_2O$ in the crystal solvent. These contrast variation data were used with the RC atomic coordinates determined by X-ray diffraction, to determine the positions of the $\beta$-OG molecules. From a consideration of the observed RC-$\beta$-OG interactions, the authors discuss the role and structural requirements of detergents in membrane protein crystallization, and obtain insight into the likely interactions existing in vivo between the RC and membrane phospholipids. These latter observations clearly have a more general significance in studies concerned with the prediction of pharmaceutically and pharmacologically relevant membrane receptor protein structures. Other similar studies have been reported for the surfactant structure in the tetragonal and trigonal crystal forms of the *Escherichia coli* ompF porin (Cowan *et al.*, 1995; Pebaypeyroula *et al.*, 1995).

With a somewhat different emphasis, Hauss *et al.* (1994) have used single crystal neutron diffraction techniques to look at the protein conformational consequences of light-induced isomerization of the (selectively deuterated) retinal ligand to bacteriorhodopsin (BR). Since the BR structure has been so extensively used in modelling proteins like the muscarinic, adrenergic, 5-HT and other neurotransmitter receptors, the data revealed by this type of study will undoubtedly help to improve our understanding of the mechanism of action of these more pharmaceutically relevant macromolecules.

Fibre diffraction is also possible using neutrons, and such studies have been particularly valuable in the determination of the solvent structure around the DNA double helix (Langan, 1997). By exploiting the ability to replace the $H_2O$ surrounding the DNA by $D_2O$, isotopic difference Fourier maps are computed in which the peaks are identified with the distribution of solvent in the unit cell. High angle neutron fibre diffraction studies have been reported on the hydration of A-DNA (Langan *et al.*, 1992, 1995) and D-DNA (Forsyth *et al.*, 1992). Used in conjunction with the information provided by X-ray studies on the locations of cations around the DNA helices, these data provide new information on the factors which help to stabilize the different conformations of the nucleic acid, and this in turn aids those concerned with the design of therapeutic agents which act by binding to DNA. Other work in the neutron fibre diffraction area includes a study of (perdeuterated) acetaldehyde binding to tendon collagen, which has yielded information relevant to the processes of cirrhosis and fibrosis of the liver (Wess *et al.*, 1994).

The use of powder neutron diffraction data in structural studies is also worthy of mention here (Wilson *et al.*, 1997), although the contributions reported that have any direct relevance to pharmaceutical scientists are currently rather scant, with the only study of note being the determination of the room temperature crystal structure of decadeuteriodopamine (Shankland *et al.*, 1996b). Such studies suffer much the same drawbacks as encountered with single crystal work, in so far as high resolution refinements are really only feasible if deuterated materials are used, with the sizes of the molecules amenable to study being rather small (of the order of 30–40 atoms), and the sample sizes required being potentially prohibitive (generally in the range 0.5–1.0 g). With these limitations set aside, however, it is important to note that the technique has much to offer pharmaceutical scientists, in particular those engaged in fundamental research on solid dosage forms. Over recent years there has been great progress made in neutron powder diffraction owing to the availability of higher neutron fluxes, improved instrumentation and the use of Rietveld's method for refining the structural data using the whole diffraction profile. Crystal structure refinement can currently be performed routinely for systems involving 100–150 structural parameters, and the prospects for the future (with the increased resolution and extension to higher and lower d-spacings provided by new instruments) will make it possible to deal with systems involving up to 500 structural parameters. There are many drugs in this size range whose structures will therefore be amenable to study using this technique, and since neutron powder diffraction (by comparison with X-ray powder diffraction) affords more straightforward investigations of any structural changes accompanying changes in temperature and pressure, it is possible to use this method to explore, for example, polymorphic phase transformations and crystal defects — phenomena that have direct bearing on the utility of solid dosage forms of drugs. Such studies are exemplified by the work of Ibberson (1996) on the temperature-induced phase transition of sulfamide, $SO_2(NH_2)_2$.

### 11.2.2  Structure of Disperse Systems

The wavelength of neutrons makes them ideal for investigating the structure of colloidal materials in the range 10–1000 Å. Since many such materials are based on organic systems, they do not scatter X-rays strongly, although the advent of intense synchrotron sources is making the use of X-rays more widespread. Neutron scattering is complementary to light scattering which provides information on larger length scales (1000 Å and greater) and thus accurately probes the overall structure rather than the fine detail. An example of this is the study of fluorocarbon emulsions (Washington *et al.*, 1996, 1997); these have a droplet diameter of 300 nm by DLS, but SANS underestimated their diameter at 140 nm. However SANS provided unparalleled detail concerning the structure of the adsorbed surfactant layer

TABLE 1. Some typical neutron scattering density lengths and $H_2O/D_2O$ matchpoints.

| Material | $b \times 10^{10}$ cm$^{-2}$ | Matchpoint (% $D_2O$) |
|---|---|---|
| Water | −0.559 | 0 |
| Deuterium oxide | 6.355 | 100 |
| Poly(ethylene oxide) | 0.637 | 17 |
| Poly(propylene oxide) | 0.343 | 13 |
| Ethanol | −0.345 | 3 |
| Perfluorodecalin | 4.183 | 72 |
| Proteins | 2.2–2.6 | 40–45 |
| Nucleic acids | 4.0–4.5 | 65–72 |
| Dimyristoyl PC | 0.31 | 12 |
| Cholesterol | 0.132 | 10.2 |
| Triglycerides | 0.247 | 10.3 |

(see below). For this reason emulsions have not been widely studied by neutron scattering but smaller colloidal systems, such as microemulsions, micellar and vesicular systems, have received more attention.

## 11.2.2.1 Microemulsions

Most microemulsion systems have structural features in the range of 100–1000 Å, and neutron scattering has contributed extensively to a knowledge of their size, interdroplet potentials (Bergenholtz et al., 1995), and shape and size fluctuations (Farago et al., 1990). Microemulsions composed of ionic surfactants have been shown to have shapes which are strongly sensitive to the counterion (Eastoe 1992, 1993). The most commonly studied microemulsion-forming surfactant is sodium di-2-ethyl-hexylsulfosuccinate (known by the tradename AOT), which forms a rich variety of normal, inverse, and bicontinuous systems (Chen et al., 1991). More complex microemulsion gels have been formed by adding gelatin, forming 4-component systems in which the microemulsion droplets are linked by surfactant-free gelatin rods (Petit et al., 1991; Atkinson et al., 1991).

## 11.2.2.2 Vesicular systems

Liposomes, the most intensively studied vesicular systems, have been examined in detail by SANS and SAXS. Despite intensive research liposomes have not lived up to their initial promise as drug carriers, but have been widely used as models of biological membranes. Recently they have been joined by a wide range of synthetic vesicular systems made from alternative surfactants. SANS has been applied to the structure of lipid bilayer systems, both transversely and laterally; for example a study of mixed chain lecithins by Lin et al. (1991) concluded that they formed discs rather than vesicular structures, and Knoll et al. (1991) found that lipids with differing chain length were in general mixed in the fluid phase, but separated into domains below the phase transition temperature. SANS studies have now been used to study the interaction of membrane-active materials with liposomes, the materials studied including clathrin (Bauer et al., 1991), annexin (Ravanat et al., 1992) and a range of anaesthetics (Ku and Tam 1997) and antimicrobial agents (Thiyagarajan 1997). Inelastic neutron scattering also provides information on the dynamic behaviour of membranes (Konig and Sackmann 1996). Finally a number of nonionic vesicle-forming surfactants have been studied both as vesicles (Oberdisse *et al.*, 1996), and in monolayers by neutron reflectometry (Barlow *et al.*, 1995, 1997).

### 11.2.2.3 Micelles

A large number of micellar systems have been studied by SANS; some of the more pharmaceutically relevant include chlorpromazine (Perezvillar *et al.*, 1993), $\kappa$-casein (Dekruif and May, 1991), gangliosides (Cantu *et al.*, 1993) and block copolymers (Mortensen, 1996). Lecithin-bile salt mixed micelles have also been extensively studied (e.g. Long *et al.*, 1994). A particularly interesting system is that of lecithin reverse micelles (Schurtenburger *et al.*, 1990) which form long polymer-like systems in nonaqueous solvents in the presence of small amounts of water. A detailed understanding of this behaviour is of considerable importance in the formulation of aerosol suspensions, in which lecithin is often used as a stabilizing surfactant in a nonpolar solvent.

### *11.2.3  Surfaces and Interfaces*

Specular neutron reflection is now a well-established technique to investigate adsorbed material at a surface or interface. It has been widely used to examine the structural properties of phospholipids at the air-water interface (see for example Vaknin *et al.*, 1991). More recently Brumm *et al.* (1994) have examined the head group conformation of the phospholipid dipalmitoyl-glycero-phosphocholine (DPPC) in monolayers at the air-water surface in the fluid-like liquid-expanded state (LE) and the crystal-like solid (S) phase. This information was obtained by contrast variation of the lipid monolayer using four differently deuterated species of DPPC; perdeuterated, chain perdeuterated, choline group perdeuterated and head group selectively deuterated. These workers were able to distinguish between changes in head group conformation and hydration moving from the LE to the S phase. The same group (Naumann *et al.*, 1995) extended these studies to examine the effect of the insertion of a nonionic surfactant ($C_{12}E_4$) into the monolayer on the hydration of the head group. Barlow and co-workers (1995, 1997) have examined the detailed structure of the monolayer formed by a dialkyl chain polyoxyethylene ether surfactant at the air-water interface by the use of selective deuteration of the surfactant. By using the information obtained from these studies it was possible to predict the molecular architecture of the vesicles formed by this surfactant, their prediction correlating well with the experimental data for the vesicles.

Lipid bilayers supported on a solid substrate have also been examined using neutron reflectivity (Krueger *et al.*, 1995). In this study the bilayer was formed by fusing single membrane vesicles (prepared from the outer membrane of *Neurospora crassa*) onto a phospholipid monolayer supported by a flat substrate. Neutron scattering studies, which were performed using both $H_2O$ and $D_2O$ as solvent, enabled the dimensions of the whole bilayer, the head group and the hydrocarbon region to be established.

The application of the technique of specular neutron reflection to the study of adsorbed protein is well illustrated in the case of $\beta$-casein at the air-water interface in the presence and absence of the nonionic water surfactant $C_{12}E_6$ (Dickinson *et al.*, 1993). Analysis of the reflectivity of pure $\beta$-casein in water at pH 7.0 gave a time-independent adsorbed amount, with the thickness increasing substantially as the pH was reduced towards the protein's isoelectric point. In the presence of surfactant, protein was displaced from the surface to the bulk aqueous phase. Indeed at a molar ratio of surfactant to protein of approximately 2.2:1 roughly half the protein was lost from the surface, although there was little change in the thickness of the adsorbed layer. At a surfactant to protein molar ratio of 10:1 protein was effectively completely removed from the surface. This work was later expanded to look at, in addition to the air/water surface, the adsorption of $\beta$-casein at the oil/water interface (Atkinson *et al.*, 1995a,b). Later studies, combining neutron reflectivity data with self-consistent-field modelling (Atkinson *et al.*, 1996) showed that the adsorbed layer of $\beta$-casein could be described by a model comprising a thin (approximately 10 Å) protein rich region at the air/water surface and a thicker (approximately 45 Å) more diffuse layer extending towards the bulk aqueous phase. The thickening observed upon reducing the pH of the solution towards the isoelectric point of the protein was attributed to secondary adsorption of protein parallel to the monolayer seen at pH 7.0. More recently these workers have extended these studies to also look at $\beta$-lactoglobulin as well as $\beta$-casein (Atkinson *et al.*, 1995b). In this study Atkinson *et al.* examined the effect of protein concentration, substrate pH, film aging and the presence of calcium ions on the adsorbed amount of these proteins at the air-water interface and were able to determine differences in the behaviour of these proteins.

Neutron reflectivity has also been used to examine the structure of bovine serum albumin (BSA) spread at the air-water interface (Eaglesham *et al.*, 1992). These workers used a combination of hydrogen-deuterium isotopic substitution in the aqueous phase and constrained model fitting to determine the distribution of protein at the interface. The resulting model suggested an upper layer containing the major portion of the molecule which lies flat along the surface (consistent with an $\alpha$-helix conformation) and a lower layer containing only 7% vol of the BSA which probably consists of a number of hydrophilic chains dangling down from the upper layer. In addition up to 30% wt of the protein was lost to the subphase on initial spreading.

The binding of the protein streptavidin to biotinylated lipid head groups has been studied at the air-water interface using $H_2O$ and $D_2O$ subphases (Vaknin *et al.*, 1993). The data were consistent with formation of a homogeneous monomolecular protein layer directly underneath the lipid. As well as examining the behaviour of the protein streptavidin at the air-water interface, the protein has also been studied at the solid-liquid surface. For example neutron reflection and surface plasmon optical experiments have been performed to evaluate structural data of the interfacial binding reaction between the protein streptavidin and a solid-supported lipid monolayer functionalized by biotin moieties (Schmidt *et al.*, 1992). By using these techniques together it was found that binding of streptavidin to the biotin groups results in an increase in layer thickness. Furthermore the results suggested the formation of a well-ordered protein monolayer, in which the biotin/spacer units of the functionalized lipids were fully embedded into the binding pockets of the proteins. Interestingly their model calculations showed that a more detailed picture of the internal structure of this supramolecular assembly could only be obtained if deuterated lipids were used to allow for a high contrast between the individual layers.

### 11.2.4  Structure of Multiphase Systems

The most commonly reported neutron scattering experiment on multiphase systems is that of small angle neutron scattering (or SANS). This powerful technique can, depending upon the design of the experiment, be used to obtain various levels of information on the system under study. For example it can be used to provide low level information such as the size and shape of particles present in a system, or much more detailed, sophisticated information, of a detail that is frequently not possible to achieve with other techniques. In addition, SANS is often combined with other techniques, to lend support for a particular interpretation of a system. For example SANS studies have been successfully used in combination with quasi-elastic light scattering and cryotransmission electron microscopy to determine the size, shape and structure of drug-containing nanoparticles prepared using an oil-in-water emulsification procedure (Sjostrom *et al.*, 1995).

Recent SANS studies by Meyer *et al.* (1995) on low-density lipoproteins (LDL) dispersed in aqueous buffer can be used as an illustration of the varying levels of information that can be gained about a system using this technique. Low level SANS studies showed that LDL (a particle composed of a single molecule of apolipoprotein in association with lipid which usually forms the core) exhibits a pronounced concentration-dependence in the radius of gyration and molecular weight when analysed using Guinier analysis, although the results are consistent with a spherical structure. More detailed analysis of the SANS data revealed that LDL could be successfully modelled by a polydisperse assembly of spheres with two internal densities and a mean radius close to 10 nm. The results of the SANS experiments were consistent with electron microscopy and ultracentrifugation data.

Another example of the sort of detailed information that can be provided by SANS studies is illustrated by the work of Jenkins and Donald (1996) and Waigh *et al.* (1996). These workers have exploited SANS using contrast variation to obtain detailed information about the distribution of water within waxy maize starch granules. On the basis of previous X-ray scattering studies, these workers had proposed that the starch granules contained three regions, namely semicrystalline stacks and amphorous lamellae embedded in a matrix of amorphous material. The results obtained from SANS studies of the starch granules dispersed in water containing different $D_2O$ contents enabled, for the first time, the quantity of water present in each of the different regions to be quantified.

The coating of particulate material with surfactants has attracted interest as a means of altering the surface characteristics of the particulate intended for use as drug delivery vehicles. SANS studies have been successfully used to study this type of system. For example Gladman *et al.* (1995) have reported the use of SANS to investigate the structure of surface adsorbed polymeric surfactant layers on colloidal particles in dispersion. Recently the structure of sodium dodecyl sulphate bound to a poly(NIPAM) microgel particle has been examined using a combination of dynamic light scattering (DLS), SANS and binding isotherm measurements (Mears *et al.*, 1997). Both DLS and SANS measurements showed that the microgel swelled in the presence of surfactant. However, the use of selective deuteration of the solvent made it possible to highlight the different components in the system, namely the poly(N-isopropylacrylamide) gel or the SDS, by rendering the other components 'invisible' to the neutrons, thereby allowing detailed information on each component in the system be obtained.

## 11.2.5  Biopolymers and Bioadhesives

A large number of biomacromolecules have been studied using neutron scattering techniques, and in particular SANS. For example the aggregation state of the enzyme lysozyme in aqueous solution has been studied as a function of temperature and pH (Giordano et al., 1993). These studies have further been extended to look at the interactions between individual lysozyme molecules, at under and super-saturation (Gripon et al., 1997). Another enzyme that has been studied by SANS is the 'blue' copper enzyme ascorbate oxidase isolated from zucchini (Maritano et al., 1996). Contrast variation using $D_2O$ and $H_2O$ yielded both the radius of gyration and the molecular weight of the enzyme. Haemcyamin from *Rapana thomasiana* has also been examined using SANS studies (Triolo et al., 1996). The results obtained from the SANS studies were in excellent agreement with those found with electron microscopy. The aggregation behaviour of zinc-free insulin has also been studied as a function of protein concentration, pH and ionic strength of the solution using SANS (Pedersen et al., 1994). The weight average molecular mass and the $z$-averaged radius of gyration were found to vary systematically with experimental conditions. All of the above studies were performed using aqueous solutions. Recently however Wang et al. (1997) have examined the aggregation behaviour of bacteriochlorophyll *a* and *c* in a non-polar benzene solution. As with the aqueous-based studies it was possible to determine a molecular weight and radius of gyration for the dimers formed, although for the oligomers it was only possible to obtain an approximation of the radius of gyration. The results from the SANS studies correlated well with other physical measurements made on the system.

In addition to examining the aggregation state of a biopolymer in solution SANS has also been successfully used to study the crystallization of lysozyme from the microscopic molecular level (Niimura et al., 1994). More recently the same group (Niimura et al., 1995) used time-resolved small-angle neutron-scattering to study the nucleation of lysozyme from a supersaturated solution. SANS has also been used to obtain a model of the extended glycoprotein region of the seven domains present in human carcinoembryonic antigen (Boehm et al., 1996). The molecular mass obtained from the neutron data agreed well with the value found using X-ray scattering techniques, and the results suggested a model of the glycoprotein that satisfactorily explained the adhesion interactions between molecules on adjacent cells and the antibody targeting of the molecule. The overall and internal structure of a recombinant yeast-derived human hepatitis B virus surface antigen vaccine particles has been investigated by SANS using the contrast variation method (Sato et al., 1995). The vaccine was established to be nearly spherical and that a large part of the particle was composed of lipids and carbohydrates from the yeast. Furthermore, these studies suggested that the surface antigens were predominately located in the peripheral region of the particle, which is favourable to the induction of anti-virus antibodies.

Bioadhesives, such as chitosan, are an increasingly important class of pharmaceutical molecules with particular potential for use in the transmucosal delivery of drugs. An important aspect of the bioadhesive action of these molecules is their interaction with water. SANS studies exploiting contrast variation have been used to investigate the changing structure of chitosan films upon swelling in water vapour which has been shown to proceed via water cluster formation within the chitosan matrix (Evmenenko and Alekseev, 1997).

### 11.2.6  Gel-Forming Systems

Gels form a very important (and indeed very diverse) class of pharmaceutical materials. Although they are routinely used for the topical application of drugs, they are being increasingly studied as vehicles for the sustained release of drugs, especially after oral or intramuscular administration. Gels that have been studied by neutron scattering techniques include natural (Kruger *et al.*, 1994) and synthetic polymers (Silberbergbouhnik *et al.*, 1995), cross-linked polymers (Evmenenko *et al.*, 1996), star polymers (Willner *et al.*, 1994) and gels formed from surfactant systems such as sodium lithocholate (Terech *et al.*, 1996a), high internal phase volume emulsions (Ravey *et al.*, 1994), cubic phases, such as those produced by monoolein (Czeslik *et al.*, 1995) and the pluronics (Mortensen, 1996, 1997), and very recently a cubic phase of densely packed monodisperse, unilamellar vesicles (Gradzielski *et al.*, 1997). To date while little neutron scattering work has examined specifically systems intended for pharmaceutical use, a significant number of the polymers investigated by neutron scattering studies are currently used pharmaceutically, for example agarose and gelatin. Furthermore, while there are no reports dealing with gelation in the presence of drugs, there is a limited amount of work concerned with the influence of additives such as the surfactant, sodium dodecyl sulphate (Cosgrove *et al.*, 1996; Hecht *et al.*, 1995) and the dye congo red (Shibayama *et al.*, 1994) on the gelation process. As pharmaceutical systems are often prepared from a mixture of ingredients, this type of information is extremely valuable. In addition, a considerable amount of work has recently been performed on polar, non-aqueous based-gels (Cohen 1996) and organogels (Terech *et al.*, 1994, 1995, 1996b) and microemulsion-based gels (Atkinson *et al.*, 1988). Although these organogels have yet to make an impact on the pharmaceutical market, they are likely to do so in the very near future. Again the effect of surfactants, in this case lecithin, on the gelation process has been investigated (Tamura and Ichikawa, 1997).

Although the majority of neutron scattering experiments examining gels are SANS studies, for example Horkay *et al.* (1994), Krueger *et al.* (1994), a small number of wide angle (Kanaya *et al.*, 1994) and QINS (Dadmun *et al.*, 1996) studies have also been undertaken. As with other systems investigated SANS is often combined with other techniques, frequently light or X-ray scattering or rheology, to lend support for a particular interpretation of a system. SANS has been extensively used to examine the structural arrangement of materials as they gel under a variety of conditions. For example, the subtle changes in gel structure due to a variation in the pre-gelation concentration of agarose have been studied (Krueger *et al.*, 1994) as has the thermoreversible gelation of agarose in water/DMSO mixtures (Rochas *et al.*, 1994). The effect of pH and salt on environmentally sensitive hydrogels consisting of poly(N-isopropylacrylamide-co-acrylic acid has been examined by Shibayama *et al.* (1996) using SANS. These workers were able to determine the effects of pH and salt concentration on the temperature of the volume phase transition and the structure factor of the hydrogels. It has also proven possible to establish the time taken for the evolution of the gel-network using time-resolved SANS measurements (Kobayashi *et al.*, 1995).

## 11.2.8  *Membrane Structure and Function*

In order to understand the effects that many drugs exert on biological membranes it is essential to understand the properties of the membranes themselves. The technology is presently available to study biological membranes using a variety of neutron scattering techniques, including SANS, neutron diffraction and neutron reflectivity. Recently, for example, a novel experimental set up has been designed to measure specularly reflected neutrons from model membranes prepared either from chain deuterated and protonated versions of dipalmitoylphosphatidylcholine (DPPC) or components of the outer membrane of *Neurospora crassa* mitochondria deposited on a flat surface to $Q$ values of 0.25 Å$^{-1}$ (Kreuger *et al.*, 1995; Koening *et al.*, 1996). These experiments enabled the determination of the overall thickness of the bilayers, as well as the thicknesses of the tail and head group regions. It was also possible to determine Ångstrom-scale thickness changes of the central membrane layer as a function of the phase state of the lipid and of the length of the hydrocarbon chains (Koening *et al.*, 1996). The advantages of using neutrons over X-rays is that it is not necessary to have heavy atoms in the molecules under study, and that the location of any water can be established. However an in-plane scattering technique has been developed by He and co-workers (He *et al.*, 1993) to look at the lateral organization of proteins and peptides in membranes. This technique, which is equally applicable to X-ray or neutron scattering, has the possible advantage of allowing the study of proteins in membranes by X-rays without the need for heavy atom labelling, as orientated proteins may provide substantial X-ray contrast against the lipid background.

SANS studies have also been widely used to gain an understanding of the properties of membranes. Recent SANS data obtained from fully hydrated, multilamellar phospholipid bilayers with deuterated acyl chains of different length, as a function of temperature, have provided a wealth of information on the properties of the bilayer (Lemmich *et al.*, 1996a). These SANS data were analyzed using a paracrystalline theory and a geometric model that allowed the determination of bilayer structure under conditions where the lamellar layers are coupled and fluctuating. Analysis of the data in this way provided indirect information on interlamellar undulation forces and bilayer bending rigidity as well as direct information on the lamellar repeat distance, hydrophobic bilayer thickness, the thickness of the aqueous and polar head group regions, the molecular cross sectional area, number of interlamellar water molecules and lateral area compressibility.

Other neutron scattering techniques such as quasi-elastic incoherent neutron scattering (QINS) have been widely used to study the local diffusion and chain dynamics of phospholipids in orientated model membranes, for example by altering the hydration and phase state of the membrane (Koning *et al.*, 1995). The technique has also been used to examine integral membrane proteins such as bacteriorhodopsin embedded in a purple membrane (Fitter *et al.*, 1996) where it has been shown that a decrease in hydration of the protein results in an appreciable decrease in internal molecular flexibility of the protein structure. This observation ties in with functional studies which show that the activity of bacteriorhodopsin is reduced if hydration is insufficient and suggests that the observed diffusive motions are essential for the function of this type of protein.

Frequently neutron scattering techniques are used in combination with other techniques to obtain very detailed information on a particular system. For example a combination of

X-ray and neutron diffraction studies have been used to determine the complete structure of a fluid phase dioleylphosphatidylcholine bilayer (Weiner and White, 1992). Using a combination of proton NMR, FT-IR, DSC and neutron specular reflection, Reinl et al. (1992) have been able to determine the changes in the physical properties in binary-mixtures of dipalmitoylphosphatidylcholine with cholesterol as a function of temperature.

It obviously makes for a more complex experiment to study the effect of drugs on these membranes, however recent work has shown that it is possible to get a considerable amount of very useful information from such experiments. For example the temperature dependence of the SANS scattering from aqueous vesicular dispersions containing small amounts of cholesterol (less than 3 mol%) has been analyzed using a simple geometric model which yields the lamellar repeat distance, the hydrophobic thickness of the bilayer and the intralamellar spacing as well as fluctuation parameters (Lemmich et al., 1996a). The same group (Lemmich et al., 1996b) have also studied the effect of incorporation of 1 mol% of amphiphilic solutes, such as cholesterol, a short chain lipid and a bola lipid into multilamellar DPPC vesicles. Small amounts of the solutes were found to soften the bilayer and thermally reduce bending rigidity. The well known rigidifying effect of cholesterol was only seen to occur at cholesterol concentrations above 3–4 mol%. Neutron diffraction studies have suggested that cholesterol sulphate has two locations in DMPC membranes (Faure et al., 1997). The interaction of volatile anaesthetics on phospholipid vesicles has also been recently investigated using SANS (Ku and Tam, 1997).

Neutron scattering studies have also been performed on dipalmitoylphosphatidylcholine (DPPC) bilayers containing as model probes, diphenyl-hexatriene (DPH) and the trimethyl-lammonium analogue (TMA-DPH) (Pebaypeyroula et al., 1994). As DPH and TMA-DPH were either protonated or deuterated in one of the phenyl rings, it was possible to locate the position of the probes in the membrane by use of contrast variation techniques. Both probes were seen to exhibit bimodal distributions in the membrane, the position, population and orientation of which depend upon the physical (fluid or gel) state of the bilayer. Furthermore, $D_2O$ exchange experiments afforded an estimate of the maximum penetration of water into the interior of the bilayer. SANS studies have been used to determine the intralamellar location of tetrahydrocannabinol (deuterated and protonated forms) in hydrated dipalmitoylphosphatidylcholine (DPPC) bilayers (Martel et al., 1993). These results were compared to previous X-ray measurements using iodine labelling. Although there were similarities in the results obtained using both techniques, a difference was found in the location of the terminal methyls on the cannabinoid side-chains. The results of the study suggested the location of the drug in relation to the active site of its transporter.

The interaction of bee venom melittin with a dioleylphosphatidylcholine bilayer has been studied using a combination of X-ray and neutron diffraction (Bradshaw et al., 1994). The results of the study showed that melittin largely incorporates into the phospholipid bilayers distributed between the surface and the centre of the bilayer and that this distribution is dependent upon pH, with a larger population of melittin in the surface when the N-terminus was unprotonated.

The effect of the presence of carbohydrates including the hydrophobically modified polysaccharide cholesterol-pullulan (CHP) on fluid phase phospholipid vesicles has been studied using a combination of SANS and X-ray scattering, both as a function of temperature and pressure (Deme *et al.*, 1996, 1997). Concomitant neutron and X-ray studies have been successfully used in accurately locating the drug amantadine in multilayers of 1,2-dioleyl-sn-glycero-3-phosphocholine (Duff *et al.*, 1993). The X-ray data were obtained by the swelling series method while the neutron data were obtained using $D_2O/H_2O$ exchange and a variation of the isomorphous replacement technique. The two sets of data complemented each other and revealed that there were two populations of amantadine within the bilayer. The major site is close to the bilayer surface, while the other, less occupied position is much deeper in the membrane.

In-plane neutron scattering has been used to study the structures of the pore of the peptide alamethicin in membranes of DLPC (He *et al.*, 1997). These investigations indicated that water was part of the high order structure of the inserted alamethicin. The studies also accurately determined the diameter of the water pore, which was found to vary with the level of hydration, and the outside diameter of the channel. In-plane neutron scattering has also been used to study the conformation of the antimicrobial peptide magainin, found in the skin of *Xenopus laevis*, in a biological membrane (Ludtke *et al.*, 1996). These workers were able to resolve the apparent discrepancies observed between other techniques used to examine its membrane activity and produced a model consistent with all published magainin data.

## 11.3 STUDIES OF VESICLE-FORMING SURFACTANTS

The last few years have witnessed a renaissance of interest in the development of surfactant and lipid vesicles for use in drug delivery (Lasic 1996). Figure 1 shows a schematic representation of a single (uni)lamellar vesicle formed from a double hydrophobic chain surfactant. Soon after it was first realized by Bangham and colleagues in 1964 that certain surfactants could form these closed structures it was proposed that vesicles could be exploited as universal drug delivery vehicles, able to incorporate both hydrophilic and lipophilic compounds in their structure. However, the initial excitement was soon tempered by the fact that after intravenous administration, vesicles were almost immediately sequestered from the blood circulation by the cells of the reticulo-endothelial system and as such did not exhibit sufficient plasma life-times to be used as drug delivery vehicles. The recent rekindled optimism in these researches has been founded upon the discovery that vesicles with hydrophilic non-ionic polymer chains expressed on their outer surface have the ability to avoid this rapid sequestration from the blood circulation by the cells of the reticulo-endothelial system and thus exhibit the much longer plasma life-times required for their successful exploitation as drug delivery vehicles. To date, the non-ionic polymeric exterior to the vesicles has been variously engineered either by incorporating into phospholipid vesicles (known as liposomes) lipids modified with polyoxyethylene glycol (POE) (Woodle *et al.*, 1992), by coating preformed liposomes with POE-containing polymers such as the pluronics (Jamshaid *et al.*, 1988), by covalently attaching POE chains to preformed liposomes (Senior *et al.*, 1991), by incorporating micelle-forming monoalkyl POE surfactants within the lipid bilayers (Kronberg *et al.*, 1990), or by using lipids with sugar-based head groups (Allen and Chonn 1987). Numerous such systems are now in various stages of clinical trials with intended applications including the delivery of doxorubicin for the treatment of solid tumours, AZT for the treatment HIV, and amphotericin B for the treatment of fungal and parasitic infections (Lasic, 1996).

Now although these Stealth® liposome preparations have properties much improved over those prepared from normal (zwitterionic) phospholipid liposomes they are not without their own serious limitations (Lawrence *et al.*, 1996a). In particular it may be noted that these systems are not at all straightforward to prepare or characterize (which presents serious difficulties in their pharmaceutical quality control), and there is evidence (Blume and Cevc, 1993) that they exhibit poor *in vitro* stability (which may necessitate their preparation immediately prior to administration, thereby increasing their cost).

Over the past few years attempts have been made to overcome these problems by the development of potentially more suitable systems prepared from a series of novel, non-ionic, synthetic dialkyl glycero polyoxyethylene ether surfactants of the general type, 1,2-di($C_nH_{2n+1}$)-$O$-glyceryl-(oxyethylene)$_m$ (henceforth abbreviated as $2C_nE_m$) (Lawrence *et al.*, 1996a). The defined chemical structure and ease of purification of these molecules will benefit their production as pharmaceuticals. It has been shown that particular surfactants in this series form micelles and others vesicles (Lawrence *et al.*, 1996b,c), and that the vesicles have a very high in vitro stability exhibiting worthwhile potential for use in drug delivery. Although these compounds are relatively straightforward to synthesize, the number of possible surfactants which might be considered for synthesis is almost limitless, and so it is essential that the design of the compounds is approached in a truly rational fashion, so that only those molecules are synthesized that can be reliably expected to give vesicles rather than some other aggregate (such as micelles) and, more importantly, to give vesicles with the desired properties. As a means to establish such predictive schema, and thereby to further our general understanding of the relationship between surfactant structures and the physico-chemical and biological properties of the resulting aggregates, systematic investigations of the 3D structures of the $2C_mE_m$ vesicles, and monolayers have been performed using neutron scattering techniques (Barlow *et al.*, 1995, 1997).

Initial neutron scattering experiments involved elementary studies of the fully protonated and alkyl chain deuterated forms of the surfactant $2C_{18}E_{12}$, dispersed to form monolayers at the air-water interface. Figure 2 shows schematically the surface pressure-area curve obtained for this surfactant at the air-water interface using a standard Langmuir trough apparatus (Lawrence *et al.*, 1996c). As can be seen from Figure 2 at low values of surface pressure the surfactant forms a very expanded (gaseous) monolayer while at intermediate surface pressures it undergoes an apparent phase transformation to a condensed (liquid) monolayer. Prior to the performance of the neutron specular reflectance experiments, it had been assumed that the collapse point of the surfactant monolayer occurred at a surface pressure around 56 mNm$^{-1}$, and an area per surfactant molecule of $\sim 50$ Å$^2$ was obtained (Lawrence *et al.*, 1996c). This area per surfactant molecule was smaller than expected from the surfactants' known ability to form vesicles. By holding the surfactant *monolayer* film at pre-determined surface pressures, it was possible to determine the structure of the film at this surface pressure using neutron specular reflectance experiments. These experiments were performed on the instrument CRISP at the Rutherford Appleton Laboratories, Didcot, UK. Rather surprisingly, the data obtained showed that the interfacial $2C_{18}E_{12}$ could only be regarded as a monolayer up to a surface pressure of $\sim 40$ mNm$^{-1}$, and that increases in the surface pressure above this point resulted in the development of a pronounced off

FIGURE 1. Schematic illustration of the structure of a vesicle formed from a double chained surfactant.

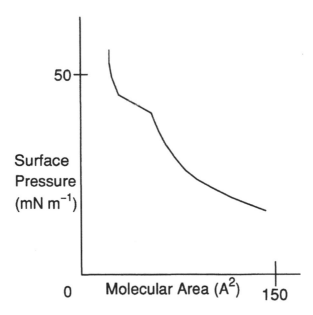

FIGURE 2. Schematic representation of the surface pressure-molecular area isotherm obtained for $2C_{18}E_{12}$ using a Langmuir trough. Surface pressure is presented in $mNm^{-1}$ and the molecular area in $Å^2$ (data taken from Lawrence *et al.*, 1996c).

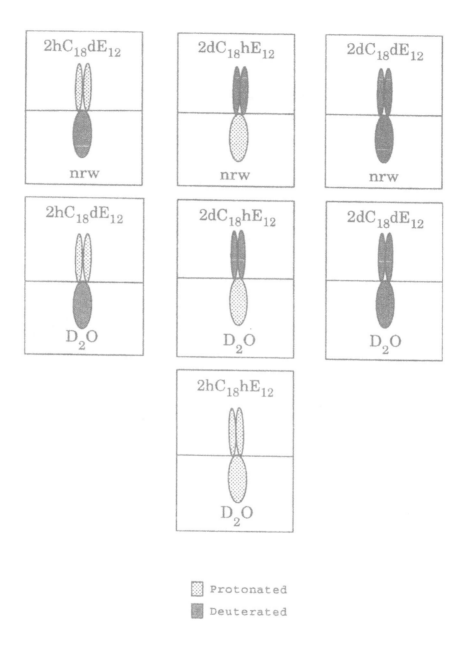

FIGURE 3. Schematic illustration showing the various contrasts employed in the seven neutron reflectivity experiments used to determine the structure of the monolayers formed by $2C_{18}E_{12}$ at an air-water interface. The prefix *h* or *d* indicates whether that portion of the $2C_{18}E_{12}$ molecule is either protonated or deterated. nrw is null reflecting water.

specular reflectance, indicative of multilayer and domain formation. This unexpected and very significant observation indicated that the novel surfactant exhibited an interfacial behaviour very different from that seen with analogous phospholipids, and suggested that there was significant mixing of the surfactant head groups and alkyl chains within the monolayers, particularly at high surface pressures. Furthermore, the results of the neutron scattering experiments showed that the $2C_{18}E_{12}$ monolayer existed only up to a surface pressure of $\sim 40$ mNm$^{-1}$ where an area per molecule of $\sim 100$ Å$^2$ was obtained. On the basis of these data it was possible to predict the size and bilayer dimensions of the vesicles formed by $2C_{18}E_{12}$, (Barlow et al., 1995) and the results obtained were found to be consistent with those obtained from photon correlation spectroscopy (Lawrence, 1992) and scanning tunnelling microscopy (Roberts et al., 1991).

More detailed studies of the $2C_{18}E_{12}$ monolayer were subsequently performed using four different isotopically labelled forms of $2C_{18}E_{12}$ including the fully protonated, fully deuterated, alkyl chain deuterated and head group deuterated species. Seven sets of reflectivity data were acquired for the monolayers formed by these surfactant species, dispersed on sub-phases of $D_2O$ and/or null-reflecting water (nrw), that is water which is neutron transparent. The seven sets of contrast examined in the study are shown in Figure 3. Each of the contrasts utilized gave different information on the monolayer. For example, when the protonated form of $2C_{18}E_{12}$ was investigated in $D_2O$, the reflectivity profile was largely determined by the solvent profile, while the reflectivity profile of the chain deuterated form of $2C_{18}E_{12}$ on nrw was predominately due to the hydrophobic portion of the chains.

The combination of the seven sets of data was then used to separate the contributions to the reflection arising from the various parts of the $2C_{18}E_{12}$ molecule and solvent using the kinematic analysis (Simister et al., 1992). Kinematic analysis of these data yielded the partial structure factors for the surfactant monolayer and gave a detailed picture of the molecular architecture of the adsorbed layer formed at an air-water interface. In each experiment the reflectivity ($R(\kappa)$) was modelled as

$$R(\kappa) = (16\pi^2/\kappa^2)[b_c^2 h_{cc}(\kappa) + b_e^2 h_{ee}(\kappa) + b_s^2 h_{ss}(\kappa)$$

$$+ 2b_c b_e h_{ee}(\kappa) + 2b_c b_s h_{cs}(\kappa) + 2b_e b_s h_{es}(\kappa)] \tag{1}$$

where $b_i$ are the sums of the empirical atomic scattering lengths for the solvent (s), $C_{18}$ chains (c) and $E_{12}$ head groups (e), $h_{ii}(\kappa)$ and $h_{ij}(\kappa)$ are the self-term and cross-term partial structure factors.

The reflectivity experiments and analyses were carried out with the interfacial $2C_{18}E_{12}$ layers maintained at surface pressures of 15 mNm$^{-1}$, 28 mNm$^{-1}$, 34 mNm$^{-1}$ and 40 mNm$^{-1}$. Figure 4 shows a computer simulation of a $2C_{18}E_{12}$ monolayer at a surface pressure of 28 mNm$^{-1}$ modelled using the constraints obtained from the reflectivity data. The analyses showed that the surfactant's two $C_{18}$ chains are highly tilted with respect to the surface normal (lying almost flat at the air-water interface) and that the $E_{12}$ head groups are far less tilted ($\sim 30°$ degrees from the vertical at 15 mNm$^{-1}$ and lying along the vertical at 40 mNm$^{-1}$). The dimensions of the $E_{12}$ head group were consistent with their behaviour as random-flight polymer chains. The analyses also revealed that there was (as suggested by the earlier studies) a very significant degree of intermixing of the $C_{18}$

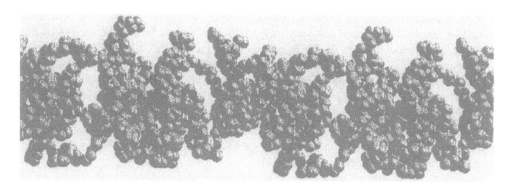

FIGURE 4. A computer simulation of a $2C_{18}E_{12}$ monolayer at a surface pressure of 28 mNm$^{-1}$ modelled using the constraints obtained from the reflectivity data.

and $E_{12}$ chains, with evidence too that segments of the $E_{12}$ chains must exist in a partially dehydrated state. With increasing surface pressure the extent of intermixing of the $C_{18}$ and $E_{12}$ chains was observed to become more pronounced, with a concomitant increase also in the proportion of the dehydrated $E_{12}$ chains. It was speculated that when the surface pressure was increased beyond 40 mNm$^{-1}$, the extent of mixing of the $C_{18}$ and $E_{12}$ chains would become so pronounced that some of the $2C_{18}E_{12}$ molecules might be extruded onto the air-side of the interface, which would give rise to 'islands' of 'bilayer' adrift in the monolayer, thus accounting for the significant level of off-specular reflection developed by the more compressed surfactant layers.

With the primary interest in these novel non-ionic surfactants lying in their potential for use in the production of drug delivery vesicles, small angle neutron scattering (SANS) experiments were also carried out using aqueous dispersions of $2C_{18}E_{12}$ vesicles, as well as those formed from $2C_{18}E_{17}$, $2C_{16}E_{12}$ and $2C_{16}E_{17}$. In these studies the aim was to obtain information directly on the size and bilayer architecture of the surfactant vesicles, so that it could be determined whether the earlier predictions of the vesicles' structures were correct, and providing empirical evidence to support or disprove the general idea of using structural data deduced from studies of *planar* surfactant *monolayers* to predict the structures of *curved* surfactant *bilayers*. All surfactants were used in their protonated form and were dispersed as vesicles in $D_2O$. These SANS experiments were performed on a diffractometer D17 at the Institute Laue-Langevin, Grenoble, France.

Photon correlation spectroscopic analyses of the $2C_nE_m$ vesicles showed that they had hydrodynamic diameters of the order of 100–120 nm and so with such low curvatures, could, in the analyses of their SANS data, be regarded as infinite planar sheets (Ma, 1997). The scattering curves (of neutron intensity $I(Q)$ vs neutron momentum transfer $(Q)$) were thus transformed to give Guinier plots of $\ln(I(Q).Q^2)$ vs $Q^2$ (Kratky, 1963) and the linear portion of these curves at low $Q$ (see Figure 5) used to determine the radius of gyration of the vesicle lamella $(R_g)$ calculated as follows:

$$R_g = (\tan\theta.\lambda^2/16\pi^2)^{1/2} \qquad (2)$$

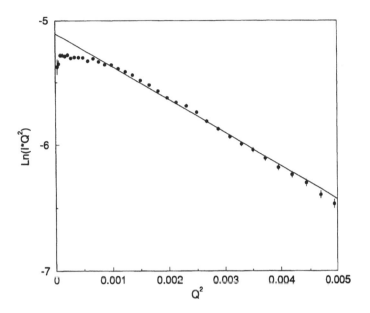

FIGURE 5. Guinier plots of $\ln(I(Q).Q^2)$ vs $Q^2$ for $2C_{18}E_{12}$ vesicles. The slight turn-down in the plot at low $Q$ is indicative of the fact that the vesicles have a finite radius and are not really infinite planar sheets.

where $\tan\theta$ is the slope of the curve in the low $Q$ domain (such that $R_g Q < 1.0$) and $\lambda$ is the wavelength of incident neutrons (in Å). The corresponding second moment thicknesses of the vesicle lamellae were then obtained as $R_g\sqrt{12}$ Å (Table 2) (Ma, 1997).

The lamellar thicknesses determined for the vesicles prepared from the $C_{18}$ chain surfactants were found to be of the order of 50–60 Å, whilst those for the $C_{16}$ chain surfactant vesicles lay in the range 90–100 Å. On the basis of these observations, therefore, it was tentatively proposed that the $C_{18}$ chain surfactant vesicles are delimited by single bilayer lamellae, and that the $C_{16}$ chain surfactants form vesicles with twin bilayers.

For $2C_{18}E_{12}$, the vesicle lamella thickness was estimated as $\sim 54$ Å. Now if it is assumed that the $E_{12}$ head groups in the bilayer adopt a random coil conformation and that they are not mixed with the $C_{18}$ chains, then the two head group layers of the bilayer will

TABLE 2. Guinier Analysis of SANS vesicular data treated as infinite planar bilayers.

| Surfactant | Radius of Gyration (Å) | Vesicle Lamellar Thickness (Å) |
|---|---|---|
| $2C_{16}E_{12}$ | $27.7 \pm 0.2$ | $96.0 \pm 0.8$ |
| $2C_{16}E_{17}$ | $31.6 \pm 0.2$ | $109.6 \pm 0.8$ |
| $2C_{18}E_{12}$ | $15.5 \pm 0.2$ | $53.8 \pm 0.5$ |
| $2C_{18}E_{17}$ | $18.9 \pm 0.3$ | $65.4 \pm 0.9$ |

have a combined thickness of $2 \times 15.6 = 31.2$ Å. Thus, if the lamella thickness of 54 Å corresponds to the thickness of a single bilayer, the two sets of $C_{18}$ chains will occupy a thickness of $54 - 31 = 23$ Å. Given that the volume of the surfactant hydrophobe is $\sim 1150$ Å$^3$ the area occupied by each $2C_{18}E_{12}$ molecule in the bilayer $(a_o)$ is obtained as $a_o = (2 \times 1150)/23 = 101$ Å$^2$. Accepting the various assumptions and approximations in these calculations, it is gratifying to observe that the area per molecule occupied by $2C_{18}E_{12}$ in their *curved* vesicle *bilayers*, is very close to the area per molecule calculated when this surfactant is dispersed as a *planar monolayer* at a surface pressure (of around $40 \text{ mNm}^{-1}$) where the molecules are beginning to be forced into a bilayer-type formation. The more general conclusion is reached, therefore that the information gleaned from neutron reflectivity studies of surfactant monolayers can be highly pertinent to modelling the structures of surfactant vesicles.

## 11.4  PERFLUOROCARBON EMULSIONS AND INTERFACIAL POLYMERS

The study of colloidal interfaces and adsorbed surfactants is of considerable interest for drug delivery. As mentioned previously, it has proven possible to prevent the rapid removal of particles injected into the bloodstream by adsorbing or grafting polymers with long polyoxyethylene chains, such as the Poloxamer PEO-PPO-PEO triblock copolymers (Illum *et al.*, 1982, 1987). We have also observed a significant targeting effect to the bone marrow when using the $A_2BA_2$ poloxamine block copolymer Tetronic 908 (BASF) on a range of microparticles and emulsions. There also appears to be an empirical relation between the *in vivo* behaviour and the structure of the adsorbed polymer layers; for example colloids with the largest adsorbed polymer hydrodynamic layer thickness exhibit the longest circulatory half-life after injection. It is likely that the biological handling of injected particles depends on the detailed structure of the interfacial layer, rather than simply its thickness, and so it is of some importance to understand the adsorbed layer structure in as much detail as possible.

We have studied the interfacial layer structure using surface potential measurements, and have obtained a direct measurement of the hydrodynamic thickness of the adsorbed layer using dynamic light scattering. This latter technique only allows the extent of the layer to be determined to a precision of $\pm 30$–40%. However, during the last decade it has been shown that SANS can provide much more detailed information about the conformations of polymer molecules adsorbed at interfaces.

Perfluorocarbon emulsions are ideal model systems for the study of polymers adsorbed at colloidal interfaces. They are of particular pharmaceutical interest, since they have been extensively studied for use as oxygen transport fluids *in-vivo*, the so-called 'artificial blood' (Johnson *et al.*, 1990a,b). Early emulsions were made from perfluorodecalin or perfluorotripropylamine, emulsified with block copolymers, and heroic attempts to use them as large-volume blood replacements were largely unsuccessful due to activation of the complement system. The droplets (of diameter 200–400 nm) are rapidly taken up into the liver and cause it to increase in mass significantly; they coalesce into larger droplets and are finally excreted as vapour in the breath over a period of several weeks. More recently the area has provoked renewed interest since the material perfluorooctyl bromide was found to accumulate in tumours, and is sufficiently radio-opaque to be used for diagnostic purposes.

From the viewpoint of the SANS experiment, perfluorocarbon emulsions have a number of useful features. A typical perfluorocarbon such as perfluorodecalin has a neutron scattering length density of $4.18 \times 10^{10}$ cm$^{-2}$; and is contrast-matched at 72% $D_2O$; thus the scattering from the core of the droplet can be minimised so that the interfacial scattering predominates. Perfluorocarbons readily emulsify with a wide range of hydrophilic polymers to make stable emulsions with droplet sizes in the range 200–400 nm (Johnson *et al.*, 1990a,b). Finally perfluorocarbons have a high density (perfluorodecalin has a density of 1.94 g cm$^{-3}$) and so they can be sedimented by gentle centrifugation, washed, and resuspended to remove unadsorbed polymer. This procedure can be carried out several times without significant emulsion coalescence due to the extreme stability of the system.

Consequently it is possible to make an emulsion of perfluorocarbon using a polymer emulsifier, wash it free of unadsorbed polymer by repeated centrifugation and resuspension, and transfer the droplets without extensive coalescence to an $H_2O/D_2O$ medium which is an exact contrast match for the oil droplet. Under these conditions the majority of the scattering in the sample will arise from the thin emulsifier layer, and the scattering pattern will contain information about the structure of the polymer layer at the interface. We have used this technique to study the structure of adsorbed layers of PPO-PEO-PPO block copolymers (the Pluronic or Poloxamer series) that have been widely used in pharmaceutical formulation in an attempt to develop targeted drug delivery systems.

## *11.4.1 Data Analysis*

In order to extract structural information from the scattering pattern, two approaches have been used:

### 11.4.1.1 Scattering from a polydisperse shell

This has been studied by Barnett (1982) and the scattering formula is given by:

$$(S/\Omega)(Q)[\text{cm}^{-1}] = 6 \times 10^{-22} \pi R_p^{-1} f_p M^2 Q^{-2} \exp(-Q^2\sigma^2)] + \text{background} \quad (3)$$

where $R_p$ is the droplet radius (in Å), $f_p$ is the volume fraction of droplets, $\sigma$ is the centre of mass of the adsorbed polymer layer, and $M$ is a parameter related to the adsorbed amount of polymer $\Gamma$ by:

$$\Gamma[\text{mgm}^{-2}] = MD/(d_p - d_s) \quad (4)$$

where $D$ (g cm$^{-3}$) is the bulk density of polymer, $d_p$ (cm$^{-2}$) is the scattering length density of the polymer and $d_s$ (cm$^{-2}$) is the scattering length density of the droplets and matched medium. The centre of mass of the adsorbed layer is a measure of central tendency, and so is expected to be of the order of half the hydrodynamic layer thickness. This approach allows the adsorbed amount $\Gamma$ to be measured; comparison of the values obtained with those calculated from measurement of the unadsorbed emulsifier (Table 3) shows good agreement, which confirms that the emulsifier was not significantly desorbed during the washing of the droplets.

TABLE 3. Structure of adsorbed layers of block copolymers on perfluorodecalin.

| Adsorbed Copolymer | $\Gamma$ (a) (mg m$^{-2}$) | $\Gamma$ (b) (mg m$^{-2}$) | $\sigma$ (a) (nm) | $l$ (c) (nm) |
|---|---|---|---|---|
| Poloxamer 188 | 1.87 | 1.85 | 2.34 | 4.8 |
| Poloxamer 188 (0.1M NaCl) | 1.94 | – | 2.40 | 8.9 |
| Poloxamer 188 (0.2M NaCl) | 1.94 | – | 2.63 | 8.7 |
| Poloxamer 407 | 3.28 | 2.93 | 3.46 | 16.0 |
| Poloxamine 908 | 1.49 | 1.18 | 3.75 | 15.2 |

(a) From SANS via adsorbed layer model
(b) By direct measurement of unadsorbed polymer in the continuous phase
(c) Effective layer thickness (95% of polymer segments)

### 11.4.1.2 Deconvolution of the segment density function

The segment density function, $\rho(z)$, is the local concentration of polymer segments as a function of distance $z$ from the droplet interface. This provides the highest detail about the layer structure, but is the most difficult description to obtain. There are two methods of calculating $\rho(z)$, by direct inversion (Crowley, 1984) using Laplace transforms, and by least-squares minimization of a parameterized profile (Barnett *et al.*, 1981), both of which require high-quality data to achieve a unique solution.

Table 3 shows the values of $\sigma$ and $\Gamma$ obtained for three block copolymers adsorbed to perfluorodecalin. As expected, they are approximately half the values obtained from measures of extent, such as the hydrodynamic thickness obtained by dynamic light scattering (Muller, 1991). As expected, the layer thickness increased with the molecular weight of the adsorbed polymer. The adsorbed amounts of the BAB polymers also increased with molecular weight, but that for Tetronic 908 was rather low considering its high molecular weight (25000). This may be due to its bulky shape and steric effects between the arms, which prevent its adsorbing efficiently.

The corresponding volume fraction profiles (Figure 6) indicate that the majority of the polymer is localised close to the droplet surface with little evidence for the extended brush-like profile which has been suggested by some authors to be important in the *in vivo* behaviour of colloids coated with these polymers (Illum *et al.*, 1982, 1987). The PEO segments must be partly looped back to the interface in order to account for these profiles.

It is important to understand the influence of factors such as ionic strength and temperature on the adsorbed layer structure since these materials are intended for intravenous use (Washington and King, 1997). Figures 7 and 8 show how the volume fraction profile and adsorbed layer centre of mass of PF68 adsorbed to perfluorodecalin change with sodium chloride concentration. Electrolyte concentrations below 0.6 M cause the polymer segments to move away from the interface, so that $\sigma$ increases to a maximum, after which the layer thickness begins to contract to near its original value. We would expect electrolytes to decrease the solvency for the PEG chains and thus to collapse the adsorbed layer (Cosgrove *et al.*, 1990). Thus the high-electrolyte concentration behaviour is understandable but the expansion at lower ionic strength is unexpected. It is possible to calculate the bound fraction of segments

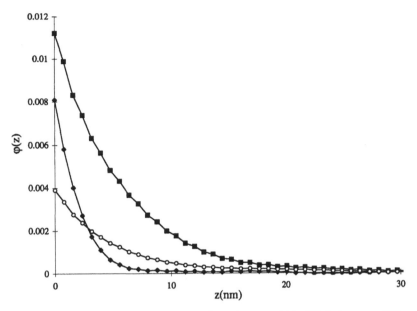

**FIGURE 6.** Volume fraction profiles, $\phi(z)$, for ($\blacklozenge$) Poloxamer 188, ($\blacksquare$) Poloxamer 407 and ($\bigcirc$) Poloxamine 908 adsorbed at the perfluorodecalin/water interface. The line joining the points is a guide to the eye.

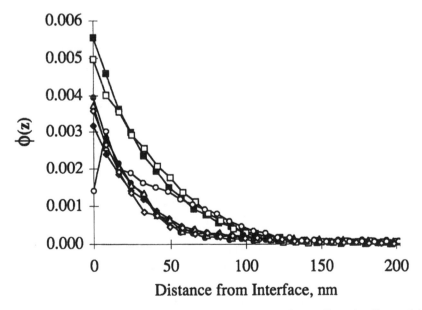

**FIGURE 7.** Volume fraction profiles for Poloxamer 188 adsorbed to perfluorodecalin emulsion in sodium chloride solutions of concentration 0–1.0 M at 25°C. 0.1 M ($\blacksquare$), 0.2 M ($\square$), 0.31 M ($\blacktriangle$), 0.48 M ($\bullet$), 0.62 M ($\bigcirc$), 0.81 M (open triangle, 1.0 M ($\lozenge$).

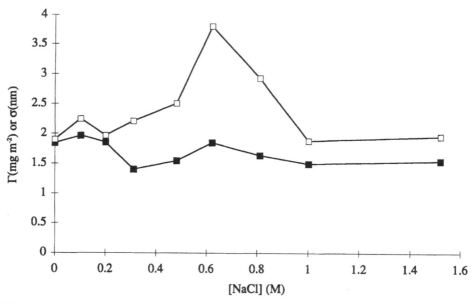

FIGURE 8. The variation of adsorbed amount Γ (■) and position of the adsorbed layer centre of mass σ (□) of adsorbed Poloxamer 188 with sodium chloride concentration.

by integrating $\rho(z)$ near the interface, and at higher electrolyte concentrations the values obtained agree with those expected on the basis that the PPO segments alone are adsorbed. The bound fraction is higher at low ionic strength, and this may be due to the adsorption of some PEO segments. It appears that the PEO segments adsorb at low ionic strength, possibly by an electrostatic interaction, and this interaction is prevented as soon as low concentrations of electrolytes are added. This allows the PEO loops to extend away from the surface and thus $\rho$ initially increases. At higher electrolyte concentration the decreased solvency causes the layer to contract as expected.

## 11.5 CONCLUDING REMARKS

The inertia on the part of pharmaceutical scientists to tap into the resource provided by neutron experiments stems partially from a general lack of awareness of the advantages offered by such studies, and partially from the perceived expense and difficulties involved in carrying out this type of experiment. The information presented above will hopefully have provided general information, and with regard to the latter points we note the following:

1. It is not always necessary to have deuterated samples in order to carry out worthwhile neutron studies of pharmaceutical systems; important information can very often be obtained by using the normal protonated systems with the required contrast(s) engineered through the use of $D_2O/H_2O$ mixtures, etc.

2. If this is not possible (for example, with single crystal or powder diffraction studies) researchers should give serious consideration to obtaining the required deuterated samples, with the time and expense devoted to this task most emphatically justified by the results obtained.

3. At present it is relatively straightforward to obtain the neutron beam time for experiments, either using the ISIS facility at the CLRC Rutherford Appleton Laboratory (RAL), Didcot, UK, or using the facilities provided at the Institut Laue-Langevin (ILL), Grenoble, France.

4. Both the ILL and RAL staff are very helpful and provide assistance in the design and on-site performance of the neutron experiments, and they also provide support in the task of data analysis and its interpretation.

Some specific problems which might benefit from neutron studies include:

1. SANS and/or single crystal neutron diffraction studies of the so-called 'DNA-liposome complexes' investigated for their potential in gene delivery.

2. SANS studies of drug distribution in drug delivery vehicles (including the more novel candidates of microemulsion-based gels, dendritic polymers and reverse vesicles).

3. Neutron powder diffraction studies of polymorphic phase transformations in drug crystals.

4. Neutron reflectivity studies of the distribution and structural perturbations caused by putative drug penetration enhancers in supported model membranes.

As drug delivery systems become more complex it is inevitable that their development will require us to understand their behaviour in unprecedented detail. Classical methods will run out of steam, and the successful pharmaceutical scientists will have to make full use of all the modern scientific tools at their disposal, including neutron scattering.

## References

Allen, T.M. and Chonn, A. (1987) Large unilamellar liposomes with low uptake into the reticuloendothelial system, *FEBS Lett.*, **223**, 42.

Atkinson, P.J., Grinson, M.J, Heenan, R.K., Howe, A.M., Mackie, A.R. and Robinson, B.H. (1988) Microemulsion-based-gels — a small-angle neutron-scattering study, *Chem. Phys. Lett.*, **151**, 494.

Atkinson, P.J., Robinson, B.H., Howe A.M. and Heenan, R.K. (1991) Structure and stability of microemulsion-based organo-gels, *J. Chem. Soc. Faraday Trans.*, **87**, 3389.

Atkinson, P.J., Dickinson, E., Home, D.S. and Richardson, R.M. (1995a) Neutron reflectivity of adsorbed protein films, *ACS Symposium Series*, **602**, 311.

Atkinson, P.J, Dickinson, E., Home, D.S. and Richardson, R.M. (1995b) Neutron reflectivity of adsorbed β-casein and β-lactoglobulin at the air-water interface, *J. Chem. Soc., Faraday Trans.*, **91**, 2847.

Atkinson, P.J., Dickinson E., Home D.S., Leermakers F.A.M. and Richardson R.M. (1996) Theoretical and experimental investigations of adsorbed protein structure at a fluid interface, *Ber. Bunsenges. Phys. Chem.*, **100**, 994.

Barnett, K.G., Cohen Stuart, M., Cosgrove, T., Sissons, D.S. and Vincent, B. (1981) Measurement of the polymer-bound fraction at the solid-liquid interface by pulsed nuclear magnetic resonance, *Macromolecules*, **14**, 1018.

Barnett, K.G. (1982), Ph.D. Thesis, University of Bristol, England.

Bauer, R., Behan, M., Hansen, S., Jones, G., Mortensen, K., Saermark, T. and Ogendal, L. (1991) Small-angle scattering studies on clathrin-coated vesicles, *J. Appl. Cryst.*, **24**, 815.

Barlow D.J., Ma, G., Lawrence, M.J., Webster, J.R.P. and Penfold, J. (1995) Neutron reflectance studies of a novel nonionic surfactant and molecular modeling of the surfactant vesicles, *Langmuir*, **11**, 3737.

Barlow D.J., Ma, G., Webster, J.R.P., Penfold, J. and Lawrence, M.J. (1997) Structure of the monolayer formed at an air-water interface by a novel nonionic (vesicle-forming) surfactant, *Langmuir*, **13**, 3800.

Bergenholtz, J., Romagnoli, A.A. and Wagner, N.J. (1995) Viscosity, microstructure, and interparticle potential of AOT/$H_2$O/n-decane inverse microemulsions, *Langmuir*, **11**, 1559.

Blume, G. and Cevc, G. (1993) Molecular mechanism of lipid vesicle longevity in vivo, *Biochem. Biophys. Acta*, **1146**, 157.

Boehm, M.K., Mayans, M.O., Thornton, J.D., Begent, R.H.J., Keep, P.A. and Perkins, S.J. (1996) Extended glycoprotein structure of the 7 domains in human carcinoembryonic antigen by X-ray and neutron solution scattering and an automated curve-fitting procedure — implications for cellular adhesion, *J. Mol. Biol.*, **259**, 718.

Bradshaw, J.P., Dempsey, C.E. and Watts, A. (1994) A combined X-ray and neutron scattering-diffraction study of selectively deuterated melittin in phospholipid-bilayers — effect of pH. *Mol. Memb. Biol.*, **11**, 79.

Brumm, T., Naumann, C., Scakmann, E., Rennie, A.R., Thomas R.K., Kanellas, D., Penfold, J. and Bayerl, T.M. (1994) Conformational changes of the lecithin head group in monolayers at the air-water interface — a neutron reflection study, *Eur. Biophys. J.*, **23**, 289.

Cantu, L., Corti, M., Zemb, T. and Williams, C. (1993) Small-angle X-Ray and neutron-scattering from ganglioside micellar solutions, *J. de Physique*, **3**, 221.

Chen, S.H., Chang, S.L., Strey, R., Samseth, L. and Mortensen, K. (1991) Structural evolution of bicontinuous microemulsions. *J. Phys. Chem.*, **95**, 7427.

Cohen, Y. (1996) The microfibrillar network in gels of poly(gamma-benzyl-L-glutamate) in benzyl alcohol, *J. Polymer Sci., Polym. Phys. Ed.*, **34**, 57.

Cosgrove, T., Crowley, T.L., Ryan, K. and Webster, J.R.P. (1990), The effects of solvency on the structure of an adsorbed polymer layer and dispersion stability, *Colloids and Surfaces*, **51**, 255.

Cosgrove, T., White, S.J., Zarbakhsh, A., Heenan, R.K. and Howe, A.M. (1996) Small-angle neutron-scattering studies of sodium dodecyl sulphate interactions with gelatin. 2. Effect of temperature and pH. *J. Chem. Soc. Faraday Trans.*, **92**, 595.

Cowan, S.W., Garavito, R.M., Jansonius, J.N., Jenkins, J.A., Karlsson, R., Konig, N., Pai, E.F., Pauptit, R.A., Rizkallah, P.J., Rosenbusch, J.P., Rummel, G. and Schirmer, T. (1995) The structure of ompf porin in a tetragonal crystal form, *Structure*, **3**, 1041.

Crowley, T.L. (1984), D.Phil. Thesis, University of Oxford, England.

Czeslik, C., Winter, R., Rapp, G. and Bartels, K. (1995) Temperature-dependent and pressure-dependent phase behaviour of monoacylglycerides monoolein and monoelaidin, *Biophys. J.*, **68**, 1423.

Dadmun, M.D., Muthukumar, M., Hempelmann, R., Schwahn, D. and Springer, T. (1996) Proton motion of poly(gamma-benzyl-L glutamate) in benzyl alcohol during gelation as measured by quasi-elastic neutron scattering, *J. Polymer Sci., Polym. Phys. Ed.*, **34**, 649.

Dekrulf, C.G. and May, R.P. (1991) Kappa-casein micelles — structure, interaction and gelling studied by small-angle neutron-scattering, *Eur. J. Biochem.*, **200**, 431.

Deme, B., Dubois, M., Zemb, T. and Cabane, B. (1996) Effect of carbohydrate on the swelling of a lyotropic lamellar phase, *J. Phys. Chem.*, **100**, 3828.

Deme, B., Dubois, M., Zemb, T. and Cabane, B. (1997) Coexistence of two lyotropic lamellar phases induced by a polymer in a phspholipid-water system, *Colloids and Surfaces A: Physicochemical and Engineering Aspects*, **121**, 135.

Dickinson, E., Home, D.S. and Richardson, R.M. (1993) Neutron reflectivity study of the competitive adsorption of $\beta$-casein and water-soluble surfactant at the planar-water-interface, *Food Hydrocolloids*, **7**, 497.

Duff, K.C., Cudmore, A.J. and Bradshaw, J.P. (1993) The location of amantadine hydrochloride and free base within phospholipid multilayers — a neutron and diffraction study, *Biochim. Biophys Acta*, **1145**, 149.

Eaglesham, A., Herrington, T.M. and Penfold, J. (1992) A neutron reflectivity study of a spread monolayer of bovine serum albumin, *Colloids and Surfaces*, **65**, 9.

Eastoe, J., Fragneto, G., Robinson, B.H., Towey, T.F., Heenan, R.K. and Leng, F.J. (1992) Variation of surfactant counterion and its effect on the structure and properties of aerosol-OT-based water-in-oil microemulsions, *J. Chem. Soc. Faraday Trans.*, **88**, 461–471.

Eastoe, J., Towey, T.F., Robinson, B.H., Williams, J. and Heenan, R.K. (1993) Structures of metal bis(2-ethylhexyl) sulfosuccinate aggregates in cyclohexane, *J. Phys. Chem.*, **97**, 1459.

Evmenenko, G.A., Budtova, T., Buyanov, A. and Frenkel, S. (1996) Structure of polyelectrolyte hydrogels studied by SANS, *Polymer*, **24**, 5499.

Evmenenko, G.A. and Alekseev, V.L. (1997) Study of swelling of chitosan films by small-angle neutron-scattering, *Vysokomol. Soedin., Ser. A & Ser. B*, **39**, 650.

Farago, B., Richter, D., Huang, J.S., Safran, S.A. and Milner, S.T. (1990) Shape and size fluctuations of microemulsion droplets — the role of cosurfactant. *Phys. Rev. Letts.*, **65**, 3348.

Faure, C., Laguerre, M., Neri, W. and Dufourc, E.J. (1997) Cholesterol sulphate has two locations in DMPC membranes. A neutron diffraction and molecular mechanics simulation, *Biophys. J.*, **72**, p.THAM7.

Fitter, J., Lechner, R.E., Buldt, G. and Dencher, N.A. (1996) Internal molecular motion of baceteriorhodopsin — hydration induced flexibility studied by quasi-elastic incoherent neutron-scattering using orientated purple membranes, *Proc. Natl. Acad Sci. USA*, **93**, 7600.

Forsyth, V.T., Mahendrasingam, A., Langan, P., Alhayalee, Y., Alexeev, D., Pigram, W.J., Fuller, W. and Mason, S.A. (1992) High angle neutron fiber diffraction studies of the distribution of water around the D-form of DNA, *Physica B*, **180**, 737.

Frampton, C.S., Wilson, C.C., Shankland, N. and Florence, A.J. (1997) Single crystal neutron refinement of creatine monohydrate at 20K and 123K, *J. Chem. Soc. Faraday Trans.*, **93**, 1875.

Freer, A.A, Bunyan, J.M., Shankland, N. and Sheen, D.B. (1993) Structure of (S)-(+)-ibuprofen, *Acta Crystallogr., Sect. C*, **49**, 1378.

Gladman, J.M., Crowley, T.L., Schofield, J.D. and Eaglesham, A. (1995) The structure of surface-adsorbed surfactants as determined by small-angle neutron scattering. *Phys. Scripta*, **T57**, 146.

Giordano, R., Wanderlingh, F., Wanderlingh, U. and Teixeria, J. (1993) Influence of pH and temperature on the structure in macromolecular solutions — a preliminary SANS study on lysozyme, *J. Phys. IV*, **3**, 237.

Gradzielski, M., Bergmeier, M., Muller, M. and Hoffmann, H. (1997) Novel gel phase: A cubic phase of densely packed monodisperse, unilamellar vesicles. *J. Phys. Chem. B.*, **101**, 1719.

Gripon, C., Legrand, L., Rosenman, I., Vidal, O., Robert, M.C. and Boue, F. (1997) Lysozyme-lysozyme interactioons in under- and super-saturated solutions: a simple relation between the second virial coefficients in $H_2O$ and $D_2O$, *J. Crystal Growth*, **178**, 575.

Hauss, T., Buldt, G., Heyn, M.P. and Dencher, N.A. (1994) Light-induced isomerization causes an increase in the chromophore tilt in the M intermediate of bacteriorhodopsin — A neutron-diffraction study, *Proc. Nat. Acad Sci. USA*, **91**, 11854.

He, K., Ludtke, S.J., Wu, Y. and Huang, H.W. (1993) X-ray scattering in the plane of membranes, *J. Phys. IV*, **3**, 265.

He, K., Ludtke, S.J., Worcester, D.L. and Huang, H.W. (1997) Neutron scattering in the plane of membranes — structure of alamethicin pores, *Biophys. J.*, **20**, 2659.

Hecht, E., Mortensen, K., Gradzielski, M. and Hoffmann, H. (1995) Interaction of ABA block-copolymers with ionic surfactants — influence of micellization and gelation, *J. Phys. Chem.*, **99**, 4866.

Horkay, F., Hecht, A.M. and Geissler, E. (1994) Small-angle neutron-scattering in poly(vinyl alcohol) hydrogels, *Macromolecules*, **27**, 1795.

Ibberson, R.M. (1996) The crystal-structure and phase-transition of sulfamide by high-resolution neutron powder diffraction, *J. Mol. Struct.*, **377**, 171.

Illum, L., Davis, S.S., Wilson, C.G., Thomas, N., Frier, M. and Hardy, J.G. (1982) Blood clearance and organ deposition of intravenously administered colloidal particles — the effects of particle-size, nature and shape, *Int. J Pharm.*, **12**, 135.

Illum, L., Davis, S.S., Muller, R.H., Mak, E. and West, P. (1987) The organ distribution and circulation time of intravenously injected colloidal carriers sterically stabilized with a block copolymer — Poloxamine 908, *Life Sciences*, **40**, 367.

Jamshaid, M., Farr, S.J., Kearney, P. and Kellaway, I.W. (1988) Poloxamer sorption on liposomes: comparison with polystyrene latex and influence on solute efflux, *Int. J. Pharm.*, **48**, 125.

Jenkins, P.J. and Donald, A.M. (1996) Application of small-angle neutron scattering to the study of the structure of starch granules, *Polymer*, **37**, 5559.

Johnson, O.L., Washington, C. and Davis, S.S. (1990a) Thermal stability of fluorocarbon emulsions that transport oxygen, *Int. J. Pharm.*, **59**, 131.

Johnson, O.L., Washington, C. and Davis, S.S. (1990b) Long term stability studies of fluorocarbon oxygen transport emulsions, *Int. J. Pharm.*, **63**, 65.

Kanaya, T., Ohkra, M., Kaji, K., Furusaka, M. and Misawa, M. (1994) Structure of poly(vinyl alcohol) gels studied by wide-angle and small-angle neutron scattering, *Marcomolecules*, **27**, 5609.

Kobayashi, M., Yoshioka, T., Imai, M. and Itoh, Y. (1995) Structural ordering on physical gelation of syndiotactic polystyrene dispersed in chloroform studied by time-resolved measurements of small-angle neutron-scattering (SANS) and infrared-spectroscopy, *Macromolecules*, **28**, 7376.

Koening, B.W., Krueger, S., Orts, W.J., Majkrzak, C.F., Berk, N.F., Silverton, J.V. and Gawrisch, K. (1996) Neutron reflectivity and atomic-force microscopy studies of a lipid-bilayer in water adsorbed to the surface of a silicon single-crystal, *Langmuir*, **12**, 1343.

Koning, S., Bayerl, T.M., Coddens, G., Richter, D. and Sackniann, E. (1995) Hydration dependence of chain dynamics and local diffusion in L-alpha-dipalmitoylphosphatidylcholine mutlilayers studied by incoherent quasi-elastic neutron scattering, *Biophys. J.*, **68**, 1871.

Knoll, W., Schmidt, G., Rotzer, H., Henkel, T., Pfeiffer, W., Sackmann, E., Mittlemeher, S. and Spinke, J. (1991) Lateral order in binary lipid alloys and its coupling to membrane functions, *Chem. Phys. Lipids*, **57**, 363.

Konig, S. and Sackmann, E. (1996) Molecular and collective dynamics of lipid bilayers, *Current Opinion In Colloid & Interface Science*, **1**, 78.

Kratky, O. (1963) X-ray and small angle neutron scattering with substances of biological interest in diluted solutions, *Prog. Biophys. & Mol. Biol.*, **13**, 107.

Kronberg, B., Dahlman, A., Carlfors, J., Karsson, J. and Arturson, P. (1990) Preparation and evaluation of sterically stabilized liposomes: colloidal stability, serum stability, macrophage uptake and toxicity, *J. Pharm. Sci.*, **79**, 667.

Krueger, S., Andrews, A.P. and Nossal, R. (1994) Small-angle neutron-scattering studies of structural characteristics of agarose gels, *Biophy. Chem.*, **53**, 85.

Krueger, S., Ankner, J.F., Satija, S.K., Majkrzak, C.F., Gurley, D. and Colombini, M. (1995) Extending the angular range of neutron reflectivity measurments from planar lipid bilayers — application to a model biological membrane, *Langmuir*, **11**, 3218.

Ku, C.Y. and Tam, C.N. (1997) The effect of volatile anesthetics on phospholipid bilayer vesicles investigated by small angle neutron scattering, *Abstracts of papers of the American Chemical Society*, **214**, 247-phys.

Langan, P. (1997) Neutron fibre diffraction: recent advances at the ILL, *Physica B*, **234**, 213.

Langan, P., Forsyth, V.T., Mahendrasingam, A., Pigram, W.J., Mason, S.A. and Fuller, W. (1992) A high angle neutron fiber diffraction study of the hydration of the A conformation of the DNA double helix, *J. Biomolecular Structure & Dynamics*, **10**, 489.

Langan, P., Forsyth, V.T., Mahendrasingam, A., Giesen, U., Dauvergne, M.T., Mason, S.A., Wilson, C.C. and Fuller, W. (1995) Neutron fiber diffraction studies of DNA hydration, *Physica B*, **213**, 783.

Lasic, D.D. (1996) Liposomes — an industrial view, *Chem. Ind.*, 210.

Lawrence, S.M. (1992) Characterization and computer modelling of novel nonionic surfactant vesicles. PhD Thesis, University of London.

Lawrence, M.J., Lawrence, S.L., Chauhan, S. and Barlow, D.J. (1996a) Synthesis and aggregation properties of dialkyl polyoxyethylene glycerol ethers, *Chem. Phys. Lipids*, **82**, 89.

Lawrence, M.J., Chauhan, S., Lawrence, S.M. and Barlow, D.J. (1996b) The formation, characterization and stability of nonionic surfactant vesicles, *STP Pharma Sciences*, **6**, 49.

Lawrence, M.J., Chauhan, S., Lawrence, S.M., Ma, G., Webster, J.R.P., Penfold, J. and Barlow, D.J. (1996c) Surfactant systems for drug delivery. In: *Chemical Aspects of Drug Delivery Systems* (Editors D.R. Karsa & R.A. Stephenson) Royal Society Chemistry, Cambridge. Chapter 5, pp. 65.

Lemmich, J., Mortensen, K., Ipsen, J.H., Honger, T., Bauer, R. and Mouritsen, O.G. (1996a) Small-angle neutron-scattering from multilamellar lipid bilayers — theory, model, and experiment, *Phys. Rev. E*, **53**, 5169.

Lemmich, J., Honger, T., Mortensen, K., Ipsen, J.H., Bauer, R. and Mouritsen, O.G. (1996b) Solutes in small amounts provide for lipid-bilayer softness: cholesterol, short chain lipids, and bola lipids, *Eur. Biophys. J. with Biophys. Lett.*, **25**, 293.

Lin, T.L., Liu, C.C., Roberts, M.F. and Chen, S.H. (1991) Structure of mixed short-chain lecithin and long-chain lecithin aggregates studied by small-angle neutron-scattering, *J. Phys. Chem.*, **95**, 6020.

Long, M.A., Kaler, E.W., Lee, S.P. and Wignall, G.D. (1994) Characterization of lecithin-taurodeoxycholate mixed micelles using small-angle neutron-scattering and static and dynamic light-scattering, *J. Phys. Chem.*, **98**, 4402.

Ludtke, S.J., He, E., Heller, W.T., Harroun, T.A., Yang, L. and Huang, H.W. (1996) Membrane pores induced by magainin, *Biochem.*, **35**, 13723.

Ma, G.K.F. (1997) Structural studies of non-ionic surfactant monolayers and vesicles. PhD Thesis, University of London.

Maritano, S., Carsughi, F., Fontana, M.P. and Marchesini, A. (1996) Study of the enzyme ascorbate oxidase by small-angle neutron-scattering, *J. Mol. Struct.*, **383**, 261.

Martel, P., Makriyannis, A., Mavromoustakos, T., Kelly, K. and Jeffrey, K.R. (1993) Topography of tetrahydrocannabinol in model membranes using neutron diffraction, *Biochem. Biophys. Acta*, **1151**, 51.

Mears, S.J., Deng, Y., Cosgrove, T. and Pelton, R. (1997) Structure of sodium dodecyl sulphate bound to a poly(NIPAM) microgel particle, *Langmuir*, **13**, 1901.

Meyer, D.F., Nealis, A.S., Bruckdorfer, K.R. and Perkins, S.J. (1995) Characterization of the structure of polydisperse human low density lipoprotein by neutron scattering. *Biochem. J.*, **310**, 407.

Mortensen, K. (1996) Structural studies of aqueous-solutions of PEO-PPO-PEO triblock copolymers, their micellar aggregates and mesophases — a small-angle neutron-scattering study, *J. Phys. Condensed Matter*, **8**, A103.

Mortensen, K. (1997) Cubic phases in a connected micellar network of poly(propylene oxide)-poly(ethylene oxide)-poly(propylene oxide) triblock copolymers in water, *Macromolecules*, **30**, 503.

Muller, R.H. (1991) Colloidal Carriers for Drug Delivery and Targeting, Wissenshaftliche Verlagsgesellschaft Stuttgart, p. 61.

Niimura, N., Minezaki, Y., Ataka, M. and Katsura, T. (1994) Small-angle neutron-scattering from lysozyme in unsaturated solution, to characterize the precrystallization process, *J. Crystal Growth*, **147**, 671.

Niimura, N., Ataka, M., Minezaki, Y. and Katsura, T. (1995) Small-angle neutron-scattering from lysozyme crystallization process, *Physica B*, **213**, 745.

Naumann, C., Brumm, C., Rennie, A.R., Penfold, J. and Bayerl, T.M. (1995) Hydration of DPPC monolayers at the air-water interface and its modulation by the nonionic surfactant $C_{12}E_4$ — a neutron reflection study, *Langmuir*, **11**, 3848.

Oberdisse, J., Couve, C. Appell, J., Berret, J.F., Ligotire, C. and Porte, G. (1996) Vesicles and onions from charged surfactant bilayers — a neutron-scattering study, *Langmuir*, **12**, 1212.

Pebaypeyroula, E., Dufourc, E.J. and Szabo, A.G. (1994) Location of diphenyl-hexatriene and trimethylammonium-diphenyl-hexatriene in dipalmitoyl-phosphatidylcholine bilayers by neutron diffraction, *Biophys. Chem.*, **53**, 45.

Pebaypeyroula, E., Garavito, R.M., Rosenbusch, J.P., Zulauf, M. and Timmins, P.A. (1995) Detergent structure in tetragonal crystals of ompf porin, *Structure*, **3**, 1051.

Pedersen, J.S., Hansen, S. and Bauer, R. (1994) Scaling in the aggregation behaviour of zinc-free insulin studies by small-angle neutron scattering, *Nuovo Cimento delta Societa Italiana Di Fisica D-Condensed Matter Atomic Molecular and Chemical Physics Fluids Plasmas Biophysics*, **16**, 1447.

Perezvillar, V., Vazqueziglesias, M.E. and Degeyer, A. (1993) Small-angle neutron-scattering studies of chlorpromazine micelles in aqueous solutions, *J. Phys. Chem.*, **97**, 5149.

Petit, C., Zemb, T. and Pileni, M.P. (1991) Structural study of microemulsion-based gels at the saturation point, *Langmuir*, **7**, 223.

Ravanat, C., Torbet, J. and Freyssinet, J.M.A. (1992) Neutron solution scattering study of the structure of annexin-V and its binding to lipid vesicles, *J. Mol. Biol.*, **226**, 1271.

Ravey, J.C., Stebe, M.J. and Sauvage, S. (1994) Fluorinated gel-emulsions, *J. Chim. Phys. Phys. — Chim. Biol.*, **91**, 259.

Reinl, H., Brumm, T. and Bayerl, T.M. (1992) Changes of the physical-properties of the liquid-ordered phase with temperature in binary-mixtures of DPPC with cholesterol — a $H^2$-NMR, FT-IR, DSC and neutron-scattering study, *Biophys. J.*, **61**, 1025.

Roberts, C.J., Davies, M.C., Tendler, S.J.B., Lawrence, S.M., Lawrence, M.J., Chauhan, S. and Barlow, D.J. (1991) Imaging of individual surfactant vesicles by scanning tunelling microscopy, *J. Pharm. Pharmacol.*, **43**(suppl), 98P.

Rochas, C., Brulet, A. and Guenet, J.M. (1994) Thermoreversible gelation of agarose in water dimethyl sulphoxide mixtures, *Macromolecules*, **27**, 3830.

Roth, M., Amoux, B., Ducruix, A. and Reisshusson, F. (1991) Structure of the detergent phase and protein-detergent interactions in crystals of the wild type (Strain Y) *Rhodobacter Sphaeroides* Photochemical Reaction Center, *Biochem.*, **30**, 9403.

Sato, M., Ito, Y., Kameyama, K., Imai, M., Ishikawa, N. and Takagi, T. (1995) Small-angle neutron-scattering study of recombinant yeast-derived human hepatitis-B virus surface-antigen vaccine particle, *Physica B*, **231**, 757.

Schmidt, A., Spinke, J., Bayerl, T., Sackmann, E. and Knoll, W. (1992) Streptavidin binding to biotinylated lipid layers on solid supports — a neutron reflection and surface-plasmon optical study, *Biophys. J.*, **63**, 1385.

Schurtenberger, P., Magid, L.J., Penfold, J. and Heenan, R.K. (1990) Shear aligned lecithin reverse micelles — a small-angle neutron-scattering study of the anomalous water-induced micellar growth, *Langmuir*, **6**, 1800.

Senior, J., Delgado, C., Fisher, D., Tilcock, C. and Gregoriadis, G. (1991) Influence of surface hydrophilicity of liposomes on their interaction with plasma protein and clearance from the circulation: studies with poly(oxyethylene glycol) coated vesicles, *Biochim. Biophys Acta*, **1062**, 77.

Shankland, N., Florence, A.J., Cox, P.J., Sheen, D.B., Love, S.W., Stewart, N.S. and Wilson, C.C. (1996a) Crystal morphology of ibuprofen predicted from single-crystal pulsed neutron diffraction data, *Chem. Comm.*, **7**, 855.

Shankland, N., Love, S.W., Watson, D.G., Knight, K.S., Shankland, K. and David, W.I.F. (1996b) Constrained Rietveld refinement of b-H-I(1) decadeuteriodopamine deuteriobromide using powder neutron diffraction data, *J. Chem. Soc. Faraday Trans.*, **92**, 4555.

Shankland, N., Florence, A.J. and Wilson, C.C. (1997a) Single-crystal neutron diffraction analysis of anion-cation interactions in perdeuteroacetylcholine bromide at 100 K, *Acta Crystallogr., Sect. B*, **53**, 176.

Shankland, N., Wilson, C.C., Florence, A.J. and Cox, P.J. (1997b) Refinement of ibuprofen at 100 K by single-crystal pulsed neutron diffraction. *Acta Crystallogr., Sect. C*, **53**, 951.

Shibayama, M., Ikkai, F. and Nomura, S. (1994) Complexation of poly(vinyl alcohol) congo red aqueous solutions. 2. SANS and SAXS studies on sol-gel transition, *Macromolecules*, **27**, 6383.

Shibayama, M., Ikkai, F., Inamoto, S., Nomura, S. and Han, C.C. (1996) pH and salt concentration dependence of the microstructure of poly(n-isopropylacrylamide-co-acrylic acid) gels, *J. Chem. Phys.*, **105**, 4358.

Silberbergbouhnik, M., Rarnon, O., Ladyzhinski, I., Mizrahi, S. and Cohen, Y. (1995) Osmotic deswelling of weakly charged poly(acrylic acid) solutions and gels, *J. Poly. Sci., Polym. Phys.*, **33**, 2269.

Simister, E.A., Lee, E.M., Thomas, R.K. and Penfold, J. (1992) Structure of a tetradecyltrimethyammonium bromide layer at the air-water interface determined by neutron reflection, *J. Phys. Chem.*, **96**, 1373.

Sjostrom, B., Kaplun, A., Talmon, Y. and Cabane, B. (1995) Structures of nanoparticles prepared from oil-in-water emulsions, *Pharm. Res.*, **12**, 39.

Tamura, T. and Ichikawa, M. (1997) Effect of lecithin on organogel formation of 12-hydroxystearic acid, *J. Am. Chem. Soc.*, **74**, 491.

Terech, P., Rodriguez, V., Barnes, J.D. and McKenna, G.B. (1994) Organogels and aerogels of racemic and chiral 12-hydroxyoctadecanoic acid, *Langmuir*, **10**, 3406.

Terech, P., Funnan, I., Weiss, R.G., Bouaslaurent, H., Desvergne, J.P. and Ramasseul, R. (1995) Gels from small molecules in organic solvents — structural features of a family of steroid and anthryl-based organogels, *Faraday Discussions*, **101**, 345.

Terech, P., Smith, W.G. and Weiss, R.G. (1996a) Small-angle scattering study of aqueous gels of sodium lithocholate, *J. Chem. Soc., Faraday Trans.*, **92**, 3157.

Terech, P., Ostuni, E. and Weiss, R.G. (1996b) Structural study of cholestryl anthraquinone-2-carboxylate (CAQ) physical organogels by neutron and X-ray small angle scattering, *J. Phys. Chem.*, **100**, 3759.

Thiyagarajan, P. (1997) In-plane small-angle neutron diffraction on the mechanism of antimicrobial peptides action on bacterial cell membranes, *Neutron News*, **8**, 16.

Triolo, F., Graziano, V. and Heenan, R.H. (1996) Small-angle neutron-scattering study of the quaternary structure of haemocyanin of *Rapana thomasiana*, *J. Mol. Structure*, **383**, 249.

Vaknin, D., Kjaer, K., Alsnielsen, J. and Losche, M. (1991) Structural properties of phosphophatidyl-choline in a monolayer at the air-water interface — neutron reflection study and reexamination of X-ray reflection measurements, *Biophys. J.*, **59**, 1325.

Vaknin, D., Kjaer, K., Ringsdorf, H., Blankenburg, R., Piepenstock, M., Diederich, A. and Losche, M. (1993) X-ray and neutron reflectivity studies of a protein monolayer adsorbed to a functionalized aqueous surface, *Langmuir*, **9**, 1171.

Waigh, T.A., Jenkins, P.J. and Donald, A.M. (1996) Quantification of water in carbohydrate lamellae using SANS, *Faraday Discussions*, **103**, 325.

Wang, Z.Y., Umetsu, M., Yoza, K., Kobayashi, M., Imai, M., Matsushita, Y., Niimura, N. and Nozawa, T. (1997) A small-angle neutron scattering study on the small aggregates of bacteriochlorophylls in solution, *Biochem. Biophys. Acta*, **1320**, 73.

Washington, C., King, S.M. and Heenan, R.K. (1996) Structure of block copolymers adsorbed to a perfluorocarbon emulsion, *J. Phys. Chem.*, **100**, 7603.

Washington, C. and King, S.M. (1997) Effect of electrolytes and temperature on the structure of a polyoxyethylene-polyoxypropylene-polyoxyethylene block copolymer adsorbed to a perfluorocarbon emulsion, *Langmuir*, **13**, 4545.

Weiner, M.C. and White, S.H. (1992) Structure of a fluid dioleoylphosphatidylcholine bilayer determined by joint refinement of X-ray and neutron-diffraction data, *Biophys. J.*, **61**, 434.

Wess, T.J., Wess, L. and Miller, A. (1994) The *in vitro* binding of acetaldehyde to collagen studied by neutron diffraction, *Alcohol and Alcoholism*, **29**, 403.

Willner, L., Jucknischke, O., Richter, D., Roovers, J. and Zhou, L.L. (1994) Structural investigation of star polymers in solution by small-angle neutron-scattering, *Macromolecules*, **27**, 3821.

Wilson, C.C., Shankland, N., Florence, A.J. and Frampton, C.S. (1997) Single-crystal neutron diffraction of bioactive small molecules, *Physica B*, **234** 84.

Woodle, M.C., Matthay, K.K., Newman, M.S., Hidayat, J.E., Collins, L.R., Redeman, C., Martin, F.J. and Papahadjopoulos, D (1992) Versatility in lipid compositions showing prolonged circulation with sterically stabilized liposomes, *Biochim. Biophys Acta*, **1105**, 193.

# Index

357